Traditional Construction Wisdom in Developing Regions: Sustainable Urbanization and Local-Eco Adaptation

Traditional Construction Wisdom in Developing Regions: Sustainable Urbanization and Local-Eco Adaptation

Guest Editors

Yin Zhang
Ahad Amini Pishro
Jin Li
Ying Huang

MDPI

Basel • Beijing • Wuhan • Barcelona • Belgrade • Novi Sad • Cluj • Manchester

Guest Editors

Yin Zhang
School of Architecture
Southwest Minzu University
Chengdu
China

Ahad Amini Pishro
School of Civil Engineering
Sichuan University of Science & Engineering
Zigong
China

Jin Li
School of Civil Engineering
Sichuan University of Science & Engineering
Zigong
China

Ying Huang
School of Fine Arts and Design
Chengdu University
Chengdu
China

Editorial Office

MDPI AG
Grosspeteranlage 5
4052 Basel, Switzerland

This is a reprint of the Special Issue, published open access by the journal *Buildings* (ISSN 2075-5309), freely accessible at: https://www.mdpi.com/journal/buildings/special_issues/DPBMYY4X37.

For citation purposes, cite each article independently as indicated on the article page online and as indicated below:

Lastname, A.A.; Lastname, B.B. Article Title. *Journal Name* **Year**, *Volume Number*, Page Range.

ISBN 978-3-7258-7116-2 (Hbk)
ISBN 978-3-7258-7117-9 (PDF)
https://doi.org/10.3390/books978-3-7258-7117-9

Contents

buildings

MDPI

Editorial

Traditional Construction Wisdom for Sustainable Urbanization and Local Ecological Adaptation in Developing Regions

Zhixiang Zuo [1], Yang Yang [1], Ahad Amini Pishro [2] and Yin Zhang [3,*]

1 School of Architecture, Southwest Minzu University, Chengdu 610225, China; 19982087276@163.com (Z.Z.); 13398432841@163.com (Y.Y.)
2 School of Civil Engineering, Sichuan University of Science & Engineering, Zigong 643000, China; ahad.a.pishro@gmail.com or ahad.ap@suse.edu.cn
3 Zoige Alpine Wetland Ecosystem National Observation and Research Station, Southwest Minzu University, Chengdu 610225, China
* Correspondence: cdzhangyin@163.com or cdzhangyin@swun.edu.cn; Tel.: +86-134-8891-8589

1. Introduction

Rapid urbanization and climate change pose severe challenges to the sustainable development of the built environment in developing regions [1], particularly in contexts where traditional building knowledge, local ecological adaptation, and modern technological systems converge [2,3]. In recent years, academia has increasingly recognized indigenous architectural wisdom, adaptive reuse of buildings, and digital design and management methods as crucial pathways toward resilient and low-carbon urbanization. This Special Issue brings together interdisciplinary research spanning architecture, environmental science, and digital technology.

Through systematic field research on traditional dwellings, combined with building modeling and performance analysis methods, relevant studies have delved into their inherent sustainability mechanisms and environmental adaptation characteristics [4,5]. These investigations systematically analyze the architectural resilience demonstrated by traditional dwellings in responding to specific regional built environments. Simultaneously, such research explores and achieves a balanced approach between enhancing energy and environmental performance while preserving the tangible and intangible cultural values of vernacular housing. Concurrently, adaptive reuse and sustainable retrofitting of existing public and healthcare buildings address population aging, public health emergencies, and evolving functional demands [6,7]. A more inclusive and environmentally sensitive built environment is possible via combining building adaptability with the preservation and rediscovery of existing architectural heritage, integrating building performance simulations, evacuation models, and age-friendly design strategies [8,9].

Furthermore, research on sustainable architectural design and community-scale interventions demonstrates the importance of integrating sustainability [10], climate change, environmental factors, and emerging technologies in achieving building sustainability. Concurrently, incorporating Building Information Modeling (BIM) into project management research expands the scope of architecture beyond spatial design and physical performance, emphasizing the significance of digital collaboration and governance in enhancing project performance and implementation efficiency [11]. In summary, these research findings reflect a shift toward integrated, multi-scale, and performance-oriented approaches that reconnect traditional architectural wisdom with contemporary sustainability agendas.

2. Overview of Contributions

The research featured in this Special Issue primarily focuses on sustainable urbanization processes and adaptive building design in developing regions. It encompasses diverse research methodologies and analytical perspectives, demonstrating significant overlap and complementarity in both content and methodological approaches. These studies effectively transcend the limitations of single-discipline perspectives, fostering interdisciplinary integration across architecture, environmental science, digital science, and socioeconomics. They provide multidimensional theoretical underpinnings and practical insights for achieving sustainable development goals and localized adaptive building design in developing regions. Based on differences in research themes and focal points, this article further categorizes the studies in this Special Issue into four directions, as shown in Table 1:

1. Subsection Analysis of Traditional Dwellings in Developing Regions.
2. Adaptive Reuse of Healthcare/Medical Buildings.
3. Architectural Sustainability Design.
4. Building Information Modeling and Project Management.
5. Industrial Heritage Preservation

Table 1. The content of the Topic "Traditional Construction Wisdom in Developing Regions: Sustainable Urbanization and Local-Eco Adaptation", and synthetic collation.

Author	Subject of the Research	Research Problem	Research Technique Instrumentality
1. Subsection Analysis of Traditional Dwellings in Developing Regions			
Zhang et al.	Thermal Comfort Analysis of Traditional Sibe Ethnic Group Dwellings	The Influence of Traditional Residential Envelope Structures and Active Heating Systems on Thermal Comfort in Living Spaces	Field Research: Sware ITES2023 and Airpak Software (Airpak 3.0)
Zhao et al.	Natural Ventilation in the Buffer Spaces of Traditional Qiang Ethnic Group Dwellings	Buffer Zones and Their Impact on Natural Ventilation Performance and Air Quality in Traditional Qiang Ethnic Group Dwellings	Field Research, Sware Software (VENT2024)
Wen et al.	Digital Analysis of Tujia Ethnic Group's "L"-Shaped Dwellings	Constructing the "L"-shaped architectural syntax of traditional Tujia dwellings and reconstructing it through transformation	Shape Grammar and Field Research
Feng et al.	Construction and Sustainable Development of Cave Dwellings	The thermal performance, energy-saving characteristics, and positive impacts of cave dwellings on efficient land use and cultural preservation	Field research
2. Adaptive Reuse of Healthcare Buildings			
Yi et al.	Adaptive Reuse of Aging Healthcare Buildings	Optimize barrier-free access and public activity spaces for the elderly	Sware Software (VENT2023 and Dali2023)
Wan et al.	Impact of Spatial Layout on Evacuation Efficiency Following the Conversion of an Exhibition Hall into a Field Hospital	The impact of evacuation time, spatial congestion characteristics, and personnel exit usage on evacuation efficiency	Pathfinder Software (2019.3.1217)

https://doi.org/10.3390/buildings16050930

Table 1. *Cont.*

Author	Subject of the Research	Research Problem	Research Technique Instrumentality
3. Architectural Sustainability Design			
Xiong et al.	Energy-Efficient Architectural Design of Banquet Halls with Integrated Tunnel Ventilation Systems	The Coordination Issue Between Indoor Thermal Comfort and Ventilation Efficiency	Field Research, DeST Software (DeST 2.0)
Zhang et al.	Multifunctional, integrated community spaces with diverse functions and sustainable development	Spatial Permeability Design within the Support Theoretical Framework	Supports Theory
4. Building Information Modeling and Project Management			
Yi et al.	Research on Project Management Under BIM Application	The Impact of Contract Governance and Relationship Governance on Project Performance	Partial Least Squares, Questionnaire Survey, and PLS-SEM Empirical Analysis
5. Industrial Heritage Preservation			
Yu et al.	Identification of Industrial Heritage Types and Research on Spatial Form	Spatial analysis of industrial heritage sites often relies heavily on qualitative descriptions, resulting in insufficient scientific rigor and comparability.	Average Nearest Neighbor Analysis, and Spatial Syntax analysis

2.1. Subsection Analysis of Traditional Dwellings in Developing Regions

Contribution 1 (Zhang et al.) conducted a systematic analysis of the thermal environmental adaptation mechanisms of traditional dwellings in cold climates, using the traditional settlement of Jianshifosi Village—home to the Sibe ethnic group in northern China—as their research subject. Combining field surveys with numerical simulations, the study first conducted an in-depth investigation and synthesis of the spatial layout, envelope structure characteristics, and typical lifestyles of local traditional buildings. Based on this foundation, representative traditional dwellings were selected as research samples. Employing Sware ITES2023 (20220401) building energy simulation software and Airpak 3.0.16 indoor airflow and thermal environment analysis software, the thermal performance of the traditional dwellings' envelope structures and the unique traditional heating facility of the Sibe people—the "Swastika Kang"—were simulated and analyzed. Analysis of simulation results compared indoor temperature variations and thermal comfort indicators under different envelope parameters and "Swastika Kang" operating conditions, focusing on the impact of traditional envelope structures and active heating systems on thermal comfort within living spaces. The research aims to reveal the thermal environment design wisdom developed by traditional Sibe dwellings through long-term adaptation to cold climates. This provides scientific basis, theoretical support, and practical reference for the renovation and energy-efficient design of contemporary traditional rural residences in cold regions.

Contribution 2 (Zhao et al.) studied the impact of architectural buffer spaces on natural ventilation performance in traditional Qiang dwellings located in high-altitude, cold regions. Based on field research, they categorized representative buffer space types found in Qiang dwellings, identifying three typical forms: courtyards, eaves spaces, and cantilevered floor slabs. Two traditional Qiang dwellings exhibiting characteristic spatial features were selected as research subjects, and corresponding architectural models were constructed. Using Sware software (VENT2024), they conducted simulation analyses and comparative studies on the configuration methods, presence/absence, and courtyard scale variations of buffer spaces under different site wind conditions. Quantitative analysis of key indicators—including wind velocity distribution, ventilation efficiency, and indoor air age—revealed that incorporating buffer spaces, particularly courtyards, significantly

https://doi.org/10.3390/buildings16050930

enhances natural ventilation performance and improves indoor air quality in traditional Qiang dwellings. Simultaneously, rational adjustment of buffer space scale parameters can further optimize internal ventilation organization. Based on simulation findings, this study proposes optimization strategies for the ventilation performance of traditional Qiang dwellings. These strategies provide a scientific basis for the design and renovation of ethnic dwellings in high-altitude cold regions, offering practical reference pathways for improving rural building ventilation environments and advancing sustainable development in rural areas.

Contribution 3 (Wen et al.) focused their research on the traditional residential architecture and construction techniques of the Tujia ethnic group. Their objective was to systematically extract spatial forms using shape grammar methodology and subsequently establish a computable digital grammar database. Conducting extensive field research primarily in southeastern Chongqing, the study employed a comprehensive approach combining map drafting, photographic documentation, interviews, and on-site sketching to systematically collect and organize data on the form characteristics and construction methods of traditional Tujia dwellings. A total of 32 representative residential cases from five districts and counties were gathered and organized, followed by classification, translation, and coded analysis of the collected materials. Integrating these findings with shape grammar methodology, a representative set of grammatical rules was derived from the stylistic characteristics of Tujia dwellings. Further leveraging the Grasshopper platform (based on Rhino 7.35), this grammar system was translated into a parametric digital model. Through parameter control and rule constraints, the model generates diverse residential forms consistent with traditional Tujia characteristics. This approach provides technical support and methodological reference for the preservation and restoration of traditional dwellings, as well as generative design of new rural residences in Tujia regions.

Contribution 4 (Feng et al.) systematically analyzed the construction characteristics and sustainable development value of traditional cave dwellings in the Loess Plateau region of Henan Province. Employing a combined approach of field research, environmental parameter monitoring, and questionnaire surveys, the study quantitatively assessed the thermal performance, energy-saving properties, and residential comfort of these structures, demonstrating their excellent thermal insulation capabilities. Further discussions explored dimensions such as efficient land resource utilization, cultural heritage preservation, and rural revitalization. These dwellings not only embody locally adapted ecological construction wisdom but also hold positive implications for promoting urban–rural integration, advancing rural tourism development, and sustaining regional cultural continuity. The findings provide theoretical foundations and practical references for the renovation and sustainable development of rural residences in cold and loess regions, while offering a model for harmonizing heritage conservation with modern residential needs.

2.2. *Adaptive Reuse of Healthcare Buildings*

Contribution 5 (Yi et al.) conducted a problem analysis of existing healthcare buildings, integrating practical needs with relevant building codes to propose optimization recommendations for two key aspects: barrier-free access for the elderly and public activity spaces. The study selected a senior care facility in Sanya, Hainan Province, China, as its research subject. By analyzing its current conditions, it identified typical issues such as insufficient public space and excessive walking distances. Through renovation strategies that increase activity space, walking convenience, and walking safety, the facility aims to provide long-term comfortable and convenient living for the elderly. Simultaneously, the study examined the building's daylighting and ventilation conditions. Using Sware software (VENT2023) for CFD calculations and Sware software (Dali2023), the indoor and

https://doi.org/10.3390/buildings16050930

outdoor airflow distribution, flow rates, and daylighting coefficients were calculated. These metrics were employed to evaluate the suitability of the renovated building for elderly residents, with corresponding optimization recommendations provided. This research offers valuable insights into the feasibility and design direction of age-friendly buildings, as well as the renovation of existing structures.

Contribution 5 (Wan et al.) addressed the post-pandemic era, where large public buildings—with their high ceilings, spacious interiors, and flexible layouts adaptable to diverse functions—are suitable for conversion into emergency field hospitals. However, their original spatial configurations, functional flow patterns, and safety evacuation routes undergo significant alterations. To prevent airborne virus transmission and meet evacuation demands during high occupancy, new spatial layout requirements emerge. Leveraging the spatial characteristics of modular emergency shelters, the study employed Pathfinder software (2019.3.1217) to simulate typical scenarios within such facilities. Key spatial factors affecting evacuation efficiency—including evacuation duration, spatial congestion patterns, and exit usage—were analyzed, leading to proposed optimization strategies. Simulation results were validated to ensure the effectiveness of these strategies. This research provides robust theoretical support and methodological guidance for the spatial layout of exhibition halls converted into prefabricated emergency shelters, thereby effectively enhancing evacuation efficiency and strengthening the resilience and safety of exhibition facilities.

2.3. Architectural Sustainability Design

Contribution 7 (Xiong et al.) conducted research on the coordination between high-standard indoor thermal comfort and ventilation efficiency. Under transitional seasonal climate conditions, they systematically evaluated the thermal and humidity environment of a banquet hall equipped with a ground-source heat exchange system. Using a banquet hall in Guanghan City, Sichuan Province, as the test subject, the study employed the YOWEXA air quality and environmental tester to collect ventilation and environmental parameter data at the bottom air intake, top exhaust outlet, and underground duct inlet of the banquet hall. Results indicate that the "bottom-underground supply air—top exhaust air" natural ventilation pattern demonstrates favorable performance in maintaining stable air intake conditions and improving indoor environments. Additionally, indoor occupancy density significantly impacts ventilation effectiveness. Furthermore, utilizing DeST software (DeST 2.0), the study simulated natural ventilation conditions in public buildings across five representative cities within China's building climate zones. Ventilation rates were calculated and residual heat exhaust efficiency evaluated, thereby validating the applicability of these findings under diverse climatic conditions. Overall, this study expands the understanding of ground-source-assisted natural ventilation systems in public buildings, providing valuable theoretical foundations and practical insights for achieving low-carbon operation and enhancing building environmental resilience.

Contribution 8 (Zhang et al.) applied the spatial permeability design framework of the support structure theory to achieve temporal adjustability and spatial adaptability in multifunctional composite community spaces. By integrating porous interfaces with modular spatial decomposition, they disrupted the closed nature of building facades, introducing permeability and variability both inside and outside the structure. This approach enhances environmental resilience, spatial adaptability, and social interaction within the architecture. Vertical stratification and horizontal interweaving create composite pathways for spatial permeability, forming a three-dimensional composite spatial system. This optimizes environmental performance (daylighting, ventilation, thermal performance), shapes social behavior (chance encounters, visual relationships), and enhances the building's spatial continuity and experiential diversity. Using a community service center in

https://doi.org/10.3390/buildings16050930

Yicheng District, Zhumadian City, China, as a case study, the paper quantitatively analyzes the post-renovation building's floor area ratio, full life-cycle costs, carbon emissions, and waste recovery under the supporting structure theory. This research provides a new theoretical framework and practical pathway for addressing the challenges of functional diversification and sustainable development in high-density urban environments.

2.4. Building Information Modeling and Project Management

Contribution 9 (Yi et al.) systematically examined the impact of contractual governance and relational governance on project performance, grounded in transaction cost theory. They introduced BIM application level as a moderating variable and constructed an analytical framework. The study combined questionnaire surveys with PLS-SEM empirical analysis, revealing that BIM application levels exert a significant moderating effect on the relationship between contract flexibility, trust, and performance. This highlights the crucial role of information exchange and coordination mechanisms in enhancing inter-organizational workflows. The findings also provide insights for transforming BIM from a technical tool into a governance support platform, promote the practical application of BIM technology, and offer actionable recommendations for fostering efficient governance practices within an environment of technological advancement.

2.5. Industrial Heritage Preservation

Contribution 10 (Yu et al.) conducted a systematic study of 39 Third Line industrial heritage sites in Guangyuan City, Sichuan Province. It identified issues such as widespread underutilization of industrial heritage and insufficient recognition of its value, and proposed adaptive mechanisms and revitalization pathways. The study employs Average Nearest Neighbor Analysis (ANN) and Spatial Syntax to establish a "type–medium–value" analytical framework and investigates spatial characteristics at the city, settlement, and building scales. At the urban level, heritage sites exhibit an overall dispersed pattern with localized clustering, while distributions within county- and district-level boundaries become more concentrated due to industrial production requirements. At the settlement level, integrating topographic and morphological indicators reveals concentrated and linear spatial patterns, mainly in plains and valley areas. At the architectural level, building layouts are shaped by production needs, daily life, and natural constraints, forming a characteristic "building–street–annex" pattern, which is further classified into three basic forms and five composite configurations through case studies. The study concludes that the spatial forms of Third Line industrial heritage are key carriers of historical, social, and environmental values. Based on this understanding, the authors propose revitalization strategies including tiered and categorized utilization, adaptive micro-scale renewal, and multi-stakeholder collaborative governance, providing methodological references for the conservation and regeneration of Third Line industrial heritage in underdeveloped regions.

3. Outlook

In the future, it will be essential to further systematize and quantify traditional construction knowledge through long-term performance monitoring and digital modeling of traditional architectural forms and structural techniques [9,12]. Translating indigenous design logic into computable parametric frameworks will enhance the applicability of traditional architectural wisdom in contemporary design [13], while promoting its transformation and application in sustainable urbanization and ecological adaptation practices across developing regions. This approach will provide innovative pathways for addressing real-world challenges such as sustainable building development and age-friendly design [14–17]. Simultaneously, the re-evaluation and revival of traditional architecture not

https://doi.org/10.3390/buildings16050930

only yield significant environmental benefits but also strengthen cultural identity, social cohesion, and community empowerment [18].

Secondly, research should deepen the adaptive reuse and resilience-oriented design of existing buildings—particularly medical, exhibition, and community facilities—under multiple scenarios including extreme weather, public health events, and demographic shifts. By integrating evacuation safety, indoor environmental quality, and flexible spatial organization early in design and renovation, while strengthening the application of multifunctional spaces and spatial permeability, combined with design concepts like modularity and adaptability, the overall resilience of buildings and their built environments in developing regions can be effectively enhanced [19].

Finally, the governance dimension of sustainable development—already well-illustrated in Building Information Modeling (BIM)-based project management research—urgently requires deep synergy between technological innovation and institutional mechanisms [20,21]. With the widespread adoption of digital tools, future research should further explore how BIM and related platforms can evolve from mere technical tools into critical enablers for fostering trust-building, collaborative governance, and cooperative decision-making [22–24].

Overall, this Special Issue emphasizes that sustainable urbanization in developing regions should not rely on a single disciplinary perspective or technological pathway. Instead, it requires integrating traditional ecological knowledge, adaptive building strategies, digital technologies, and governance innovations to form systemic solutions. It is hoped that the perspectives presented here will stimulate more interdisciplinary research and provide valuable references for context-sensitive, resilience-oriented, and culturally rooted practices within the global sustainable development process.

Funding: This research was funded by Fundamental Research Funds for the Central Universities, Southwest Minzu University (No. ZYN2026253), and Humanities and Social Sciences Research of the Ministry of Education of China (No. 25YJC760118).

Acknowledgments: We sincerely thank the authors who contributed to this Special Issue and submitted high-quality manuscripts. We also thank the reviewers for their responsible and rigorous feedback, which has greatly improved each manuscript. Finally, we sincerely thank the editorial team of *Buildings* for their professional support, which ensured the success of this Special Issue.

Conflicts of Interest: The authors declare no conflicts of interest.

List of Contributions:

1. Yi, B.; Nie, N.L.S. Effects of Contractual and Relational Governance on Project Performance: The Role of BIM Application Level. *Buildings* **2024**, *14*, 3185. https://doi.org/10.3390/buildings14103185.
2. Xiong, J.; Wu, J.; Cao, J.; Tian, Z.; Yang, Q. Energy-Efficient Architectural Design of a Banquet Hall with Integrated Tunnel Ventilation: Monitoring Performance During the Transitional Season in China. *Buildings* **2024**, *14*, 3335. https://doi.org/10.3390/buildings14113335.
3. Yi, C.; Han, W.; Liu, Y.; Lin, Y.; Qi, Y. Aging Adaptation Transition of Health Care Buildings for Accessibility Optimization for the Elderly. *Buildings* **2025**, *15*, 379. https://doi.org/10.3390/buildings15030379.
4. Zhang, M.; Shang, Z.; Luo, K.; Xie, K. New Insights into Traditional Construction Behind Sibe Dwellings with Swastika Kang for Space Heating in North China. *Buildings* **2025**, *15*, 795. https://doi.org/10.3390/buildings15050795.
5. Zhao, Y.; Li, K.; Han, M.; Xiong, J.; Zhang, Y. Natural Ventilation in Building Buffer Spaces of Traditional Qiang Dwellings: Field Study in Western China. *Buildings* **2025**, *15*, 794. https://doi.org/10.3390/buildings15050794.
6. Wen, Q.; Zhao, Y.; Huang, X.; Wang, G. A Digital Analysis of the "L"-Shaped Tujia Dwellings in Southeast Chongqing Based on Shape Grammar. *Buildings* **2025**, *15*, 900. https://doi.org/10.3390/buildings15060900.

7. Feng, W.; Meng, Y. Research on the Construction and Sustainable Development of Cave Dwellings in Mogou Village, Mengzhou City. *Buildings* **2025**, *15*, 1460. https://doi.org/10.3390/buildings15091460.
8. Wan, Z.; Yang, F.; Zhou, T.; Li, C. The Effects of Spatial Layout on Efficiency of Safe Evacuation After Conversion of an Exhibition Building to a Fangcang Shelter Hospital. *Buildings* **2025**, *15*, 2880. https://doi.org/10.3390/buildings15162880.
9. Zhang, M.; Yang, Y.; Yi, C.; Jia, M.; Zhang, M.; Yang, Q. Application of Supports Theory in Building Design: Multi-Dimensional Permeability and Spatial Structure in Versatile Community Centers. *Buildings* **2026**, *16*, 92. https://doi.org/10.3390/buildings16010092.
10. Yu, H.; Xiang, M.; Yang, Q.; Qi, Y.; Xiong, J.; Tang, Y.; Huang, X.; Yang, J.; Dong, X. Typological Identification and Revitalisation Strategies for Third Front Industrial Heritage: A Case Study of Guangyuan. *Buildings* **2026**, *16*, 446. https://doi.org/10.3390/buildings16020446.

References

1. Qiao, W.; Huang, X. Assessment the urbanization sustainability and its driving factors in Chinese urban agglomerations: An urban land expansion-Urban population dynamics perspective. *J. Clean. Prod.* **2024**, *449*, 141562. [CrossRef]
2. Sanagustín-Fons, V.; Stavrou, P.; Moseñe-Fierro, J.A.; Escario Sierra, F.; Castrolla, G.; Rocha, C.; Bazco Nogueras, E. Cultural Heritage Architecture and Climate Adaptation: A Socio-Environmental Analysis of Sustainable Building Techniques. *Land* **2025**, *14*, 1022. [CrossRef]
3. Akbar, N.; Abubakar, I.R.; Bouregh, A.S. Fostering Urban Sustainability through the Ecological Wisdom of Traditional Settlements. *Sustainability* **2020**, *12*, 10033. [CrossRef]
4. Philokyprou, M.; Michael, A. Environmental sustainability in the conservation of vernacular architecture. The case of rural and urban traditional settlements in Cyprus. *Int. J. Archit. Herit.* **2021**, *15*, 1741–1763. [CrossRef]
5. Luyao, H.; Ling, L.; Wenjun, Z. Winter Thermal Comfort in Zhuang Ethnic Stilt-Style Dwellings: A Case Study of the Longsheng Region with Field Measurement Validation. *Case Stud. Therm. Eng.* **2025**, *76*, 107412. [CrossRef]
6. Wu, S.; Chen, X.; Ma, C.; Wu, D.; Xu, Y.; Xiong, Y. Vertical Transportation and Age-Friendly Urban Renewal: A Systematic Framework for Sustainable and Inclusive Communities. *Sustainability* **2025**, *17*, 9594. [CrossRef]
7. Chen, Q.; Sun, Z.; Li, W. Effects of COVID-19 on Residential Planning and Design: A Scientometric Analysis. *Sustainability* **2023**, *15*, 2823. [CrossRef]
8. Cao, Y.; Yang, A. Spatial syntax analysis on the renewal of the external space of mining heritage buildings—A case study of Zollverein Coal Mine Industrial Complex. *Archit. Eng. Des. Manag.* **2025**, *21*, 136–153. [CrossRef]
9. Jo, H.H.; Yuk, H.; Kang, Y.; Kim, S. Conservation of architectural heritage: Innovative approaches to enhance thermal comfort and promote sustainable usage in historic buildings. *Case Stud. Therm. Eng.* **2023**, *51*, 103500. [CrossRef]
10. Hauashdh, A.; Nagapan, S.; Jailani, J.; Gamil, Y. An integrated framework for sustainable and efficient building maintenance operations aligning with climate change, SDGs, and emerging technology. *Results Eng.* **2024**, *21*, 101822. [CrossRef]
11. Abuhussain, M.A.; Waqar, A.; Khan, A.M.; Othman, I.; Alotaibi, B.S.; Althoey, F.; Abuhussain, M. Integrating Building Information Modeling (BIM) for optimal lifecycle management of complex structures. In *Structures*; Elsevier: Amsterdam, The Netherlands, 2024.
12. Hou, C.; Gao, W.; Gao, Y. Winter indoor thermal environment and livability analysis of traditional residential houses in northeast Sichuan, China. *Case Stud. Therm. Eng.* **2024**, *60*, 104636. [CrossRef]
13. Wu, Z.; Li, J.; Zhang, Y.; Zhang, Z.; Xu, G.; Zhao, H. Learning sustainable wisdom from the past: A generative framework for extracting climate-responsive design strategies from vernacular architecture. *Energy Build.* **2025**, *350*, 116669. [CrossRef]
14. Hsu, C.-H.; Wu, C.C.; Liao, K.-M. Fire evacuation strategies for traditional public markets in an aging society. *Case Stud. Therm. Eng.* **2025**, *71*, 106156. [CrossRef]
15. Monika, S.; Sundaram, A.M. Efficient and optimum design of native architecture–A means for sustainability. Case study of Residential units in Kottar, Kanyakumari. *Energy Build.* **2023**, *298*, 113586. [CrossRef]
16. Santos, M.M.; Ferreira, A.V.; Lanzinha, J.C. Sustainable vernacular architecture to improve thermal comfort in African countries. In *International Conference on Recovery, Maintenance and Rehabilitation of Buildings*; Springer: Berlin/Heidelberg, Germany, 2023.
17. Hassas, N. Investigation of the concept of sustainability of the nomad vernacular architecture, a case study of the Qashqai tribe in Iran. *Built Herit.* **2025**, *9*, 47. [CrossRef]
18. Jiménez Rios, A. Cob Vernacular Architecture Conservation and Its Contribution Towards a Sustainable Built Environment. In *Sustainable Materials for the Built Environment*; Springer: Berlin/Heidelberg, Germany, 2025; pp. 89–109.
19. Hamida, M.B.; Jylhä, T.; Remøy, H.; Gruis, V. Circular building adaptability and its determinants–A literature review. *Int. J. Build. Pathol. Adapt.* **2023**, *41*, 47–69. [CrossRef]

20. Croce, V.; Caroti, G.; Piemonte, A.; De Luca, L.; Véron, P. H-BIM and Artificial Intelligence: Classification of Architectural Heritage for Semi-Automatic Scan-to-BIM Reconstruction. *Sensors* **2023**, *23*, 2497. [CrossRef]
21. Panah, R.S.; Kioumarsi, M. Application of Building Information Modelling (BIM) in the Health Monitoring and Maintenance Process: A Systematic Review. *Sensors* **2021**, *21*, 837. [CrossRef]
22. Li, J.; Liu, Z.; Han, G.; Demian, P.; Osmani, M. The relationship between Artificial Intelligence (AI) and Building Information Modeling (BIM) technologies for sustainable building in the context of smart cities. *Sustainability* **2024**, *16*, 10848. [CrossRef]
23. Gutiérrez, V.; Theodoridis, E.; Mylonas, G.; Shi, F.; Adeel, U.; Diez, L.; Amaxilatis, D.; Choque, J.; Camprodom, G.; McCann, J.; et al. Co-Creating the Cities of the Future. *Sensors* **2016**, *16*, 1971. [CrossRef]
24. Lagsaiar, L.; Shahrour, I.; Aljer, A.; Soulhi, A. Modular Software Architecture for Local Smart Building Servers. *Sensors* **2021**, *21*, 5810. [CrossRef]

https://doi.org/10.3390/buildings16050930

buildings

MDPI

Article

Typological Identification and Revitalisation Strategies for Third Front Industrial Heritage: A Case Study of Guangyuan

Hongcheng Yu, Mingming Xiang *, Qianru Yang *, Yicong Qi, Jianwu Xiong, Yao Tang, Xinyi Huang, Jiefeng Yang and Xinyi Dong

School of Architecture, Southwest Minzu University, Chengdu 610225, China; yhcwyh999999999@sina.com (H.Y.); 80300254@swun.edu.cn (Y.Q.); 80300151@swun.edu.cn (J.X.); 23500011@swun.edu.cn (Y.T.); auzny@icloud.com (X.H.); 532930a276z.cdb@sina.cn (J.Y.); boketto@sina.com (X.D.)

* Correspondence: xiangmm@swun.edu.cn (M.X.); yangqianru@swun.edu.cn (Q.Y.); Tel.: +86-159-2854-4674 (M.X.); +86-156-8001-6301 (Q.Y.)

Abstract

The industrial heritage of the Third Front construction (hereafter referred to as Third Front industrial heritage) serves as a significant physical manifestation of China's urban society, economy, and culture during a unique historical period. Its widespread abandonment not only constitutes a waste of social resources but also accelerates the erosion of collective memory surrounding the Third Front initiative. As one of Sichuan Province's (including present-day Chongqing) key Third Front construction regions during that era, Guangyuan City possesses a substantial legacy of Third Front industrial heritage sites. These sites are predominantly idle and face ongoing risks of deterioration, necessitating comprehensive and systematic research into their classification, protection, and regeneration. This paper focuses on 39 Third Front industrial heritage sites in Guangyuan City, employing architectural typology to construct a 'type-medium-value' research framework integrating field research with strategic distribution analysis at the urban level, spatial form analysis at the settlement level, and spatial combination analysis at the building level to quantitatively identify and qualitatively deconstruct the spatial logic of these sites. This enables the analysis of the functional characteristics, structural logic, and spatial intent embodied by different types, thereby exploring the multidimensional value implications of Third Front industrial heritage through this value medium. Ultimately, this research proposes targeted adaptive mechanisms and revitalisation pathways for Third Front industrial heritage. It aims to promote the cultural legacy of this heritage and perpetuate the Third Front spirit within the context of strengthening the Chinese national community consciousness in the new era, while aligning with the Party and state's development strategies. This approach aims to provide a reference for revitalising and utilising Third Front industrial heritage in other underdeveloped regions.

Keywords: third front construction; industrial heritage; typology; type identification; revitalisation and utilisation

1. Introduction

1.1. Research Status

Academic inquiry into the Third Front construction has yielded substantial outcomes both domestically and internationally, encompassing fields such as history, sociology, humanities, economics, and architecture. Scholars across disciplines have actively engaged in this study, examining diverse sources including official documents, personal biographies, and memoirs. However, research specifically addressing the industrial heritage of the Third Front remains relatively scarce. Research into Third Front industrial heritage originated in 2006 when Chen Donglin of the Chinese Academy of Social Sciences first linked the concepts of 'Third Front Construction' and 'industrial heritage' in his article 'Third Front Construction: The Industrial Heritage Closest to Us'. Since then, research into the industrial heritage of Third Front construction has gradually gained prominence [1]. Domestic research on Third Front industrial heritage has primarily focused on foundational theory, conservation, renewal applications, and value assessment, with relatively little attention given to spatial morphology.

Regarding foundational theory, scholars such as Lü Jianchang define Third Front industrial heritage as the cultural remnants left behind by military–industrial and related supporting enterprises constructed in western China during the Third Front construction period. This encompasses both tangible forms (buildings, sites, products, archives, etc.) and intangible forms (strategic thinking, management wisdom, the 'Third Front spirit', collective memory, craft knowledge, etc.) [2,3]. Chen Donglin, through analysis of the scope, industrial sectors, and scientific achievements of the Third Front construction, emphasises the rich history, social value, and era-specific memory embodied by Third Front industrial heritage within China's industrial development history [1]. Zhang Fengqi focuses on the relationship between industry and urbanisation during the Third Front period, noting that Chongqing's industrial development outpaced its urban construction at the time. Under the influence of the planned economy system and administrative divisions, leading to diminished economic influence and an intensified urban–rural dual structure, serving as a significant historical lesson in the urbanisation process [4]. In summary, Third Front industrial heritage is defined as the cultural remains of military–industrial and supporting enterprises established in western China during the Third Front construction period, encompassing both tangible and intangible forms. It bears significant historical and social value within China's industrial and urbanisation history, while also reflecting the urban–rural developmental relationships and collective memory of that specific era.

In the field of conservation and renewal applications, Ding Xiaoshan, drawing upon social memory and place theory, explores the connection between the transformation of Third Front industrial heritage and the extraction of its spiritual essence. She proposes fostering a sense of place through the creation of natural and man-made environments, enabling visitors to immerse themselves, develop value recognition, and achieve the regeneration of Third Front memory [5]. Zhang Yuming examines Third Front industrial heritage in the Sichuan–Chongqing region, analysing its characteristics of extensive quantity, widespread distribution, and diverse current typologies. Integrating symbiotic philosophy, he proposes renewal strategies across cultural, urban, and spatio-temporal dimensions, demonstrating feasibility through the practical case of Mianyang Chaoyang Machinery Factory [6]. Zhang Pan introduced catalyst theory into the preservation of Third Front industrial heritage. By analysing the characteristics of heritage in the Mianyang region, he constructed an activation mechanism and selected typical cases for practical verification, achieving the integration of theory and application [7]. The scholars mentioned above have explored the theoretical construction and practical application of Third Front industrial heritage in conservation and regeneration through various approaches, such as the regeneration of social memory, symbiotic renewal, and catalytic revitalisation. Their work has advanced the field of heritage conservation from a focus on material preservation toward a multidimensional integration of spiritual inheritance and functional rejuvenation.

In the realm of value assessment, scholars such as Fu Yubing have referenced methodologies for evaluating industrial heritage value. By integrating the foundational context of

https://doi.org/10.3390/buildings16020446

typical Third Front cities in Sichuan Province, they have established a value assessment framework for Third Front industrial heritage, thereby providing a reference for subsequent research into the valuation of such heritage sites [8]. Scholars, including Zhao Li, have investigated assessment models for the conservation and development of industrial heritage architecture from the Third Front construction. They propose employing probabilistic processes to enhance reliability assessments of industrial heritage buildings, constructing an evaluation framework for the conservation and development of Third Front industrial heritage around triple indicators of economic, social, and environmental factors [9]; Liu Hanxi developed a value and risk assessment system for Third Front industrial heritage based on the Analytic Hierarchy Process (AHP) and incorporating risk factors. This system was applied to conduct comprehensive value and risk assessments for selected heritage sites in the former eastern Sichuan and northern Guizhou regions [10]. The aforementioned studies have systematically advanced research on the value identification and assessment of Third Front industrial heritage primarily by constructing a value evaluation framework, introducing probabilistic processes and multi-criteria indicators, and integrating methods such as the Analytic Hierarchy Process (AHP) and risk assessment. These approaches provide a quantitative foundation and decision-making reference for subsequent conservation and development efforts.

While research in the aforementioned areas is well-established, studies on spatial morphology lack systematic typological categorisation and quantitative analysis, often exhibiting considerable subjectivity. For instance, Li Tingting employed morphological and typological analysis to study Third Front industrial remnants in northern Sichuan but failed to summarise typologies of such heritage [11]; Wang Xinyi employed morphological typology to analyse the factory-city architecture of Puji Textile General Factory, categorising it into industrial, residential, and public building forms, yet failed to elucidate the relationship between spatial morphology and architectural typology [12]. Huang Zhihan utilised urban morphology to examine the built environment characteristics of Hubei's Third Front construction, outlining its built environment and spatial form, but her spatial morphological analysis lacked scientific rigour [13].

Simultaneously, as a distinct category within industrial heritage, research into Third Front industrial heritage fundamentally constitutes research into industrial heritage itself. Research on industrial heritage has reached considerable maturity, employing diverse perspectives, methodologies, and disciplines. For instance, scholars including Jing Lv pioneered a multi-level genealogical analysis framework at the provincial level. Utilising GIS spatial analysis techniques such as kernel density analysis and mean-centred analysis, they systematically traced and visually represented the evolutionary trajectory of industrial heritage, providing systematic theoretical underpinnings and practical references for regional industrial heritage conservation and renewal [14]. Through historical archival research and interdisciplinary collaboration, scholars including Judit Ramírez-Casas employed integrated geographic information technology and materials science methodologies to complete a systematic documentation and value assessment of this cement factory. This work provides a referenceable technical pathway and collaborative paradigm for the salvage research, archiving, and revitalisation of similarly 'forgotten' industrial heritage sites [15]. Scholars, including Zhuoran Jiang, employed multidimensional quantitative spatial syntax analysis to systematically reveal issues in spatial structural organisation, traffic flow patterns, and visual perception within such industrial estates. They subsequently proposed specific, actionable spatial optimisation strategies [16]. Hui Tao et al., framing their work within Lefebvre's theory of 'spatial production,' conceptualised industrial heritage as a dynamic field for the reproduction of cultural meaning. Employing semi-structured interviews and an 'author-text-reader' narrative analysis model, they sys-

https://doi.org/10.3390/buildings16020446

tematically deconstructed its narrative mechanisms, revealing how power, materiality, and identity interact to construct a dynamically evolving narrative ecology [17]. Contemporary industrial heritage research has matured into a multidisciplinary methodology. Approaches such as genealogical analysis, GIS quantification, spatial syntax, interdisciplinary archival construction, and spatial narrative theory provide crucial methodological frameworks for systematically revealing the spatial characteristics of Third Front industrial heritage and scientifically advancing its revitalisation and utilisation.

1.2. Research Background

The Third Front construction was a national defence economic initiative launched in the early 1960s to prepare for potential conflict [18]. Centred on the defence science and technology industry, it established a relatively comprehensive industrial system grounded in railways, coal, power, petroleum, steel, non-ferrous metals, and raw materials. The scale of its construction and the scope of its coverage were unprecedented in history [19]. Third Front construction projects were concentrated in the Sichuan Province (including present-day Chongqing), Guizhou, and Gansu regions. Among these, Guangyuan became an essential location for Third Front development due to its significant strategic position and unique geographical and natural conditions. Clusters of electronic industries, including radar, command instruments, and electronic components, were established in remote mountainous areas far from urban centres [20]. By the 1980s, as the Third Front construction entered a phase of transformation and restructuring, numerous Third Front enterprises (hereafter referred to as Third Front enterprises) declared bankruptcy or relocated after failed attempts at military-to-civilian conversion. This gradually left behind vast quantities of idle factory buildings, forming a distinctive industrial heritage of the era. However, current developmental lag and limited economic capacity render heritage preservation challenging. For Guangyuan's Third Front industrial heritage, which urgently requires revitalisation and utilisation, comprehensively and systematically understanding its intrinsic qualities and identifying its value constitutes the foremost task in its revitalisation.

2. Research Methods

2.1. Typological Research Approach

In studying Third Front industrial heritage, it is widely acknowledged in heritage studies that the subject of heritage research belongs to the past, its value is assessed in the present, and its purpose, according to Wang Lijun's interpretation, serves the future [21]. However, due to its unique historical context, the historical, social, and other values of Third Front industrial heritage are predominantly embodied in intangible forms, resulting in an incomplete understanding of its worth. This necessitates a critical inquiry: how should the material form of Third Front industrial heritage be valued? How should its significance evolve with shifting human understanding? The typology's holistic view of cities and architecture, its historical perspective blending tradition and modernity, and its ecological vision of urban–natural relationships [22] offer fresh insights for valuing Third Front industrial heritage. For instance, Rossi's architectural typology primarily elucidates the dialectical relationship between form and meaning (type and value): familiar forms or archetypes retain constancy, yet designers may imbue these fixed forms with new significance. Thus, form acquires fresh meaning through adaptation to context, while context and object transform with the addition of new meaning [22]. Applying Rossi's typology allows the abstraction of 'prototypes' from historical and regional model forms, whose formation hinges on 'function,' 'image,' and 'structure.' Drawing on Rossi's typology, prototypes are abstracted based on the following: function primarily concerns utilitarian and social functions; imagery relates to spatial value and meaning; and structure principally involves

https://doi.org/10.3390/buildings16020446

cognitive schemata of spatial and existential awareness, spatial syntactic organisation, and textual spatial composition [22].

2.2. Quantitative Research Methods for Spatial Patterns

2.2.1. Average Nearest Neighbour (ANN) Analysis

The nearest neighbour index primarily analyses the spatial distribution characteristics of heritage sites—whether randomly dispersed or clustered—by comparing the average distance between the centre of a heritage site and its nearest neighbouring site against the expected average distance under a hypothetical random distribution. The calculation is expressed as follows:

$$\mathrm{ANN} = \frac{\overline{D}_o}{\overline{D}_e} = \frac{\sum_{i=1}^{n} d_i/n}{\sqrt{n/A}\,/2} = \frac{2\sqrt{\lambda}}{n}\sum_{i=1}^{n} d_i \tag{1}$$

where D_o denotes the observed mean distance between each heritage site and its neighbouring sites; D_e represents the expected mean distance of heritage sites under a hypothetical random pattern; n is the total number of heritage sites; d denotes distance; and A is the area of the minimum bounding rectangle encompassing heritage sites within the study area [23]. ANN < 1 indicates a clustered distribution pattern; ANN > 1 suggests a tendency towards random distribution.

2.2.2. Shape Ratio (PAR)

The shape ratio represents the ratio of the perimeter of a closed boundary to the perimeter of a circle of equal area. It reflects the deviation of settlement morphology from an equi-area circle and indicates the smoothness of the site boundary, serving as a key indicator for assessing the complexity of site boundaries (Figure 1) [24]. Its calculation formula is as follows:

$$\mathrm{PAR} = \frac{L}{2\sqrt{\pi S}} \tag{2}$$

where L denotes the perimeter of the old site boundary (m) and S denotes the area of the old site boundary (m^2). Simultaneously, settlements with a PAR value exceeding 2 exhibit linear development, whereas those with a PAR value below 2 indicate clustered development [25].

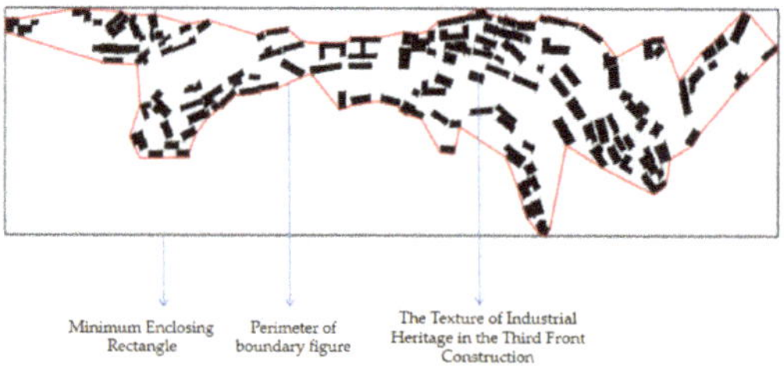

Figure 1. Schematic diagram of quantitative metrics. (Source: Drawn by the author).

2.2.3. Terrain Position Index (TPI)

The Terrain Position Index calculates the difference between a point's elevation and the average elevation within a specified surrounding area, using the following formula:

$$\mathrm{TPI} = z_i - \overline{z} \tag{3}$$

$$\overline{z} = \frac{1}{n_R}\sum_{i\in R} z_i \tag{4}$$

 https://doi.org/10.3390/buildings16020446

where z_i denotes the elevation of the centre point of the study area, while z represents the average elevation of the surrounding area.

2.2.4. Spatial Syntax

Spatial syntax constitutes a series of theories and techniques concerning space and society. It abstracts the interconnections between spaces into a network diagram, then employs fundamental principles of graph theory to conduct topological analyses of the spatial accessibility of axes or features. This ultimately yields a set of morphological analysis variables. Among these, the depth value reflects whether a particular area has been deliberately designed for low accessibility; integration indicates the degree to which a space within a system clusters with or disperses from other spaces; and traversability denotes the likelihood that a space within a system is traversed by other shortest paths.

2.3. The Holistic Research Framework of 'Type-Medium-Value'

Typological research is inherently qualitative. To enhance scientific rigour, this study adopts quantitative methodologies employed in analysing traditional village spatial forms. By incorporating nearest neighbour analysis, morphological indices, topographical position indices, and spatial syntax, a multi-tiered scientific classification system for Third Front industrial heritage is established. This approach facilitates a paradigm shift from qualitative description to quantitative analysis, effectively overcoming the limitations of subjectivity and ambiguity inherent in traditional typologies.

Secondly, the spatial form of third-tier enterprises is shaped to varying degrees by 'function', 'structure', and 'intent', while simultaneously exerting a reciprocal influence on these elements. This dynamic interaction manifests distinct value connotations, rendering 'function, imagery and structure' more pronounced. Consequently, this study employs a combined qualitative and quantitative methodology to construct a 'type-medium-value' analytical framework (Figure 2). This framework first draws upon the methodology of industrial archaeology to systematically identify and spatially quantify the material form of heritage, transforming complex spatial types into analysable data systems. Subsequently, it incorporates the 'medium' perspective from Lefebvre's theory of 'spatial production,' using 'function,' 'image,' and 'structure' as dynamic hubs connecting material 'type' with abstract 'value.' This reveals how material space bears and continuously produces multiple values, including specific social meanings and historical significance. Ultimately, the research focuses on revitalisation practices. By precisely identifying the value mediators dominant in different heritage types, it provides theoretical grounding for formulating differentiated conservation and regeneration strategies. This shifts heritage protection from universal principles towards value-logic-based precision decision-making.

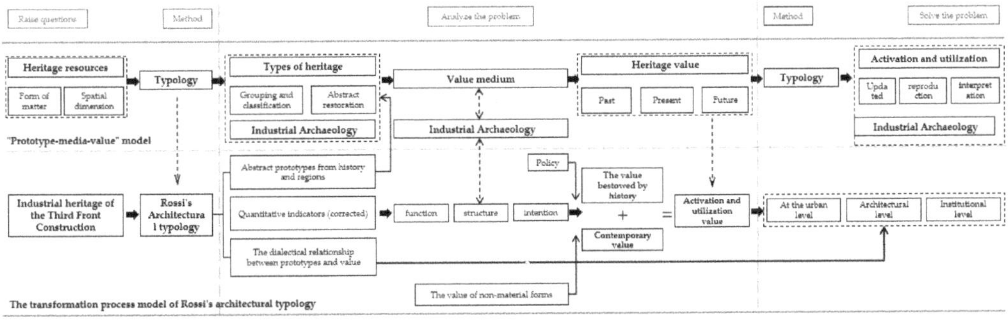

Figure 2. Research framework for Third Front industrial heritage from a typological perspective. (Source: Drawn by the author).

https://doi.org/10.3390/buildings16020446

2.4. Data Source

The scope of information collection for the research subjects was determined based on the spatial dimensions examined in this study, primarily encompassing the following: information on Guangyuan City's Third Front industrial heritage, chiefly derived from field investigations; secondly, documentary research covering the specific locations, scale, and functional uses of Third Front industrial heritage sites. Latitude and longitude coordinates were obtained through field surveys supplemented by Baidu Picker [https://lbs.baidu.com/maptool/getpoint (accessed on 10 November 2024)], while scale and architectural texture data were derived from rough measurements using Global Mapper GM23 and AutoCAD2014 drafting. The digital elevation model data for Guangyuan City was sourced from the Chinese Academy of Sciences Geospatial Data Cloud Platform [https://www.gscloud.cn/search (accessed on 12 March 2025)].

3. Results

3.1. Analysis of Urban-Level Types

During the Third Front construction period, factories adhered to the policy guidelines of 'backing onto mountains, dispersal, and concealment' in their deployment, forming a construction layout characterised by 'large-scale dispersal and small-scale concentration'. The 'extensive dispersion' aimed to mitigate the risk of concentrated destruction, embodying an 'anti-agglomeration' strategy. For instance, the Second Automobile Manufacturing Plant in Shiyan, Hubei, centred its main assembly plant with three subsidiary clusters distributed around it at intervals of approximately 5 km [26]. Such Third Front enterprises featured dispersed layouts between main plants and supporting sub-plants, forming a 'one factory, multiple sites' distribution structure. While all Third Front enterprises followed identical construction principles for wartime preparedness, regional variations led to distinct layouts. In Guangyuan's mountainous terrain with limited flat land, this resulted in a concentrated distribution pattern.

This study employs ArcGIS 10.6 software to calculate the ANN values of Guangyuan's Third Front industrial heritage sites, thereby revealing their wartime distribution characteristics. Given that the statistical significance of average nearest neighbour analysis is strongly influenced by the size of the study area, this paper employs two different area definitions—the minimum bounding rectangle (Table 1) and the administrative boundary (Table 2)—as the study area to compute ANN values, thereby conducting a sensitivity assessment for the ANN analysis in Guangyuan (Cite from ANN Analysis Report in Supplementary Materials).

Comparative analysis using the two different study area definitions reveals that at the municipal level (Guangyuan City), the changes in Z-score and p-value are minimal, indicating that these metrics are insensitive to area changes at this scale. Conversely, at the county/district level, both Z-score and p-value show variations, indicating that these metrics are sensitive to area changes. Notably, Jiange County and Qingchuan County exhibit substantial changes in both Z-score and p-value, signifying that these metrics are extremely sensitive to area changes in these two counties.

Simultaneously, using the administrative boundary area as the study area yields ANN values generally less than 1 across most regions (Table 2), which would imply a clustered pattern inconsistent with the 'mountainous, dispersed, and concealed' principle of the Third Front construction. Analysis suggests two primary reasons for this discrepancy: the heritage sites in the current study do not represent the complete original deployment, as the locations of some critical enterprises remain classified, and other known sites have been completely lost due to transformation or relocation; therefore, analysing the spatial dispersion of the currently surviving sites using vast administrative areas introduces

https://doi.org/10.3390/buildings16020446

significant error. It must be acknowledged that for Jiange County and Qingchuan County, with only two and three known sites, respectively, results will inherently contain a degree of error regardless of whether the administrative boundary or the MBR is used as the area definition.

Table 1. Nearest neighbour analysis of Third Front industrial heritage in Guangyuan City. The study area is defined as the minimum bounding rectangle. (Data source: Field survey and calculation by the authors).

District/ County	Number of Points	Study Area (km^2)	Mean Expected Distance (km)	Mean Observed Distance (km)	ANN Ratio	Z-Score	*p*-Value	Type
Guangyuan Municipality	39	10,382.9	8.158	3.632	0.445	−6.626934	0.000000	Clustered
Lizhou District	22	405.28	2.146	1.814	0.845	−1.384200	0.166297	Clustered
Wangcang County	11	233.237	2.302	2.334	1.01	0.088211	0.929709	dispersed
Jiange County	3	2.604	0.465	10.671	22.9	72.591202	0.000000	dispersed
Qingchuan County	2	0.037929	0.068	18.965	275	742.473941	0.000000	dispersed
Zhaohua District	1	/	/	/	/	/	/	/

Table 2. Nearest neighbour analysis of Third Front industrial heritage in Guangyuan City. The study area is defined by the administrative boundary. (Data source: Collected and compiled by the author).

District/ County	Number of Points	Study Area (km^2)	Mean Expected Distance (km)	Mean Observed Distance (km)	ANN Ratio	Z-Score	*p*-Value	Type
Guangyuan Municipality	39	163,312.86	10.232	3.632	0.354948	−7.706510	0.000000	Clustered
Lizhou District	22	1533.24	4.174	1.816	0.434966	−5.070103	0.000000	Clustered
Wangcang County	11	2996.11	8.252	2.326	0.281892	−4.556347	0.000000	Clustered
Jiange County	3	3203.83	16.340	10.676	0.653350	−1.148636	0.250706	Clustered
Qingchuan County	2	3219.26	20.060	18.972	0.945783	−0.146683	0.883382	Clustered
Zhaohua District	1	/	/	/	/	/	/	/

To account for the non-contiguous distribution caused by the 'concealment' policy, the minimum bounding rectangle (MBR) of heritage clusters was utilised as the reference area, rather than administrative boundaries. This adjustment corrects for the ecological fallacy where vast administrative areas dilute the density of specific strategic clusters (Table 1).

In summary, to enhance the scientific rigour and representativeness of the analysis, this study adopts the area of the smallest bounding rectangle encompassing heritage sites within each district or county as the study area when assessing the spatial dispersion of

https://doi.org/10.3390/buildings16020446

heritage. This approach enables an examination of its morphological characteristics at a spatial scale more closely aligned with the actual distribution of heritage. Measurement indicates an average nearest neighbour index of 0.445 at the municipal level, classifying the distribution as clustered. Furthermore, the Z-value of −6.627 indicates only a 1% or lower probability that this clustering pattern resulted from random processes. At the county level, Lizhou District exhibited an average nearest neighbour index of 0.845, indicating a clustered distribution pattern. Conversely, Qingchuan County, Wangcang County, and Jiange County recorded average nearest neighbour indices of 0.845, 1.01, and 22.9, respectively, all reflecting dispersed patterns (Table 1).

It should be clarified that although the ANN values for Jiange and Qingchuan counties appear elevated, this does not stem from computational error. Rather, it arises from the extremely limited number of heritage sites within these counties, resulting in relatively small minimum bounding rectangle areas. These results should be interpreted as exploratory analysis. This outcome precisely reflects the objective constraints imposed on studying extant heritage by the Third Front's stringent secrecy requirements and complex historical transformations. While subject to certain limitations, this methodology nevertheless enables the analysis of known remnants to reveal structural characteristics of spatial distribution through the principle of 'seeing the big picture through the small'.

The findings reveal that at the municipal level, Guangyuan's Third Front industrial heritage exhibits clustered distribution patterns. While seemingly at odds with the 'dispersed' construction policy, this merely represents a difference in perspective. As the industrial heritage sites currently under study do not constitute the full deployment of the Third Front construction era, the specific reasons for which have been outlined above. Therefore, while the clustered distribution of extant Third Front construction industrial heritage within Guangyuan's municipal boundaries is understandable, it also indicates that typological analysis at the urban level should focus on the district and county level. At the district and county level, this reflects a locally adapted layout formed by integrating the topographical conditions and production collaboration requirements, while adhering to strategic principles. With the exception of Lizhou District, heritage sites across other districts and counties exhibit dispersed characteristics, widely distributed among mountain valleys. To balance production coordination and strategic security, each factory complex achieved relative internal concentration within an overall dispersed framework. This ultimately crystallised into two typical wartime operational models: the 'one factory, multiple sites' type for cross-valley collaboration and the 'one factory, one site' type for concentrated production.

Among these, Third Front enterprises adopting the 'one factory, multiple sites' type dispersed and concealed their facilities within different mountain valleys to mitigate the risk of military strikes, maintaining a certain distance between sites. Simultaneously, to ensure production coordination and component assembly, the sites required relative proximity to one another. Taking Base 081 as an example, within its structure of 'eight factories, one warehouse, one institute, and one school', the main plant—Chuanbei Electronics Industrial Company—was responsible for research and development. The remaining factories were dispersed across the mountain valleys of Lizhou District, undertaking functions such as radar assembly, component production, medical services, and technical personnel training, thereby forming a complete system of collaborative production (Figure 3).

The 'one factory, multiple sites' type prioritises security considerations above production efficiency. Conversely, the 'one factory, one site' type approach represents two extremes in balancing production efficiency and security: one exhibits extremely high secrecy, prioritising security above all else, with subsidiary plants deployed solely for efficiency; the other exhibits lower secrecy, prioritising production efficiency first and security secondarily.

https://doi.org/10.3390/buildings16020446

For instance, the Wuzhou Chemical Plant, operating under the highest security protocols, is deployed as a standalone facility. Its subsidiary plants are distributed in a linear formation along mountain valleys, maintaining close coordination to collectively undertake fuel production and processing. Conversely, the Zhaohua Paper Mill, operating under lower security protocols, is centrally located in a relatively flat terrain, backed by forested hills. This allows for the local sourcing of materials, with the various mills collaborating to produce paper.

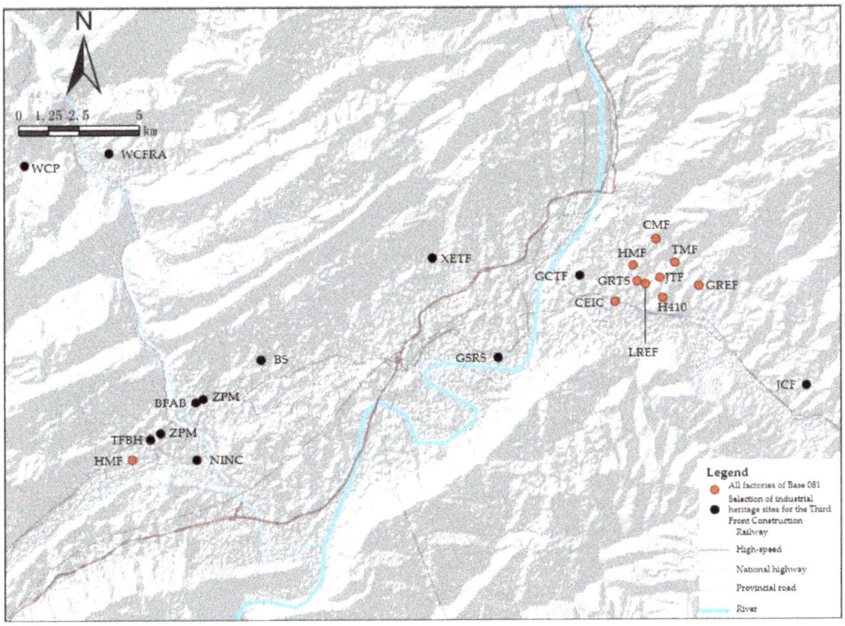

Figure 3. Dispersed distribution structure of the 'one factory, multiple sites' type (Base 081) for war preparedness. (Source: Drawn by the author).

3.2. Settlement-Level Typological Analysis

The deployment of Third Front enterprises was influenced not only by construction policies but also by the integration of production processes with topographical features. To ensure production continuity and minimise environmental impact, distinct layout patterns emerged within different mountain valleys. For instance, Hubei Huaguang Equipment Factory and Hongwei Machinery Factory adopted a sequential valley-aligned configuration, while the China Aero-Engine Corporation Sichuan Gas Turbine Research Institute's High-Altitude Simulation Test Base in Jiangyou City, Sichuan, adopted a dispersed layout within a valley. Enterprises in Guangyuan's Third Front initiative predominantly served defence and military industries, where specialised production processes combined with complex, varied topography generated more intricate spatial configurations.

This study utilised AutoCAD to compile foundational morphological index data. GIS was employed to calculate Terrain Position Indices for heritage sites, enhancing spatial precision at the meso-scale. The average elevation of areas within a 2 km radius of each heritage site's central point was computed. Finally, all calculation results were standardised using Z-scores. This process generated a scatter plot illustrating the morphological typology of heritage sites (Figure 4).

 https://doi.org/10.3390/buildings16020446

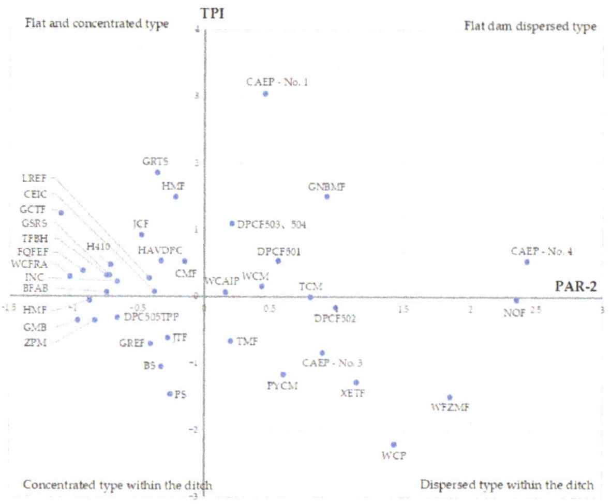

Figure 4. Scatter plot of morphological types of heritage sites. (Source: Drawn by the author).

Flat-plain dispersed production workshops are interconnected and arranged linearly along rivers and roads, such as Guangyuan City's First Building Materials Factory. Concentrated flat-plain layout consists of production workshops that are centrally arranged for coordinated operations, e.g., Honglun Machinery Factory is situated on the flat plains of Baolun Town, with its various production workshops operating in concert to carry out specialised, fixed-location manufacturing of electronic equipment vehicles. Concentrated gully layout consists of workshops that are densely arranged to minimise mutual interference, e.g., Guangming Radio Equipment Factory, nestled amidst mountains on all sides, engages in the production of highly confidential servo systems, terminal display systems, and precision machining. Gully dispersed layout comprises the following two types: Single-linear serial gully arrangement consisting of smaller-scale workshops with minimal interconnectivity and low transport volume, e.g., Donghe 502 Factory, concealed within a mountain valley along the Donghe River in Wangcang County, independently completed the printing of banknotes. And larger-scale, closely interconnected workshops with substantial transport volumes that adopt a multi-linear layout in parallel along the gully, exemplified by Wuzhou Chemical Works with their independently arranged yet closely interconnected sub-facilities that carried out the production and processing of fuel within the mountainous valley of Sandui Town (Table 3).

It should be noted that this study employs a circular area with a radius of 2 km, centred on the heritage site's focal point, to calculate the average elevation of its surrounding region. The 2 km radius was determined through iterative sensitivity testing based on the study area's actual conditions and multiple trials. Guangyuan's topography is complex, predominantly hilly and mountainous. Employing a uniform average elevation within administrative boundaries would render the Terrain Position Index (TPI) ineffective for distinguishing between different heritage sites. Conversely, arbitrarily defining the scope would fail to objectively reflect the actual terrain conditions. Therefore, this study comprehensively considered the spatial scale characteristics of the 39 heritage samples (with maximum long sides ranging from 3459 m to 110 m, as detailed in Appendix A.1) and the surrounding topographical undulations. Circular areas with radii of 0.5 km, 1 km, 2 km, 2.5 km, and 3 km were tested, with their average elevations calculated, respectively. The results indicate (Appendix A.3) that for the majority of heritage sites, the average elevation increases with expanding radius, consistent with Guangyuan's predominantly

 https://doi.org/10.3390/buildings16020446

mountainous and hilly terrain. Simultaneously, the 2-kilometre radius circular area fully encompasses the longest boundaries of all heritage sites while adequately representing the surrounding topographical features. It thus possesses favourable spatial representativeness and computational stability, leading to its final selection as the analytical unit. This choice represents an empirical calibration developed through multi-scale comparisons, aimed at enhancing the scientific rigour and explanatory power of topographical factors within heritage environmental analysis (Cite from TPI Analysis in Supplementary Materials).

Table 3. Spatial structure of the Third Front industrial heritage in Guangyuan City. (Source: Compiled and drawn by the author).

Spatial Type	Archetypal Representative
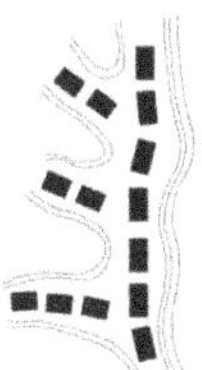 Flat-area dispersed	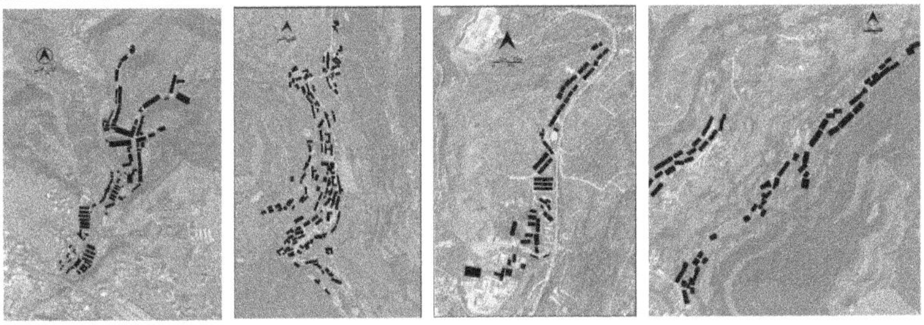 Donghe Printing Plant No. 503 and No. 504; Institute No. 1 of the China Academy of Engineering Physics; Guangyuan First Building Materials Factory; Institute No. 4 of the China Academy of Engineering Physics
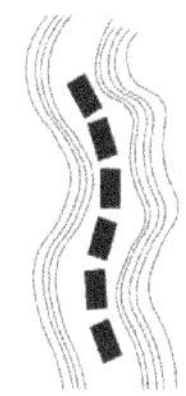 Gully dispersed Gully dispersed	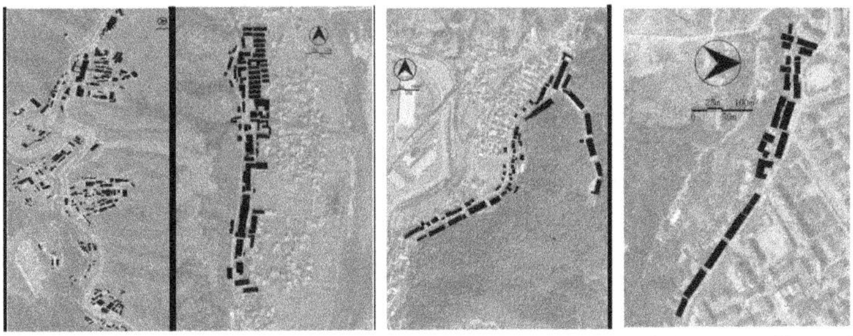 Wuzhou Chemical Works; Donghe Printing Plant 502; Wanzhong Machinery Works; Xinguang Electronics Factory
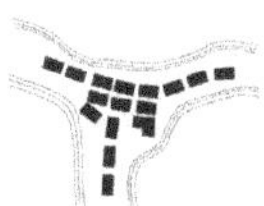 Gully centralised Gully centralised	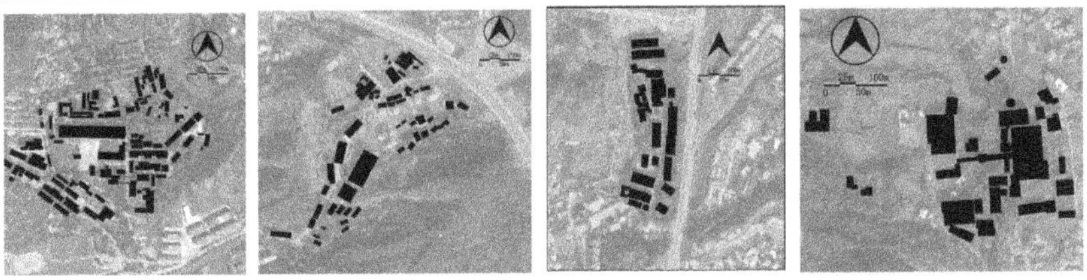 Zhaohua Paper Mill; Guangming Radio Equipment Factory; Jiangling Cable Factory; Donghe Printed Circuit Factory 505
 Flat-area centralised	 Jianping Tool Factory; China Water Resources and Hydropower Engineering Bureau No. 5; Honglun Machinery Factory; Guangyuan South Railway Station

https://doi.org/10.3390/buildings16020446

3.3. Typological Analysis at the Architectural Level

The Third Front enterprises in Guangyuan possess self-contained social functions, with buildings primarily comprising production and living facilities. Influenced by industrial requirements and complex topography, their spatial arrangements exhibit diverse characteristics (Figure 5). Firstly, the architectural plan layout exhibits rich textures, encompassing various forms such as linear, 'L'-shaped, 'O'-shaped, 'H'-shaped, and 'U'-shaped configurations, each corresponding to distinct functions. For instance, L-shaped and enclosed layouts predominantly serve as office buildings, while O-shaped structures are primarily water towers and chimneys. Secondly, street and lane spaces are categorised into internal lanes and boundary roads. Internal lanes follow a 'building-street-building' sequence, exhibiting strong enclosure reflecting process flow influences. Boundary roads follow a 'building-street-annex' sequence, flanked by buildings on one side and natural elements on the other, balancing stability with openness while serving composite functions of transport, production, and culture. Both types form a grid-like street system through spatial continuity. Finally, ancillary spaces encompass farmland, water features, vegetable plots, and plazas. Within Third Front enterprises, these spaces predominantly feature small-scale, diverse forms, while peripheral areas exhibit greater extensibility, forming an interlocking texture. Surrounding farmland spreads out in a parallel arrangement, imbuing the entire landscape with an agrarian, rural character.

(**a**) (**b**)

Figure 5. Spatial arrangement of buildings at Guangyuan Third Front enterprises. Image source: Photographs by the authors. (**a**) Partial spatial configuration diagram for Plants 503 and 504. (**b**) Partial spatial configuration diagram for Plant 502.

The architectural spaces, connecting spaces, and ancillary spaces of Third Front enterprises exhibit diverse forms while maintaining close interconnections. Given the intimate relationship between these three spatial categories, might their boundaries dissolve as scale increases, yielding interwoven and integrated configurations that form higher-level composite spatial arrangements?

Based on the preliminary cluster analysis, three representative samples were purposively selected to correspond to the identified morphological clusters (e.g., concentrated-valley, dispersed-plain). Criteria for selection included spatial integrity, data availability, and typological distinctiveness. Three representative cases were purposively selected based on the cluster analysis results to exemplify the following distinct morphological categories: (1) integrated (e.g., Factory 502), (2) dispersed (e.g., Factory 501), and (3) mixed. This selection ensures typological coverage across the identified spectrum. Employing depth map X to generate depth value and integration analysis diagrams (Figure 6). For Donghe Printing Plant 501, the analysis commences from the office zone, while for Donghe Printing Plant 502, both office and production zones serve as starting points for comparative depth value analysis.

 https://doi.org/10.3390/buildings16020446

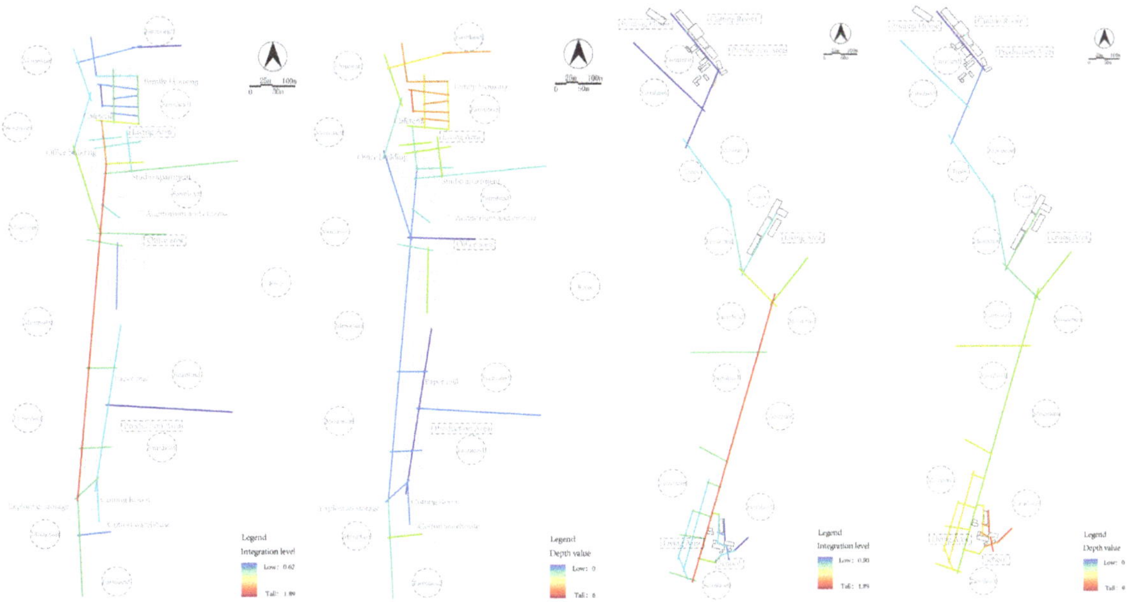

(a) (b) (c) (d)

Figure 6. Comparative analysis of depth and integration values in representative cases. (Source: Drawn by the author). (**a**) Integration level of Plant 502. (**b**) Depth value of Plant 502. (**c**) Integration Level of plant 501. (**d**) Depth value of Plant 501.

The results indicate (Table 4) that accessibility between the residential quarters and production or administrative areas at Plant 502 is relatively low, suggesting a high level of confidentiality within the facility, coupled with a high degree of integration and compact layout between zones. Conversely, accessibility between the production, residential, and administrative areas at Donghe Printing Plant 501 is also low, yet these zones exhibit lower integration and a more dispersed layout. Consequently, Donghe Printing Plant 501 presents a 'dispersed factory-compound model' with dispersed management across zones; Donghe Printing Plant 502 exhibits a 'unified management model' with integrated management across all zones.

Table 4. Statistics of space syntax data for Factory 501 and Factory 502. (Source: Compiled by the author).

	Factory 501 Living Quarters	**Factory 501 Production Zone**	**Factory 502 Living Quarters**	**Factory 502 Administrative Zone**	**Factory 502 Production Zone**
Integration (min)	0.59	0.50	0.85	0.82	0.62
Integration (max)	1.89	0.62	1.89	1.89	1.89
Clustering density	spatially dispersed			spatially integrated	
Depth (is origin?)	no	yes	no	yes	yes
Depth (min)	4	0	1	0	0
Depth (max)	9	2	6	3	3
Accessibility from the Living Quarters to This Area	/	low	/	low	low

https://doi.org/10.3390/buildings16020446

Moreover, Third Front enterprises exhibit a typical integrated industrial–agricultural model, where industrial production and farmland cultivation are geographically contiguous and functionally complementary, collectively forming self-sufficient community units. Analysis of the Zhaohua Paper Mill reveals (Table 5) that the connecting spaces between residential and industrial zones exhibit high integration and concentrated layout; depth analysis originating from farmland indicates high accessibility from residential and office areas to farmland, while accessibility from production zones to farmland is low; and choice analysis further indicates that residential zones and the plaza spaces within these zones exhibit significant potential for traversal (Figure 7). Consequently, the aforementioned analysis permits the identification of three spatial patterns for the Third Front industrial heritage: courtyard–field combinations, factory–field combinations (as exemplified by Factory 502), and street–factory combinations.

Table 5. Statistical table of space syntax analysis for the Zhaohua Paper Mill. (Source: Compiled by the author).

	Living Quarters	Production Zone	Administrative Zone	Farmland	Square
Integration (min)	0.41	0.76	0.56	/	/
Integration (max)	1.32	1.32	1.22	/	/
Clustering Density		spatially integrated		/	/
Depth (Is Origin?)	no	no	no	yes	no
Depth (min)	0	2	2	0	/
Depth (max)	3	5	5	0	/
Zones to Farmland	tall	low	low	/	/
Choice (min)	430	0	0	0	567
Choice (max)	795	694	694	306	795
Integrated Choice Potential	tall	low	low	low	tall

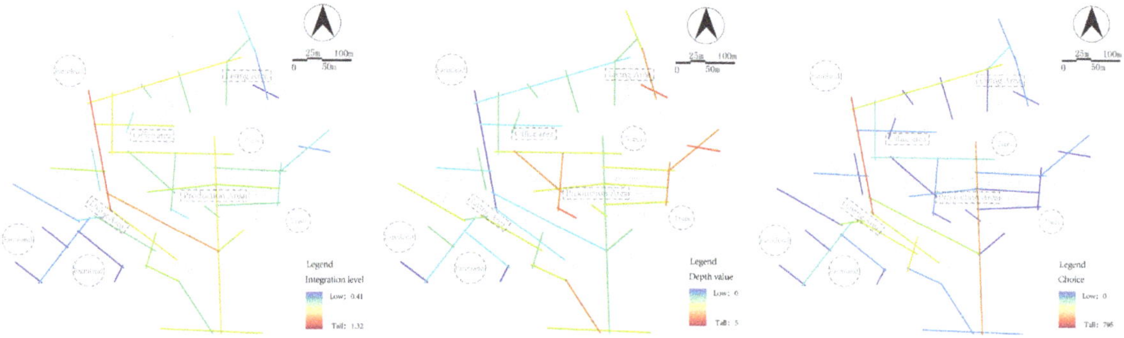

(a) (b) (c)

Figure 7. Comparative analysis of depth, integration, and intelligibility values in representative cases. (Source: Drawn by the author). (**a**) Integration level of Zhaohua Paper Mill. (**b**) Depth value of Zhaohua Paper Mill. (**c**) Penetration level of Zhaohua Paper Mill.

Based on the above analysis, the spatial morphology of Third Front enterprises at the architectural level transcends typological boundaries through scaled elevation, inter-

 https://doi.org/10.3390/buildings16020446

weaving, and integrating the foundational 'architectural space', 'connective space', and 'ancillary space' (Cite from Space Syntax in Supplementary Materials). This evolution manifests as higher-order composite spatial configurations, encompassing the following five combinations (Figure 7):

(1) 'Integrated Factory-Courtyard' Type: Suitable for facilities with moderate security requirements and ample land, typically utilising natural topography for concentrated layout under unified military-style management. For instance, Guangming Radio Equipment Factory occupies a low-lying mountainous area, using the hillside as a boundary and tunnels as access points, with living and production functions centrally arranged.
(2) 'Decentralised Factory-Courtyard' Type: Common in enterprises with stringent confidentiality requirements. Production zones requiring flat terrain are situated in gently sloping valley areas, while residential quarters are dispersed across nearby hillsides to minimise accessibility between the two. For instance, the residential area of Wuzhou Chemical Factory is concealed in a valley west of Sandui Town, with the production zone located on the eastern hillside, approximately four kilometres apart.
(3) 'Courtyard-Field Combination' Type: A typical arrangement for self-sufficiency, where living quarters adjoin farmland to facilitate agricultural labour by staff. For instance, the residential area of the First Institute of the Ninth Academy is surrounded by cultivated fields, which provided crucial logistical support for research during periods of material scarcity.
(4) 'Factory-Field Combination' Type: Also serving self-sufficiency objectives, this configuration prioritises production zones due to limited flat land, resulting in direct spatial integration between production areas and farmland. Such arrangements are relatively uncommon.
(5) 'Street-Square Combination' Type: As a prevalent composite form, this layout addresses diverse employee needs by arranging public spaces like squares along highly accessible streets, creating clusters of communal functions that effectively foster social interaction and leisure activities.

3.4. Typological Atlas of Third Front Industrial Heritage in Guangyuan City

Summarising the above analysis reveals the following: at the urban level, ANN analysis indicates that Third Front industrial heritage sites at the district/county level exhibit a dispersed distribution pattern, specifically categorised as 'one factory, multiple sites' type and 'one factory, one site' type; at the settlement level, combining morphological index analysis and topographical location index, the spatial form of Third Front construction industrial heritage is classified into 'flat-area dispersed', 'gully dispersed', 'gully centralised', and 'flat-area centralised'; at the building level, spatial syntax analysis was employed to examine the spatial configurations of former sites. Through case studies, Guangyuan's Third Front industrial heritage was categorised into five spatial patterns: 'integrated factory-courtyard' type, 'decentralised factory-courtyard' type, 'courtyard-field combination' type and 'factory-field combination' type, and 'street-square combination' type.

As a critical period embodying the nation's collective memory, the Third Front construction left profound imprints across the vast expanse of the land. Guangyuan's Third Front journey, a distinctive chapter within this narrative, crystallises both the collective will of its era and a unique regional adaptability. Its spatial form emerged as a complex product of strategic imperatives, geographical constraints, and practical adaptation.

https://doi.org/10.3390/buildings16020446

To systematically decipher this 'spatial text,' this study deconstructs various spatial elements, mapping a typology that spans from macro to micro scales (Figure 8). This typology is structured through a linguistic metaphor: establishing the tone with 'a Network' of unfolding layers via 'four Systems' enriches its texture with 'five Clusters,' and ultimately rests upon a solid foundation of 'Combinations and Elements,' transforming the physical space into a readable textual system.

This diagram serves not merely as a descriptive tool but as an interpretative methodology. Through its grammatical structure of 'Networks-Systems-Clusters-Combinations-Elements', it reveals how the Third Front architecture responds to topography, organises functions, and mediates transitions between public and private spaces. This approach offers fresh perspectives for understanding heritage value while establishing a theoretical foundation for conservation and renewal, enabling the spatial forms to convey their historical significance.

Urban-level spatial typology (localised: Lizhou District)

Networks

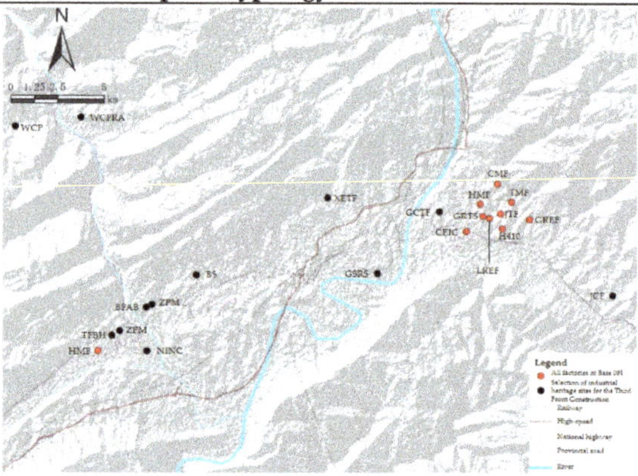

'one factory, multiple sites' type and 'one factory, one site' type

Spatial Types at the Settlement Level

Systems

Flat-area dispersed

Gully dispersed

Flat-area centralised

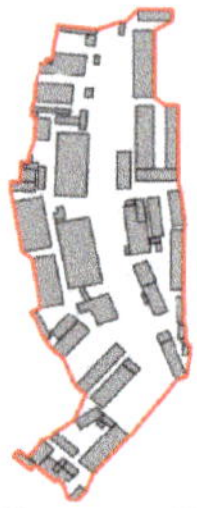

Gully centralised

Figure 8. *Cont.*

https://doi.org/10.3390/buildings16020446

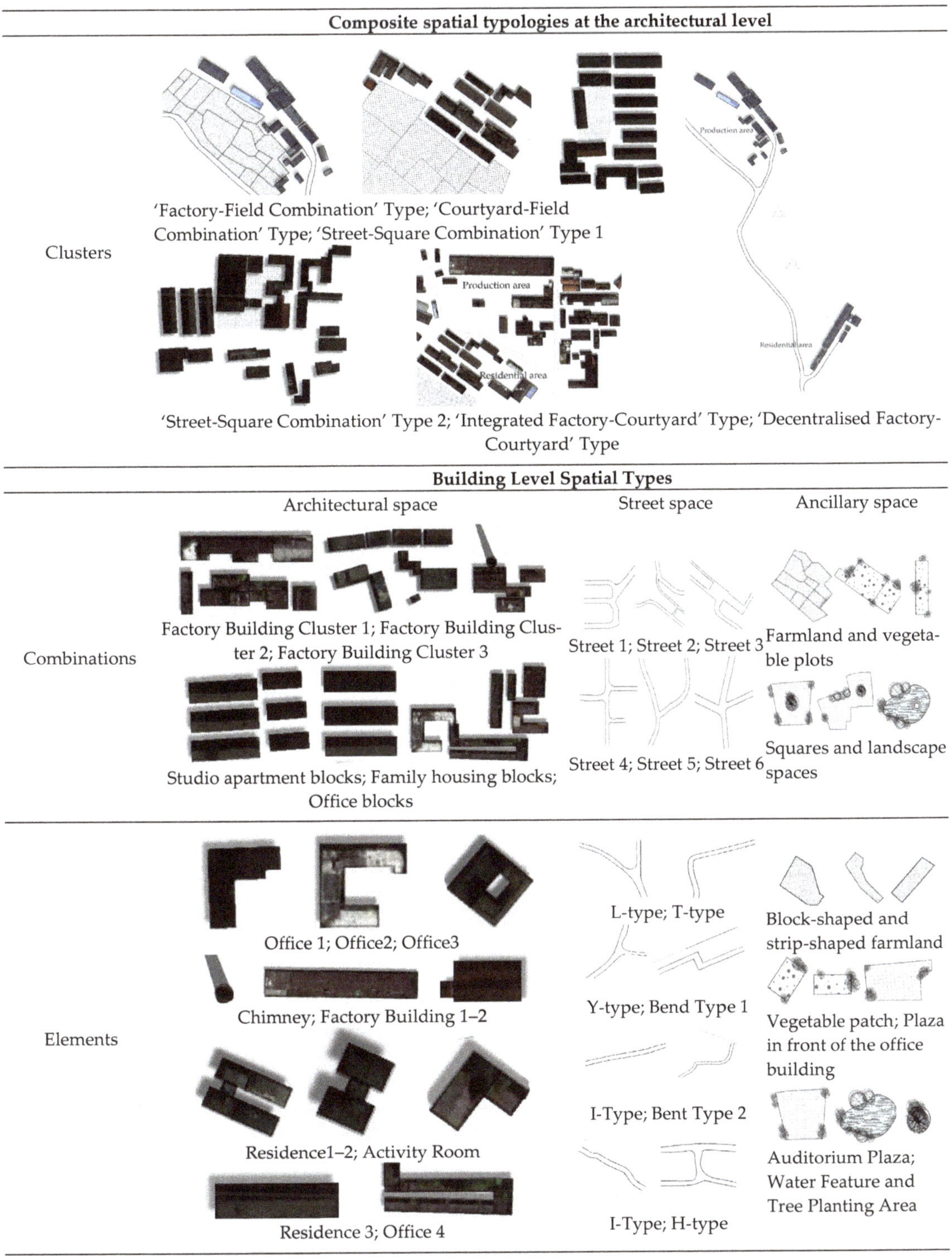

Figure 8. Summary of the spatial vocabulary of constituent elements in Guangyuan's Third Front industrial heritage. (Source: Drawn by the author).

https://doi.org/10.3390/buildings16020446

4. Discussion

4.1. Assessment of Guangyuan's Third Front Industrial Heritage Value

Through typological research across three spatial dimensions of Guangyuan's Third Front industrial heritage, its material form systematically articulates multidimensional value through the mediating elements of 'function', 'structure', and 'intent' (Figure 9).

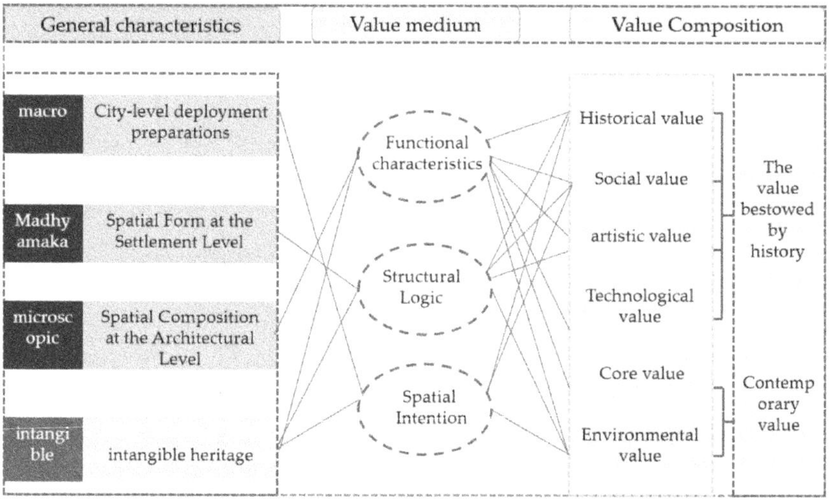

Figure 9. Relationship between typology and value of Guangyuan's Third Front industrial heritage. (Source: Adapted from literature [13]).

At the macro-urban level, 'spatial intent' reflects the differing emphases on production and security under wartime preparedness policies, holding significant historical and social value [27]. At the meso-settlement level, 'structural logic' reveals the organic integration of diverse spatial patterns, production processes, and topographical conditions, possessing notable historical and environmental value. At the micro-architectural level, 'functional characteristics' demonstrate spatial compounding tailored to varying demands, exhibiting rich historical and intrinsic value.

Moreover, the intangible aspects of Third Front heritage encompass corporate history, construction techniques, production processes, the 'Third Front Spirit,' and collective memory. Corporate archives comprehensively document its construction, development, and transformation, holding significant historical and social value. Construction and production techniques achieved local innovation while building upon Soviet learning, possessing both technological and social value. The 'Third Front Spirit' and collective memory crystallise the patriotic sentiments of a specific era, carrying profound historical and cultural significance [13,28].

4.2. Strategies for Revitalising Guangyuan's Third Front Industrial Heritage

Current national policies present a significant opportunity for revitalising Third Front industrial heritage. The 14th Five-Year Plan for Cultural Development explicitly advocates promoting high-quality cultural industry development and integrating culture with tourism. The 2024 Decision of the Central Committee of the Communist Party of China on Further Comprehensively Deepening Reform and Advancing Chinese Modernisation further emphasises 'building strategic hinterlands and critical industrial backups,' reinforcing the significance of preserving and researching Third Front heritage. Against this backdrop, Guangyuan City has completed surveys of 30 heritage sites, transforming the Guangming

https://doi.org/10.3390/buildings16020446

Radio Equipment Factory into the 'Third Front Memory' Museum and repurposing the Guangyuan Cotton Textile Mill into a commercial and cultural exhibition zone. However, current conservation efforts remain fragmented and monolithic in approach, facing multiple challenges including complex ownership structures, insufficient recognition of value, lack of community engagement, and overly generalised technical strategies. Therefore, based on a typological analysis and value assessment of Guangyuan's Third Front industrial heritage, this paper proposes the following three systematic strategic recommendations:

(1) Multidimensional Coordination: Region-specific coordination based on classification activation

The author recommends transcending the limitations of individual heritage sites by integrating the conservation and utilisation of Third Front heritage within the broader frameworks of urban renewal, rural revitalisation, and regional development. This approach would facilitate systematic planning and the implementation of tiered, categorised strategies.

Firstly, establish a framework for the classification, grading, protection, and revitalisation of heritage. Adopt differentiated strategies based on heritage characteristics: develop distinctive industries for sites with unique features, while applying similar approaches to ordinary heritage sites to control costs. At the macro level, the 'one factory, multiple sites' type could be developed into cultural touring routes, linking sites to form closed circuits; the 'one factory, one site' type could be integrated into this system. At the meso level, dispersed sites within valleys and on flat plains could be developed as patriotic education bases; concentrated sites within valleys and on flat plains would be suitable for industrial heritage parks. At the micro level, the 'integrated factory-courtyard' type could collaborate with rural development to create 'Third Front Military Industry Towns'; while 'decentralised factory-courtyard' type sites could transfer residential areas to community use and convert production zones into museums [29–31] (Figure 10). For instance, the former site of Donghe Printing Plant 502, now listed as a provincial cultural heritage site, possesses an intact spatial structure and distinctive spatial imagery. It should be designated as a core conservation and exhibition node, adopting a 'museum plus' model. Within the factory buildings, digital technologies should be employed to recreate the printing process flow—such as using AR/VR to restore the 'printing-cutting-quality inspection' work scenarios. This should be integrated with the development of the neighbouring Xiahetan Village community to offer in-depth research and cultural–creative experiences, thereby ensuring its historical and social value is accurately interpreted and conveyed.

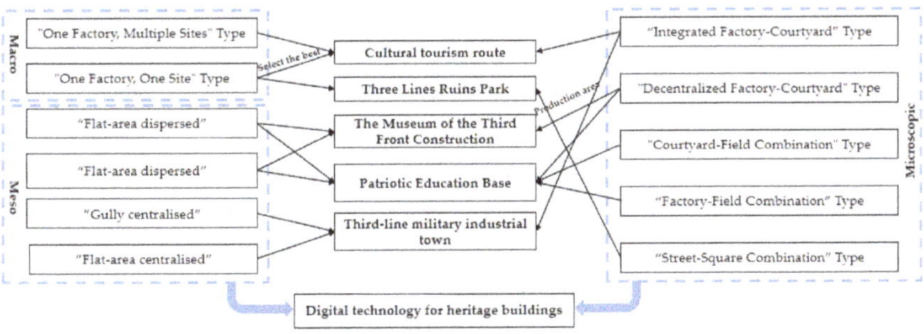

Figure 10. Schematic diagram of graded and classified adaptive reuse. (Source: Drawn by the author).

Secondly, regional synergy should be fostered through heritage corridors and thematic clusters based on classification. For instance, the 'one factory, multiple sites' type heritage base 081 could develop a 'Third Front Military Industry Cultural Theme Route' for its

 https://doi.org/10.3390/buildings16020446

various factories. Drawing inspiration from the 'anchor-link' model of Britain's Ironbridge Gorge, this route would connect sites along railways, rivers, and roads to form a cohesive narrative loop.

Finally, deeply integrate heritage revitalisation into the rural revitalisation process. For heritage sites located in townships, collaborate with surrounding villages to establish 'Third Front Memory' military–industrial towns. For instance, the 'integrated factory-courtyard' type of the Zhaohua Paper Mill could develop distinctive guesthouses, local cuisine, and agricultural experiences, transforming heritage into a catalyst for local development.

(2) Micro-space renovation: Adaptive reuse based on environmental and performance enhancement

The conservation of Third Front industrial heritage necessitates the renewal and regeneration of physical structures.

Firstly, conduct comprehensive condition assessments of significant architectural heritage. Systematically evaluate existing buildings for structural safety, thermal performance of envelope structures, and natural daylighting and ventilation efficiency to identify physical deficiencies and energy-saving potential. Given Guangyuan City's limited economic resources, such assessments shall be prioritised for principal buildings only. Subsequently implement performance-based green refurbishment. Based on the 'decentralised factory-courtyard' type identified in Section 3.3, which naturally facilitates cross-ventilation, renovation strategies should priorities passive cooling techniques similar to Italy's 'Piano Casa' energy retrofit programmer. Streamlined approval processes will facilitate upgrades such as natural ventilation optimisation in ageing industrial buildings, achieving energy savings and efficient utilisation.

Secondly, employing typological approaches to achieve innovative transformation. For buildings that are severely damaged, difficult to restore to their original state, yet possess significant historical value, one may draw upon Aldo Rossi's typological principles. This involves extracting their spatial archetypes—such as U-shaped courtyards, L-shaped workshops, materials, and decorative details—and reconfiguring them through contemporary functional requirements. This approach facilitates the contemporary translation of historical DNA. This strategy of integrating old and new both perpetuates the Third Front ethos and collective memory while achieving innovative spatial transformation.

Finally, the preservation of industrial heritage must shift from a paradigm of 'static conservation' to one of 'dynamic adaptation' [32], thereby enhancing ecological restoration and environmental performance. Emphasis should be placed on preserving the historical characteristics of the surrounding environment, restoring the distinctive ambience of various Third Front-era settings, and creating authentic, tangible 'Third Front construction experience sites' [33,34]. Ecological restoration of disused sites and water systems within the industrial estate shall be undertaken, incorporating sponge city principles to manage surface runoff while preserving and enhancing surrounding ecosystems. This approach integrates the revitalisation of Third Front industrial heritage with the optimisation of regional ecological security patterns.

(3) Multi-stakeholder collaborative governance: Establishing an implementation mechanism based on clearly defined responsibilities and authorities

A critical barrier to revitalisation is the fragmented ownership structure resulting from the military-to-civilian transition [27]. Most such assets are state-owned. Following the relocation of Third Front enterprises to developed cities, their former sites have long remained idle, neglected, and gradually fallen into disrepair. Effective revitalisation and utilisation depend upon clear ownership arrangements, inclusive participation mechanisms

 https://doi.org/10.3390/buildings16020446

and pragmatic governance models. Clarifying and rationalising ownership relationships is therefore a prerequisite for effective conservation and utilisation.

Firstly, establish a differentiated mechanism for the transfer of ownership rights and the sharing of benefits. High-value, well-preserved defence and military–industrial sites may be transferred in their entirety to government enterprises or those for unified planning and development. Sites of average quality may involve social groups or organisations. Military support facilities, typically better preserved, may have ownership divided among government, enterprises, social groups, or even individuals. Public participation in revitalising spaces like workshops and dormitories is particularly encouraged. Transportation heritage, owing to its specialised functions, should be directly transferred to local governments or relevant enterprises for continued operation (Figure 11). For instance, regarding core heritage sites within the defence and military industries, it is recommended to adopt a model combining state ownership with professional operational management. For underutilised ancillary factory complexes and residential areas near city centres, exploring a model of partial property transfer or long-term leasing coupled with community cooperative operation could be considered. Drawing inspiration from the 'temporary use' strategy employed in Germany's Ruhr IBA initiative, this approach would attract cultural and creative institutions, social enterprises, and local residents for phased occupancy, thereby revitalising idle assets. For significant heritage sites situated further from city centres, initial phases may require government or professional institution-led investment and platform development to cultivate fundamental spatial conditions and foster core activities. Once conditions mature, a gradual transition towards community-led 'temporary use' models should be pursued.

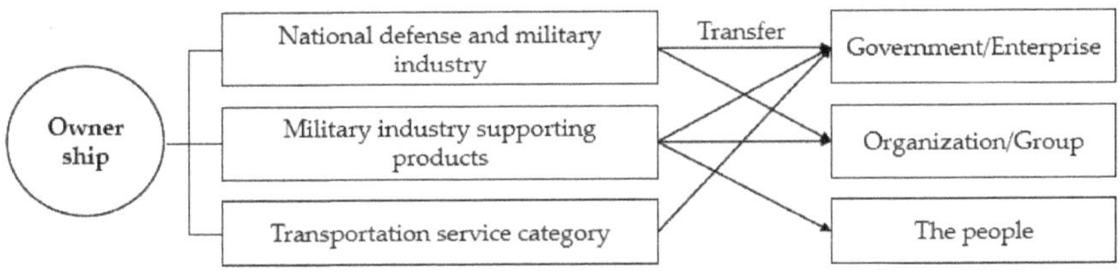

Figure 11. Diagram of ownership transfer models. (Source: Drawn by the author).

Secondly, the community's pivotal role in heritage narration and revitalisation must be reinforced. The sustainable regeneration of industrial heritage necessitates community engagement and intergenerational dialogue. Through bottom-up collaborative mechanisms, economic, social, and cultural imperatives must be balanced, transforming heritage into a dynamic resource that bridges past and future while shaping inclusive collective identities—rather than a static spectacle solely serving capital appreciation [35]. Therefore, systematic oral history collection, community archiving, and initiatives such as 'memory workshops' are required to transform the individual recollections and collective narratives of the 'Third Front People' into core elements of heritage interpretation and integral components of the visitor experience. This approach ensures the socio-cultural authenticity of revitalisation efforts.

Finally, by embedding new functions within the industrial framework, a narrative tension of symbiosis between structure and substance is created, achieving a transformation from a 'place of production' to a 'place of experience' [17]. This approach not only safeguards physical spaces but also integrates cultural experiences and social memory, enabling sustainable development. Future implementation must delineate clear property rights and transfer protocols specifically for the 'grey zones' between military management and local municipal jurisdiction. Preliminary cost estimates, market analyses, and stakeholder impact assessments should be conducted for various functional replacement or revitali-

 https://doi.org/10.3390/buildings16020446

sation schemes. This process should identify potential conflicts—such as property rights disputes or community exclusion—and design negotiation pathways to ensure strategic recommendations are grounded in operational and manageable realities.

In summary, the subject of Third Front industrial heritage research pertains to the past, its value recognition is grounded in the present, while its revitalisation and reuse look towards the future. Throughout the process of conservation and utilisation, it is imperative to coordinate the relationship between past, present, and future, thereby advancing the sustainable development of this precious heritage through enduring preservation.

5. Conclusions

With the conclusion of the Third Front construction, Guangyuan City has gradually accumulated a substantial quantity of Third Front industrial heritage possessing multiple values. This paper, grounded in relevant theories and incorporating typology, constructs an analytical framework of 'type-medium-value' to systematically dissect the value connotations of Guangyuan's Third Front industrial heritage in its material form.

The findings reveal the following: (1) At the urban level: Guangyuan's Third Front industrial heritage exhibits distinct spatial characteristics. Employing 'spatial imagery' as its value medium, it manifests multifaceted value dimensions, profoundly reflecting the strategic wisdom of adapting construction to local conditions during the Third Front era. (2) At the settlement level: Utilising 'structural logic' as its medium, it demonstrates the organic integration of national strategic directives, complex topographical conditions, and production processes during the Third Front period. (3) At the architectural level: The diverse spatial forms of the heritage collectively document historical information from this unique period of military preparedness. Through 'functional characteristics,' they bear historical value and express the agency of industrial architecture.

In terms of academic contribution, this study not only completed the typological identification and value analysis of regional Third Front industrial heritage but also advanced methodology in two respects: firstly, it established a mediated interpretative pathway between material form and intangible value, enhancing the systematic and hierarchical nature of heritage value recognition; secondly, it combined typological analysis tools with spatial quantification methods, providing a transferable and verifiable analytical framework for Third Front industrial heritage research.

Nevertheless, the present study retains certain limitations: its focus remains on typological identification and revitalisation research at the macro-urban level of wartime deployment, the meso-settlement level of spatial morphology, and the micro-architectural level of spatial composition. More granular dimensions—such as structural performance, material craftsmanship, and post-occupancy evaluation at the individual building level—have not been thoroughly explored. This stems both from research constraints imposed by the classified nature of Third Front industrial heritage and represents avenues for future research. Furthermore, owing to the constraints of available research materials and the unique secrecy protocols of the Third Front construction, the author acknowledges limitations in data granularity due to the analysis and discussion. Readers' understanding and valuable feedback are sincerely appreciated.

Looking ahead, the author envisions establishing a clear implementation mechanism with defined responsibilities to achieve multi-stakeholder governance. Integrating heritage characteristics should drive graded and categorised management, incorporating these sites into urban renewal and rural revitalisation processes to foster regional coordination. Building renovations should prioritise environmental enhancement and performance upgrades to promote organic heritage revitalisation. Concurrently, future research will incorporate dimensions such as environmental performance, community participation, and long-term

 https://doi.org/10.3390/buildings16020446

usage assessments, propelling the conservation of Third Front industrial heritage towards a more refined, inclusive, and sustainable trajectory.

Supplementary Materials: The following supporting information can be downloaded at: https://www.mdpi.com/article/10.3390/buildings16020446/s1.

Author Contributions: Conceptualisation, H.Y. and M.X.; methodology, H.Y., Q.Y. and Y.Q.; software, H.Y., X.H. and J.Y.; validation, J.X. and Y.T.; formal analysis, H.Y. and X.D.; investigation, H.Y., M.X., X.H., J.Y. and X.D.; resources, M.X. and Q.Y.; data curation, H.Y. and X.H.; writing—original draft preparation, H.Y., M.X. and Q.Y.; writing—review and editing, M.X. and Q.Y.; visualisation, H.Y., J.Y. and X.D.; supervision, M.X.; project administration, J.X. and Y.T.; funding acquisition, M.X., Q.Y. and Y.Q. All authors have read and agreed to the published version of the manuscript.

Funding: This research was funded by the Humanities and Social Science Fund of Ministry of Education of China (No. 23XJAZH008); Sichuan Center for Strategic Studies of Integrated Military and Civilian Development Research Project (No. JMRH-2401); Fundamental Research Funds for the Central Universities, Southwest Minzu University (No. RQD2022031); National Natural Science Foundation of China (No. 52308038); Sichuan Science and Technology Program (No. 2025ZNSFSC1309).

Data Availability Statement: All data generated or analysed during this study are included in this published article.

Acknowledgments: The authors would like to thank the anonymous reviewers for their constructive comments. Gratitude is also extended to financial support throughout the research process. The tutors promptly identified shortcomings in the research and imparted invaluable methodologies and techniques. Under their guidance, the author has broadened their horizons, accumulated extensive knowledge, and achieved continuous growth on the path of scientific inquiry. Simultaneously, heartfelt gratitude is extended to the junior colleagues within the research group for their solidarity and collaboration. With their assistance, multiple industrial heritage site surveys were successfully completed, ultimately enabling the research to be concluded more effectively and leaving behind a splendid conference experience.

Conflicts of Interest: The authors declare no conflicts of interest.

Abbreviations

The following abbreviations are used in this manuscript:

ANN	Average nearest neighbour
TPI	Terrain Position Index
NOF	New Optoelectronic Factory
H410	Hospital 410
HMF	Huachang Machinery Factory
TMF	Tianyuan Machinery Factory
JTF	Jianping Tools Factory
ZPM	Zhaohua Paper Mill
WCP	Wuzhou Chemical Plant
INC	Industry No. 23 Company
JCF	Jiangling Cable Factory
GSRS	Guangyuan South Railway Station
GRTS	Guangyuan Radio Technical School
PS	Puji Station
GCTF	Guangyuan Cotton Textile Factory
TFBH	The Fifth Bureau of Hydropower
BS	Baishiyan Substation
DPCF501	Donghe Printing Company, Factory 501
DPCF502	Donghe Printing Company, Factory 502
DPC505TPP	Donghe Printing Company 505 Thermal Power Plant

https://doi.org/10.3390/buildings16020446

DPCF503, 504	Donghe Printing Company, Factory 503 and 504
PAR	Shape Ratio
WZMF	Wan Zhong Machinery Factory
XETF	Xuguang Electron Tube Factory
CMF	Changsheng Machinery Factory
LREF	Liyuan Radio Equipment Factory
GREF	Guangming Radio Equipment Factory
HMF	Honglun Machinery Factory
CEIC	Chuanbei Electronic Industry Company
WCFRA	Wuzhou Chemical Factory Residential Area
CAEP-No. 1	Institute No. 1 of the China Academy of Engineering Physics
CAEP-No. 3	Institute No. 3 of the China Academy of Engineering Physics
CAEP-No. 4	Institute No. 4 of the China Academy of Engineering Physics
GNBMF	Guangyuan No. 1 Building Materials Factory
WCAIP	Wangcang Coal and Iron Plant
PYCM	Pick Yinyan Coal Mine
WCM	Wangcang Coal Mine
TCM	Tangjiahe Coal Mine
BFAB	The Bailongjiang Forestry Administration Bureau of the Forestry Department
HAVDPC	The headquarters and vault of Donghe Printing Company
FQFEF	The family quarters of Fenglei Equipment Factory
GMB	The former site of Guangwang Mining Bureau

Appendix A

Appendix A.1

Table A1. Main information table of Third Front construction industrial heritage in Guangyuan City. (Source: field survey and calculation by the authors).

Heritage Site Name	Military Project	Internal Mail Drop	Longitude	Latitude	Length of the Long Axis of the Site (m)
WZMF	789	789	105.224466	32.575201	483.08
NOF	879	/	105.410936	32.641223	679.09
XETF	779	109	105.769358	32.455977	1132.31
CEIC	081	122	105.849204	32.438810	249.09
CMF	787	116	105.867362	32.462266	286.35
HMF	885	105	105.857084	32.452567	144.48
LREF	4961	118	105.862494	32.445294	406.4
TMF	4190	112	105.875603	32.453366	546.94
GREF	4130	120	105.885826	32.444456	486.36
JTF	4520	110	105.868889	32.447689	428.41
HMF	4192	102	105.637354	32.381011	609.51
GRTS	/	/	105.85873	32.446372	362.05
H410	410	/	105.870000	32.440000	278.67
WCP	821	/	105.590843	32.492041	3459.32
WCFRA	/	/	105.628121	32.496390	889.09
INC	/	/	105.665426	32.380786	622.54
JCF	608	213	105.932808	32.406662	662.66
CAEP-No. 1	/	/	105.301291	31.830349	1543.04
CAEP-No. 3	/	/	105.360000	31.770000	1575.27
CAEP-No. 4	/	/	105.200000	31.930000	1663.27
GSRS	/	/	105.797824	32.418308	627.78
PS	/	/	106.459584	32.237761	119.05
GNBMF	/	/	105.649906	32.39083	825.56
ZPM	/	/	105.668397	32.403623	561.05
BFAB	/	/	105.665147	32.402396	376.6

https://doi.org/10.3390/buildings16020446

Table A1. *Cont.*

Heritage Site Name	Military Project	Internal Mail Drop	Longitude	Latitude	Length of the Long Axis of the Site (m)
GCTF	/	/	105.833885	32.448751	219.22
TFBH	/	/	105.645490	32.388563	705.82
BS	/	/	105.693849	32.418178	213.76
DPCF501	501	/	106.022106	32.266605	1764.3
DPCF502	502	/	106.303232	32.246781	1055.4
DPCF503, 504	503, 504	/	106.268393	32.235912	1086.47
DPC505TPP	505	/	106.302334	32.258349	422.99
HAVDPC	507	/	106.307858	32.239683	209.18
FQFEF	756	/	106.308442	32.238331	175.94
GMB	/	/	106.29136	32.233913	110.65
WCAIP	/	/	106.219892	32.209875	550.06
PYCM	/	/	106.012444	32.321317	988.21
WCM	/	/	106.027453	32.262982	930.96
TCM	/	/	106.236688	32.218429	1056.71

Appendix A.2

Table A2. Statistical table of PAR and TPI data for Third Front construction industrial heritage in Guangyuan City. (Source: field survey and calculation by the authors).

Heritage Site Name	Perimeter (m)	Area (m^2)	PAR	PAR-2	PAR-2 (Z-Score Normalisation)	Elevation of the Centroid	Mean Elevation	TPI	TPI (Z-Score Normalisation)
WZMF	1200.93	10,734.24	3.27	1.27	1.85	767	884.45	−117.45	−1.5
NOF	2997.09	54,177.58	3.63	1.63	2.35	640	670.41	−30.41	−0.05
XETF	3807.95	151,090.94	2.76	0.76	1.15	502	606.15	−104.15	−1.28
CEIC	684.71	13,900.55	1.64	−0.36	−0.38	484	507.61	−23.61	0.07
CMF	858.94	17,847.79	1.81	−0.19	−0.15	604	600.90	3.10	0.52
HMF	362.86	3386.69	1.76	−0.24	−0.22	569	507.61	61.39	1.49
LREF	1367.58	57,489.72	1.61	−0.39	−0.42	496	507.61	−11.61	0.27
TMF	1875.28	66,217.53	2.06	0.06	0.2	533	600.90	−67.90	−0.67
GREF	1595.025	77,010.77	1.62	−0.38	−0.41	530	600.03	−70.03	−0.71
JTF	1110.69	33,660.5	1.71	−0.29	−0.28	536	600.90	−64.90	−0.62
HMF	1680.42	138,999.21	1.27	−0.73	−0.88	480	511.24	−31.24	−0.06
GRTS	947.76	26,073.81	1.66	−0.34	−0.36	590	507.61	82.39	1.85
H410	1006.95	41,518.86	1.39	−0.61	−0.72	508	507.61	0.39	0.47
WCP	12,060.19	1,325,109.95	2.96	0.96	1.43	506	665.29	−159.29	−2.21
WCFRA	2559.51	381,371.48	1.17	−0.83	−1.03	588	598.58	−10.58	0.29
INC	2032.34	160,585.4	1.43	−0.57	−0.67	496	510.29	−14.29	0.22
JCF	1750.35	99,053.3	1.57	−0.43	−0.48	562	534.80	27.20	0.92
CAEP-No. 1	4211.77	280,164.75	2.25	0.25	0.46	831	677.58	153.42	3.04
CAEP-No. 3	5596.77	376,308.62	2.57	0.57	0.89	463	540.44	−77.44	−0.84
CAEP-No. 4	5751.6	192,983.07	3.69	1.69	2.43	655	650.92	4.08	0.53
GSRS	1680.98	117,865.15	1.38	−0.62	−0.73	494	503.37	−9.37	0.31
PS	381.85	3868.68	1.73	−0.27	−0.26	488	603.27	−115.27	−1.47
GNBMF	2296.11	62,815.8	2.59	0.59	0.92	573	511.24	61.76	1.5
ZPM	1617.39	124,100.36	1.30	−0.70	−0.84	512	560.94	−48.94	−0.36
BFAB	1179.86	59,034.33	1.37	−0.63	−0.75	507	531.22	−24.22	0.06
GCTF	638.56	26,272.05	1.11	−0.89	−1.1	554	507.61	46.39	1.24
TFBH	1951.91	162,563.32	1.37	−0.63	−0.75	502	511.24	−9.24	0.31
BS	579.67	9448.96	1.68	−0.32	−0.33	554	644.04	−90.04	−1.05
DPCF501	3893.03	222,482.42	2.33	0.33	0.56	598	593.81	4.19	0.53
DPCF502	2883.3	94,812.58	2.64	0.64	0.99	471	508.90	−37.90	−0.17
DPCF503, 504	3195.63	189,981.28	2.07	0.07	0.21	541	503.53	37.47	1.09
DPC505TPP	1288.08	65,005.64	1.43	−0.57	−0.67	462	508.90	−46.90	−0.32
HAVDPC	701.94	13,863.96	1.68	−0.32	−0.33	513	508.90	4.10	0.53
FQFEF	605.02	18,827.95	1.24	−0.76	−0.93	504	508.90	−4.90	0.38
GMB	342.25	6347.79	1.21	−0.79	−0.97	460	508.90	−48.90	−0.36
WCAIP	1637.41	51,905.8	2.03	0.03	0.16	460	484.19	−24.19	0.06
PYCM	3142.75	141,317.44	2.36	0.36	0.6	714	811.66	−97.66	−1.17
WCM	2513.71	99,918.66	2.24	0.24	0.44	575	593.81	−18.81	0.15
TCM	3333.8	141,025.04	2.50	0.50	0.8	456	484.19	−28.19	−0.01

https://doi.org/10.3390/buildings16020446

Appendix A.3

Table A3. Comparison of Multi-Radius Average Elevations in TPI Analysis (Source: Field survey and calculation by the authors).

Heritage Site Name	Elevation of the Centroid	Mean Elevation (500 m)	Mean Elevation (1000 m)	Mean Elevation (2000 m)	Mean Elevation (2500 m)	Mean Elevation (3000 m)
WZMF	767	788.26	819.72	884.45	899.77	916.44
NOF	640	631.98	622.22	670.41	698.62	717.14
XETF	502	524.70	566.81	606.15	614.43	632.02
CEIC	484	506.87	504.93	507.61	515.26	522.19
CMF	604	608.53	585.80	600.90	625.81	648.16
HMF	569	529.37	526.29	507.61	515.26	522.19
LREF	496	520.26	526.29	507.61	515.26	522.19
TMF	533	543.83	553.50	600.90	625.81	522.19
GREF	530	546.09	544.25	600.03	504.77	648.16
JTF	536	536.99	520.73	600.90	515.26	522.19
HMF	480	483.14	483.89	511.24	524.45	537.35
GRTS	590	520.26	526.29	507.61	515.26	522.19
H410	508	521.41	520.73	507.61	515.26	522.19
WCP	506	536.78	575.48	665.29	695.80	711.52
WCFRA	588	560.90	553.54	598.58	624.60	659.25
INC	496	476.52	478.11	510.29	514.05	537.35
JCF	562	535.99	539.78	534.80	541.00	556.76
CAEP-No. 1	831	750.10	713.02	677.58	659.97	640.52
CAEP-No. 3	463	479.16	491.58	540.44	564.90	578.11
CAEP-No. 4	655	654.69	664.17	650.92	631.08	621.67
GSRS	494	491.25	486.37	503.37	518.20	541.94
PS	488	513.94	545.26	603.27	626.29	646.16
GNBMF	573	552.90	527.25	511.24	524.45	537.35
ZPM	512	514.41	513.71	560.94	582.70	519.59
BFAB	507	514.41	513.71	531.22	514.05	519.59
GCTF	554	533.99	520.98	507.61	515.26	522.19
TFBH	502	552.90	527.25	511.24	524.45	537.35
BS	554	598.18	603.99	644.04	679.31	646.67
DPCF501	598	581.15	574.33	593.81	621.99	645.31
DPCF502	471	475.98	493.41	508.90	519.20	538.86
DPCF503, 504	541	510.42	497.71	503.53	516.22	530.99
DPC505TPP	462	482.95	519.27	508.90	519.20	538.86
HAVDPC	513	493.11	493.41	508.90	519.20	538.86
FQFEF	504	493.11	497.10	508.90	519.20	538.86
GMB	460	470.51	487.24	508.90	519.20	538.86
WCAIP	460	462.44	463.74	484.19	504.79	526.41
PYCM	714	712.06	745.69	811.66	817.23	807.21
WCM	575	558.11	574.33	593.81	621.99	645.31
TCM	456	462.00	473.04	484.19	504.79	526.41

References

1. Chen, D.L. Three-line construction—the nearest industrial heritage to us. *Chin. Natl. Geogr.* **2007**, 43–44. Available online: https://kns.cnki.net/kcms2/article/abstract?v=VYuoLtjwl8O4c03GZqOcDldafc5mGeIU4mkrsJKc1WMPq2AuuHD5wjWbKoQ

https://doi.org/10.3390/buildings16020446

Fs-sG3Wgs2LcDIq9wr2XtXpNwB27WblRJa0muei-BZYJ5rRQkwtzBT2u2b9q_C2nDychFI7P5mwTvfaoruR5L5f-GmYReJTjyhNPx kvCCXRKW0CNb7pOnyjCtpXjYF8R0jwG1&uniplatform=NZKPT&captchaId=ba9f6d98-a17a-48a1-9d87-7879963be68f (accessed on 16 May 2024).
2. Lv, J.C.; Yang, R.M.; Li, S.T. A preliminary study on the concept of Three-line industrial heritage. *Soc. Sci. Ningxia* **2020**, *10*, 139–146. [CrossRef]
3. Lv, J.C. On the concept of Three-line construction and Three-line industrial heritage. *Acad. China* **2023**, 124–131. [CrossRef]
4. Zhang, F.Q. On Three-line construction and Chongqing's urban modernization. *Chongqing Soc. Sci.* **2007**, 79–83. [CrossRef]
5. Ding, X.S. Research on the regeneration path of the cultural memory of Third-line industrial heritage. *Soc. Sci. Res.* **2021**, 198–206. [CrossRef]
6. Zhang, Y.M. The Study on Renewal Strategy of Third-Front Industrial Heritage in Sichuan-Chongqing Region Under the Symbiosis Theory. Master's Thesis, Chongqing University, Chongqing, China, 2015.
7. Zhang, P. Research on the Protection and Activation of the Industrial Heritage of the Third-Line Construction in the Former Mianyang Area from the Perspective of Catalysts. Master's Theses, Southwest University of Science and Technology, Mianyang, China, 2022.
8. Fu, Y.B. Study on the Value Evaluation System of the Third-Line Industrial Heritage in Sichuan. Master's Thesis, Southwest University of Science and Technology, Mianyang, China, 2019.
9. Zhao, L.; Tang, T.; Du, S.M.; Zhou, J.; He, C.H.; Wu, C. Study on evaluation model of protection and development of industrial heritage buildings in Three-line construction. *Mod. Prop. Manag.* **2020**, 4–5. [CrossRef]
10. Liu, H.X. The Research on Value Assessing and the Feasibility of Protection-Reuse of the Industrial Heritage of the Strategic Construction—Illustrated by Cases of the Industrial Heritages Located in the Former East Sichuan and the North Guizhou. Master's Thesis, Huazhong University of Science and Technology, Wuhan, China, 2012.
11. Li, T.T. Research on Spatial Morphology of Industrial Remains of the Third Front Construction in Northern Sichuan. Master's Theses, Southwest University of Science and Technology, Mianyang, China, 2023.
12. Wang, X.Y. Case Study of Puqi Textile Factory City as a Typical Third-Front Movement Enterprise. Master's Thesis, Huazhong University of Science and Technology, Wuhan, China, 2023.
13. Huang, Z.H. The Research on the Classification and Activation Strategy of Hubei's Third Front Construction Remains Based on Value Evaluation. Master's Thesis, Huazhong University of Science and Technology, Wuhan, China, 2023.
14. Lv, J.; Yang, H.; Zhang, X.; Wang, Y.; Zhang, H.; Zhao, Y. Spatial and Temporal Distribution Characteristics and Genealogical Framework of Shaanxi's Industrial Heritage. *Buildings* **2024**, *14*, 2527. [CrossRef]
15. Ramírez-Casas, J.; Gómez-Val, R.; Buill, F.; González-Sánchez, B.; Navarro Ezquerra, A. Forgotten Industrial Heritage: The Cement Factory from La Granja d'Escarp. *Buildings* **2025**, *15*, 372. [CrossRef]
16. Jiang, Z.; Qi, Z.; Chen, L.; Xu, L.; Wan, D.; Burak-Gajewski, P.; Zawisza, R.; Liu, L. External Spatial Morphology of Creative Industries Parks in the Industrial Heritage Category Based on Spatial Syntax: Taking Tianjin as an Example. *Buildings* **2024**, *14*, 559. [CrossRef]
17. Tao, H.; Wen, Y.; Liu, M.; Wu, Y. Industrial Heritage Protection from the Perspective of Spatial Narrative. *Land* **2025**, *14*, 1105. [CrossRef]
18. Lv, J.C. Characteristics, value and protection and utilization of Third-Front construction industrial heritage. *Urban Rural Plan.* **2020**, 54–62. [CrossRef]
19. Ren, J.; Chen, M. Look back and ponder on 'the three line buildings'. *Mil. Hist.* **2001**, 27–31. [CrossRef]
20. Party History Research Institute of the CPC Sichuan Provincial Committee. *A Brief History of the Third Front Construction in Sichuan*; Party History Publishing House: Beijing, China, 2009; pp. 280–281.
21. Xu, S.B. Reflections on deepening research on industrial heritage. *Stud. Cult. Soft Power* **2024**, *9*, 113–116. [CrossRef]
22. Wang, L.J. *Architectural Typology*; Tianjin University Press: Tianjin, China, 2005; ISBN 978-7-5618-2198-5.
23. Li, H.B.; Zhang, X.L.; Wu, J.G.; Zhu, B. Spatial pattern and its driving mechanism of rural settlements in southern Jiangsu. *Sci. Geogr. Sin.* **2014**, *34*, 438–446.
24. Chen, Y.X.; Liu, D.Y. Study on the construction of the evaluation model of village development potential under the land and space planning system: A case of Xinzhan Town in Zhaoyuan County, Daqing, Heilongjiang. *Urban. Archit.* **2022**, *19*, 50–57. [CrossRef]
25. Qian, Y.X.; Du, Y.J.; Xu, C.B. A three-dimensional quantitative study on the spatial morphology of rural settlements on mountainous islands: A case study of Nantian Island in Xiangshan County, Ningbo City. *Dev. Small Cities Towns* **2024**, *42*, 100–110.
26. Tan, G.Y.; Gao, Y.Z.; Xu, L.Q. Research on the Third Front construction heritage based on industrial archaeology. *Time Archit.* **2019**, *6*, 44–51.
27. Yu, H.C.; Xiang, M.M.; Wang, S.Q.; Yang, J.F. Study on industrial heritage value of Guangyuan Third Front construction from typology perspective. In *Proceedings of the Toward China Modernization: Value and Action of Planning—Proceedings of China Urban Planning Annual Conference 2025 (10 Urban Cultural Heritage Protection)*; Urban Planning Society of China: Shenyang, China, 2025; pp. 652–664.

28. Tan, G.Y.; Xu, L.Q.; Huang, Z.H. Value evaluation of Hubei's Third Front construction remains under the guidance of activation and reuse. *Architect* **2025**, 25–34. [CrossRef]
29. Lv, J.C. Current status and future solutions: The protection and utilization of the Third Front construction industrial heritage. *Southeast Cult.* **2019**, 6–12. [CrossRef]
30. Jing, S.Y. Thoughts on protection and reuse of chemical industrial heritage museum. *Constr. Sci. Technol.* **2022**, 83–85. [CrossRef]
31. Xu, S.B.; Lu, Z.C.; Wang, R.R.; Nobuo, A. Research on the optimisation path of industrial heritage regeneration from the perspective of actor network theory: Taking Tianjin central city as an example. *Urban Dev. Stud.* **2023**, 65–74. [CrossRef]
32. De Gregorio, S.; De Vita, M.; Paris, A. Industrial Heritage Rethinking: Flexibility Design for Eco-Friendly Environments. *Buildings* **2023**, *13*, 1048. [CrossRef]
33. Tan, G.Y.; Geng, X.C.; Cao, X.M.; Gao, Y.Z. The production and reproduction of space: Activation strategies for Third Front industrial remains. *Int. Soc. Sci. J.* **2025**, *42*, 78–93.
34. Jia, Y.F.; Zhang, M.W.; Tan, G.Y.; Liu, H.L. A study on the conservation of industrial heritage-based historic area from the perspective of production unit: Taking Dongfang Steel Plant in Huangshi City as an example. *City Plan. Rev.* **2023**, *47*, 101–109.
35. Chiodi, C.A.M.; De Lucia, R.; Giunchi, C.; Molinari, P. Looking for a Balance Between Memories, Patrimonialization, and Tourism: Sustainable Approaches to Industrial Heritage Regeneration in Northwestern Italy. *Sustainability* **2025**, *17*, 905. [CrossRef]

https://doi.org/10.3390/buildings16020446

buildings

MDPI

Article

Application of Supports Theory in Building Design: Multi-Dimensional Permeability and Spatial Structure in Versatile Community Centers

Mingrui Zhang, Yang Yang, Chang Yi, Mingxuan Jia, Menglong Zhang * and Qianru Yang *

School of Architecture, Southwest Minzu Uiniversity, Chengdu 610225, China; 19836639583@163.com (M.Z.); 13398432841@163.com (Y.Y.); isyichang@163.com (C.Y.); jiamingxuan_0813@163.com (M.J.)

* Correspondence: z18768972650@163.com (M.Z.); yangqianru@swun.edu.cn (Q.Y.); Tel.: +86-19822990400 (M.Z.); +86-15680016301 (Q.Y.)

Abstract

With the evolution of social structure and the intensification of population aging, traditional community service centers struggle to meet residents' complex needs due to their functional singularity and spatial rigidity. In response to the continuously evolving social structure and functional requirements, this research proposes a strategy based on the "Separation of Support and Infill," distinguishing between the building's permanent Support Structure and its replaceable Infill Components. These two parts are combined with modularization to achieve long-term spatial adaptability and sustainability throughout the entire life cycle. In terms of functional space, through the combination of vertical stratification, horizontal staggering and spatial permeability, a three-dimensional composite space system is constructed, which not only enhances the functional flexibility but also improves the environmental performance. Taking a design case in Yicheng District, Zhumadian City as an example, through a comparative analysis with the traditional building model, the comparative analysis demonstrates that this framework increases the Floor Area Ratio (FAR) by approximately 0.15 compared to traditional models. Furthermore, the modular characteristics significantly enhance demountability and reusability, reducing construction and demolition waste while lowering life-cycle costs by an estimated 15% to 25%. These studies show that the support structure and the composite functional space system can not only promote social interaction and community cohesion but also reduce the life-cycle cost and carbon emissions. The framework proposed in this paper constructs a theoretical and practical system for sustainable community buildings from the perspectives of functional compounding and low-carbon community development. Its innovation lies in its flexible spatial organization mode and the enhancement of the sustainability of community buildings.

Keywords: support structure theory; community service center; modular design; spatial permeability; sustainable architecture

1. Introduction

1.1. Research Background

With the continuous evolution of social structure and population characteristics, the process of population aging is accelerating, and the birth rate shows a continuous downward trend [1]. These demographic shifts have exacerbated the disparity between the supply and demand of community public service facilities. Meanwhile, traditional universal provision models and static planning strategies are no longer adequate to meet diverse societal needs [2]. Community embedded service facilities have become the core carriers for undertaking governance transformation. They need to integrate diverse functions such as elderly care, childcare, cultural and entertainment services, and medical care within a limited space [3]. These factors collectively drive the transformation and upgrading of community service facilities. Their functions have progressively expanded from single-purpose cultural and recreational venues to encompassing diverse spaces for education, exhibitions, leisure, office use, and ecological services. However, most of the existing community service facilities are configured with a single function, and their spatial organization mode is relatively fixed. They lack the flexibility to cope with function adjustment and service expansion, and it is difficult to meet the compound needs of residents arising from changes in the life cycle and family structure.

From the perspective of public service provision and facility development, community service centers are not only service delivery platforms but are increasingly being regarded as "social infrastructure" or "service hubs". Their characteristics—functional integration, spatial consolidation, green and low-carbon design, and digital transformation (digital empowerment)—are becoming increasingly significant [4,5]. Among these characteristics, functional diversity, spatial efficiency, and digitalization represent the prevailing trends, as evidenced by the concepts of 'community-embedded service facilities' and 'digitalization empowering urban community-embedded services. Against this backdrop of developmental trends and policy support, community service centers as public buildings must also incorporate these principles. Within their effective spatial boundaries, they should achieve functional diversity, flexibility, sustainability, and low-carbon development.

Consequently, overcoming the traditional paradigm of "static planning and single-service provision" to construct community centers that are functionally integrative and spatially adaptable has become a critical imperative in current architectural research. This study aims to explore spatial organization and construction for community service centers that cater to the multi-level needs of residents and maintain openness and sustainability in the time dimension, so as to respond to the continuously evolving social life scenarios.

1.2. Research Status

With the rapid development of urbanization, according to the prediction of the OECD, China's urbanization rate will reach 75% by 2050. Among them, as the basic constituent unit of modern cities, urban communities are not only the main space and key support for the daily production and life of urban residents, but also the intersection of political, economic, cultural and other issues in urban society [6]. Concurrently, in response to the national 'dual carbon' objectives, low-carbon communities (or sustainable/eco-communities) are the most basic components of a low-carbon society and models for low-carbon cities. Moreover, in a sense, they can be regarded as the foundation for building a low-carbon society and realizing the low-carbon development strategy [7]. Low-carbon communities represent a new paradigm for sustainable urban development, serving as an essential pathway towards achieving the carbon peaking and carbon neutrality targets [8]. The green and low-carbon retrofitting of existing community service centers, as well as new construction projects, has emerged as a research and piloting priority. Key research areas include green community design strategies, the application of low-carbon and zero-carbon building technologies within community facilities, and the integration of community energy management with renewable energy sources [9]. The concept of the community center as a collection of physical facilities, services, and social resources has received significant attention [10], with a growing emphasis on "transitioning from single-purpose community centers towards

 https://doi.org/10.3390/buildings16010092

multiservice facilities" [11]. Consequently, the diversification of community functions and spatial integration has progressively gained importance. In terms of functional integration (mixed-use), Jacobs proposed the concept of mixed-use in the 1960s, opposing the modernist practice of dividing cities into single-function zones [12], While contemporary research predominantly focuses on the community level, the applicability of mixed-use at the building scale is now widely recognized [13].

In addition to the above policy requirements, the current design requirements and research gaps of community centers can also be seen in some practical cases. For example, the overall design of the Suk Agawa Community Center emphasizes alignment with community needs and flexible use. It is mainly characterized by open stepped terraces, cantilevered floors, and an activity-oriented floor plan. At the same time, it also takes into account the integration of the building with the surrounding environment and the city, successfully connecting the city and the residents [14]. Chen et al. (2025) [15] took the renovation of an abandoned boiler room in Shenyang as an example. The research shows that transforming old spaces into community service centers with composite functions can greatly improve the site utilization efficiency (the utilization rate increases by 2.2 times) and vitality (the daily activity duration is extended by 12 h, and the cross-age interaction frequency increases by 43%). The research proposes the mechanism of "functional hybridization–spatial permeability–usage sustainability" and recommends community service centers as the preferred model for urban renewal to promote multi-functional spaces and low-carbon utilization (reduce large-scale demolition) [15]. In the design of the Shepherd Park Community Center, two modes are integrated: an educational mode for daily use and a community mode for night use. During the day, the annex building serves as an activity area for students, and in the evening, it is "flipped" for community use. The design exceeds the LEED Silver standard. Through intelligent technology and an efficient electromechanical system, the building's energy consumption is close to net-zero, achieving low-carbon operation. The "day–night flip mode" significantly saves land use and improves the functional utilization rate.

These three cases each have different focuses and have some deficiencies. For example, in the future, as the population structure of the community changes, the functional requirements will also change accordingly. Then, how to carry out the renovation with the lowest loss cost and carbon emissions? Under this consideration and combined with policy requirements, this study proposes a theoretical and practical system for sustainable community buildings. Taking the support structure theory as the main body, extend the entire life cycle of buildings and reduce the construction waste generated during building demolition and renovation, to achieve the goal of low carbon. Habraken's "support–infill" concept and Open Building have laid the theoretical foundation for "permanent structure and variable function". The "time-based architecture" proposed by Leupen et al. further incorporates the time dimension into design. These theories support the application of modular and support strategies in the sustainable design of public buildings. Habraken's statement is that "the supporting structure belongs to the public domain and is permanent; while the fillers belong to individuals and can be changed. Public participation and user freedom of choice are the key goals" can also serve as theoretical support for the characteristic that "the community service center has diverse functions, and its functional structure is not stable, requiring changes according to the needs of residents." Therefore, in terms of mixed functions, a composite space system is adopted to increase the diversity and real-time practicality of functions in the limited land area. Meanwhile, supported by the "support structure–Infill System" structure, when the functional requirements change in the subsequent stage, not only can the functions be replaced conveniently, but also the cost and carbon emissions can be reduced.

https://doi.org/10.3390/buildings16010092

1.3. Summary of the Current Situation and Research Objectives

With the acceleration of the urbanization process and the proposal of the "dual carbon" goal, the design of community service centers is moving towards the direction of functional compounding and low-carbon development. The construction of the "15-min living circle" driven by policies is committed to enhancing the functional diversification of community service centers through diversified service facilities, and the construction of low-carbon communities will also become an important path to achieve sustainable development. Meanwhile, functional mixing and community resilience have gradually become the research focus, emphasizing enhancing the long-term adaptability of communities through flexible spatial design. Theoretically, the "Support-Infill" theory proposed by Habraken and the "Time-Based Architecture" theory by Leupen provide support for modular and functionally variable design and promote the long-term adaptability of building structures and functions. The community service center, a public building, aims to meet the diverse needs of local people and the requirements for indoor and outdoor activities. Considering the different behavior patterns of the crowd, functionally independent spaces are designed to meet different needs. These independent spaces also penetrate each other to form a complete space.

In this design, the mutual Permeability between spaces of different forms and spaces with different functions is studied. By overlapping and staggering blocks, various forms of spaces such as set-back terraces, outdoor activity platforms, and gray spaces under the eaves, as well as outdoor and semi-outdoor activity spaces are constructed to meet the needs of activity participants for gathering spaces, enhancing the interest and architectural vitality. Thus, a public building is created where the spaces have rich and diverse morphological changes, and these changing spaces blend with each other to form an integrated whole. At the same time, as a public activity platform that promotes communication among neighbors and provides comfortable and comprehensive services, the community service center also has important significance for improving the community management level and promoting the comprehensive development of the community. By staggering, overlapping, and stacking spaces with different functions or forms vertically or horizontally, a space with a sense of hierarchy and mutual integration is created. This design concept not only enriches the form of the space but also improves the utilization efficiency of the space and promotes communication and interaction between different functional areas. On the other hand, the design of community service centers urgently needs to explore an integrated path of "high density, multi-function, low energy consumption, and strong sense of belonging" to solve problems such as space efficiency, functional diversity, and the entire building life cycle.

2. Methodology

As shown in Figure 1, the research based on the Supports Theory as the structural framework and integrating the composite functional space system, a sustainable design methodology system for community service centers is constructed. This framework aims to achieve the structural variability, spatial complexity, and architectural resilience of buildings, and respond to the flexibility and sustainability requirements of community public buildings in a high-density urban environment.

https://doi.org/10.3390/buildings16010092

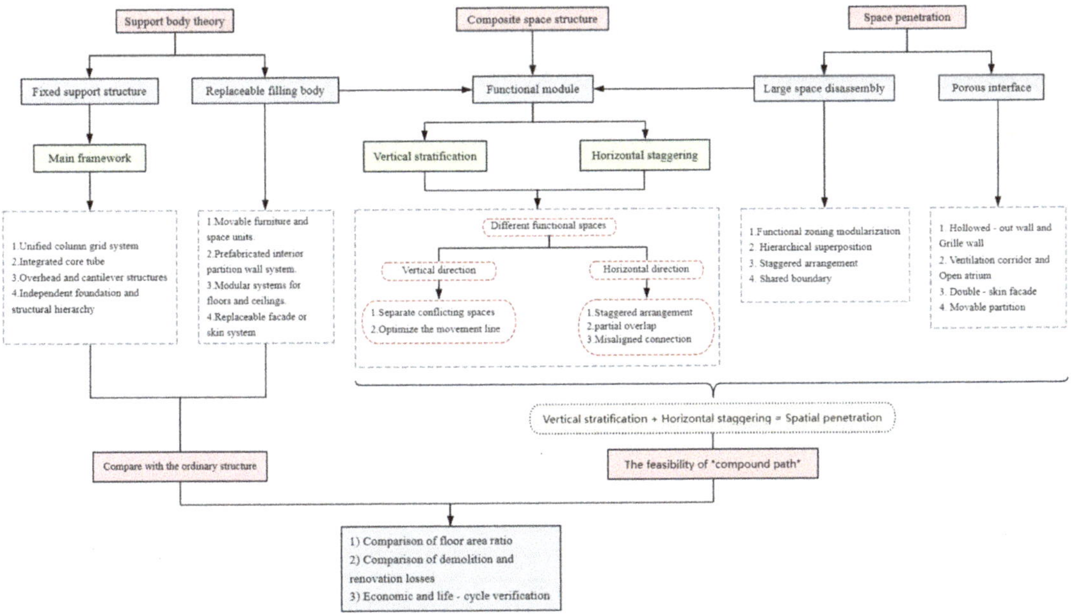

Figure 1. Method framework diagram.

2.1. Supports Theory

Community service centers serve as multifunctional integrated facilities designed to meet the diverse needs of residents. Consequently, they demand high levels of functional versatility, adaptability, and architectural sustainability. Conventional community public buildings, however, typically feature fixed spatial layouts and rigid structures, rendering them ill-equipped to respond effectively to the rapid evolution of community requirements. This often results in structures becoming functionally obsolete before reaching their physical lifespan. Dutch scholar Jan Habraken's Supports Theory offers a novel approach to resolving this issue. As illustrated in Figure 2, this theory emphasizes dividing buildings into long-term, stable 'public domain' and adaptable 'infill elements'. By separating permanent structural supports (such as load-bearing walls and column grids) from replaceable infill elements (such as partition walls and equipment modules), it enables temporal adjustability and spatial adaptability of the building [16]. This concept aligns closely with the contemporary architectural emphasis on "design for adaptability".

Concurrently, the global construction industry faces challenges including excessive energy consumption, significant waste of building materials, and pressure to reduce carbon emissions [17]. As community service functions frequently evolve with demographic shifts and policy directives (e.g., converting elderly care facilities into childcare centers), the use of support structures with pre-set standardized interfaces (such as unified column grids and centralized utility shafts) enables rapid replacement of infill modules (prefabricated partition walls, intelligent storage systems). This approach avoids extensive demolition and reconstruction work [18]. This approach aligns closely with the Sustainable Development Goals outlined in the United Nations' 2030 Agenda for Sustainable Development [19], and resonates with the life-cycle management and circular construction models advocated by the European Union's Circular Economy Action Plan [20], offering a viable pathway for the green, flexible, and sustainable development of future community service centers. By integrating diverse functional requirements with sustainability imperatives, the supporting

https://doi.org/10.3390/buildings16010092

structure theory combines with building life-cycle management and circular economy principles to explore forward-looking design models for community service centers.

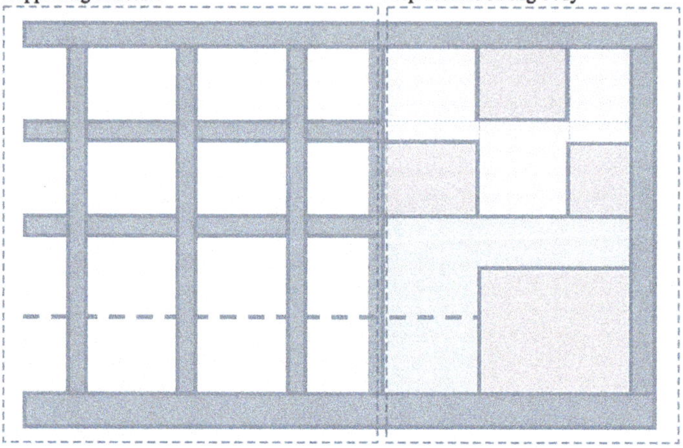

Figure 2. Fixed Support Structure and Replaceable Infill System.

2.2. Composite Spatial Structure

Facing the diverse functional requirements of community service centers, the traditional single layout often struggles to meet the needs of the compound use of space. From the renovation of small public spaces to the design of complex community centers, they all need to provide spaces for leisure activities, social interaction, and enhancing community awareness. The most typical design concept of community architecture is the community center, which provides a place for meetings and interactions, and its functions stem from the needs of the specific community [21]. Based on this requirement, the architectural design achieves an efficient combination and dynamic balance of space by staggered superimposing and organically integrating different functional spaces in the vertical or horizontal direction.

Specifically, this approach employs a three-dimensional layout to segregate conflicting functions. For instance, modules exhibiting markedly contrasting characteristics—such as noisy versus quiet, public versus private, clean versus polluting—are separated across the vertical plane, thereby effectively preventing mutual interference. Simultaneously, utilizing vertical stratification allows additional functions to be stacked upon a limited building footprint, maximizing spatial resource utilization while preserving the building's overall fluidity and coherence. On the "plan" level, horizontal interweaving techniques are employed to arrange different functions or spaces in an interlocking pattern, with partial overlaps or offset connections creating a visual permeability and functional interaction.

As shown in Figure 3, "vertical stratification" rationally organizes functions; "horizontal interlacing" emotionally creates communication. This method not only enhances the complexity and utilization efficiency of the building space at the physical level but also promotes communication and interaction between different groups of people and functional modules at the social level. In a complex spatial environment, residents can not only conveniently access diverse services but also enhance the overall vitality and cohesion of the community during cross-functional communication processes [22]. Especially under the current development requirements of "high density, multiple functions, low energy consumption, and strong sense of belonging", functional diversification has become the core strategy for the design of community service centers.

https://doi.org/10.3390/buildings16010092

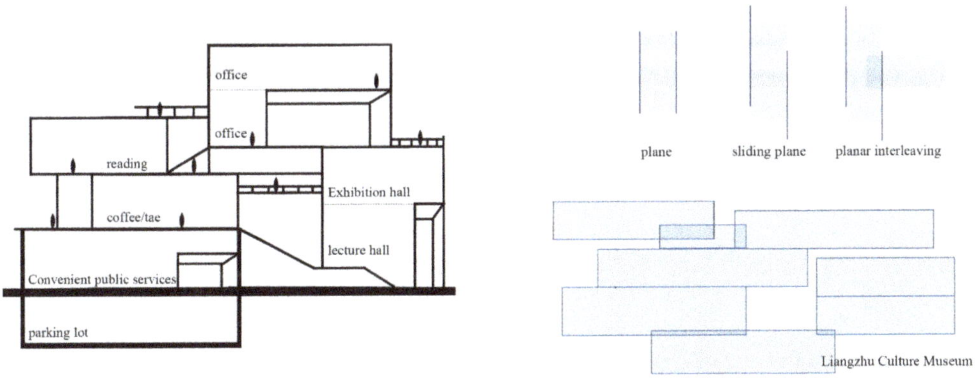

(a) (b)

Figure 3. (**a**) Vertical stratification; (**b**) Horizontal staggering.

With the continuous concentration of urban populations and the increasing scarcity of land resources, the rational integration of diverse functions—including public services, cultural and recreational activities, healthcare and wellness, as well as education and training—within a single building system can effectively alleviate land constraints and resource fragmentation. This multifunctional layering approach not only optimizes service efficiency but also enhances social value, transforming community service centers into comprehensive platforms and spiritual anchors for urban residents' lives. More significantly, this design approach responds to the strategic imperative of modernizing China's governance systems and capabilities. As the smallest unit of social governance, communities benefit from enhanced capacity and flexibility through integrated design, facilitating the establishment of convenient service networks that support the '15-min living circle' concept [23].

2.3. Spatial Permeability

Spatial Permeability proposes an adaptive design methodology within the framework of the "Supporting Structure Theory", integrating porous interfaces with modular spatial decomposition. Its core objective lies in enhancing a building's environmental resilience, spatial adaptability, and social interactivity by introducing permeability and variability simultaneously to both its exterior and interior. By deconstructing large-scale service spaces into smaller-scale units, the spatial arrangement becomes more flexible and user-friendly. These smaller units are then staggered and layered to form a 'porous interface', enabling spatial permeability, communication, and interoperability. The specific parameters of this design approach are detailed in Table 1:

Table 1. Composition of spatial permeability.

Dimension	Porous Interface	Large Space Disassembly	Space Permeability
Core goal	Break the closure of the spatial interface	Break the integrity of the spatial volume.	Let the spaces flow and penetrate each other.
Implementation methods	Semi-transparent interfaces such as grilles and glass partitions	Modular space, box-in-box, staggered layout	Maintain the continuity of vision and circulation.
Sensory experience	Light, air, and sight lines can penetrate.	Human activities and lines of sight can interact and extend.	Blur the boundaries between "inside and outside", "big and small", and "front and back".
Design effect	Soften the interface and make the space transparent.	Intimate scale and flexible layout	The fluidity and complexity of space are enhanced.

 https://doi.org/10.3390/buildings16010092

(1) "Porous" interface: Spaces are not entirely sealed or fragmented, but rather allow light, sightlines, air and activity to permeate and flow through porous interfaces, layered variations or scale divisions, thereby enhancing spatial openness and continuity. Brzezicki (2019) notes that 'translucent facades' represent a significant contemporary architectural trend balancing privacy and openness, achieving visual and environmental permeability through material porosity and layered construction [24]. Oyedeji (2022) further proposes that building facades should be conceptualized as 'ecological interfaces', facilitating the reciprocal flow of energy, air, and social activity between space and environment [25]. Regarding ventilation, empirical research by Shen et al. (2025) demonstrates that porous facades can significantly reduce wind pressure and improve the ventilation efficiency of double skin curtain walls [26].

(2) Large-scale spatial decomposition: This approach aims to deconstruct expansive continuous spaces into independent, reconfigurable small-scale modular units. It achieves spatial flexibility and functional versatility while preserving overall continuity, aligning with Habraken's theory of 'separating structure from infill'. Leuven proposed the concept of 'time-based architecture', emphasizing that buildings should possess structurally stable frameworks with adaptable functions [27]. Kendall & Teicher, in Open Building, note that integrating modularity with the supporting structure enables long-term renewal without altering the building's fundamental framework [28].

The coupling of permeability with modularity, combined with porous interfaces and the deconstruction of modular spaces, achieves dual-level adaptability: Physical Adaptability: The building forms a "breathing" system through permeable facades, optimizing ventilation and daylighting. Functional Adaptability: Modular layouts enable functional transitions and shifts in usage patterns without altering the structural framework. As illustrated in Figure 4, these elements collectively constitute a dynamic spatial system under the 'supporting structure-Infill System's logic, enabling the building to maintain sustainable vitality and dynamic renewal capacity amidst societal and environmental shifts.

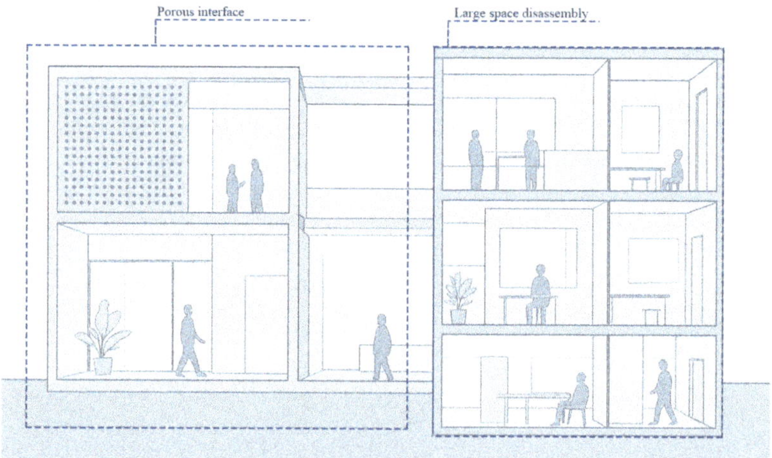

Figure 4. Diagrammatic language of spatial permeability.

(3) Space Permeability: Space Permeability refers to breaking the enclosure of building interfaces and using design techniques and structural layouts to allow light, air, sightlines, and human activities to penetrate and flow, thereby enhancing the openness, continuity, and interactivity of space. Different from the traditional division of enclosed functional areas, space Permeability emphasizes the interaction and Permeability between the interior

https://doi.org/10.3390/buildings16010092

and exterior of space, promoting the formation of an organic connection between functional modules and between people and space. This design concept is often realized by using permeable interfaces (such as perforated walls, double facades, ventilation corridors, etc.) or through the interlacing and overlapping of space layouts. While enhancing the fluidity and permeability of the space, it also improves the interactivity and diversity of the space. Spatial permeability design is not only for the optimization of functional zoning but also to promote interaction between different functional areas. Through staggered layouts and the overlapping of functional modules, visual and circulation Permeability and interaction are formed. For example, multiple functional areas are connected by a shared atrium or corridor, so that each space is not only independent but also can interact with the surrounding spaces, enhancing the sociality and dynamism of the space. This design method enables the space to more flexibly meet the changing usage needs and at the same time, enhance the spatial experience.

2.4. Three-Dimensional Composite Spatial System

Guided by the concept of 'spatial permeation', vertical stratification and horizontal inter-weaving jointly constitute the composite logic of spatial organization within the community service center. The former achieves spatial continuity and interaction through the vertical layering and stacking of distinct functional units. The latter breaks the closed nature of spatial boundaries by interweaving and shifting functional zones across the plane, and by coupling this with the installation of semi-transparent interfaces, as illustrated in Table 2. The combined action of these two approaches in both vertical and horizontal directions creates multidimensional pathways for light, airflow, and pedestrian circulation. This enhances the building's spatial openness, continuity, and complexity, as illustrated in Figure 5. That is, 'Vertical Stratification + Horizontal Inter-weaving = The Composite Pathway for Spatial Permeability'. The former creates permeability in the three-dimensional plane, while the latter extends it across the horizontal plane. Together, they elevate the building's spatial continuity, ventilation, and experiential diversity.

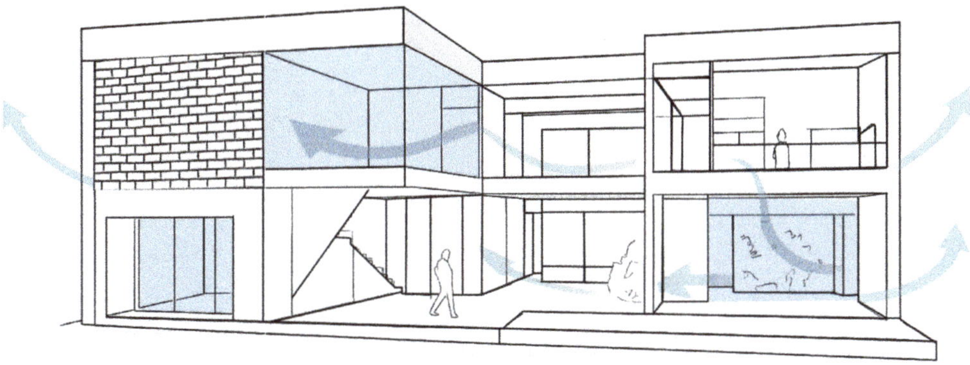

Figure 5. Schematic diagram of spatial permeability.

https://doi.org/10.3390/buildings16010092

Table 2. Implementation path of three-dimensional composite space system.

Dimension	Vertical Stratification	Horizontal Staggering	Space Permeability
Space organization	Vertical superposition of different functions	Interlaced arrangement of functions on the same floor	Visual and streamline integration occurs in the space.
Interface relationship	Floor openings and atriums connect the upper and lower spaces.	Semi-transparent partition, interlayer dislocation	Form multi-level visual and air permeability.
Usage experience	The functions are clearly defined, but the upper and lower spaces can still be perceived.	The plane is connected while keeping the boundary blurred.	Multi-dimensional interaction, continuous experience

(1) Vertical stratification: Resolves functional conflicts whilst accommodating multifunctional integration and segregating noise levels and privacy requirements; vertical stratification achieves spatial continuity and interaction through the vertical layering of distinct functional units along the height dimension [29].

(2) Horizontal interweaving: Promotes functional interaction and enhances spatial efficiency; for instance, Mies van der Rohe's Barcelona Pavilion employs interwoven walls to construct spaces that not only fulfill diverse functional requirements but also offer users an enhanced spatial experience [30], thereby improving spatial utilization and fostering serendipitous social encounters. This approach is frequently combined with spatial syntax, accessibility analysis, and behavioral observation to quantify the impact of staggered layouts on circulation patterns and congregational behavior [31].

(3) Spatial Permeability: Breaking down enclosed interfaces to enhance visual and airflow connectivity, emphasizing the shift from "isolation" to "connection": Strategies such as grilles, semi-transparent partitions, ventilated corridors, and atriums enable light, air currents, sightlines, and human activity to permeate and circulate between spatial units. This ensures functional differentiation while enhancing environmental connectivity and social dynamics [32]. Spatial permeability serves both as a means to optimize environmental performance (daylighting, ventilation, thermal behavior) and as a strategy to shape social behavior (chance encounters, visual relationships).

3. Results

The construction site is located in Yicheng District, Zhumadian City, Henan Province. The building type is a community service center. The main purpose of the design is to provide a sustainable public building that is connected to the urban space. Based on the urban spatial structure around the construction site and the functional needs of community residents, the building functions and the social and cultural values that the building must possess are reasonably planned.

As shown in Figure 6, in the design of the general plan, considering the surrounding environment of the site and the road grade, the entrances and exits of the site are first determined. The west side of the site is a city arterial road, and there is a green buffer zone that cannot be demolished. Therefore, the main entrance of the site is set on the south side, the secondary entrance on the north side, the main entrance of the building on the west side, and the secondary entrance on the east side. The southwest side of the site is open to the outside, and landscape steps are set up to correspond to the entrance square on the south side and provide a place for people to visit and rest. Since the north side of the site is a planned green space, functions such as catering and entertainment are set on the north side. According to the needs of the users, an outdoor activity square is set in the northeast

https://doi.org/10.3390/buildings16010092

corner of the site, making full use of the surrounding landscape for the activity and leisure places. The south side of the site is adjacent to the Yicheng District Government, so the exhibition-style spaces and the lecture hall are set on the south side of the site. In addition, the entrance of the underground garage is set on the north side of the site because the road adjacent to the north side of the site is a city branch road with a relatively low road grade.

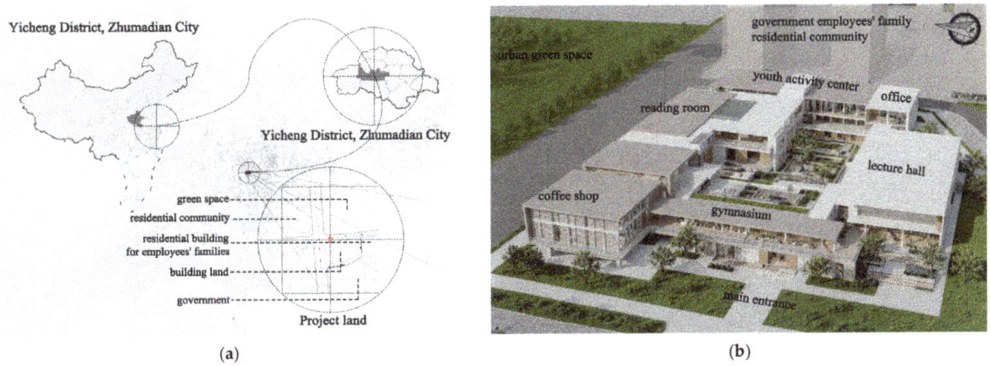

Figure 6. (**a**) Location analysis map; (**b**) Aerial view of the building.

3.1. Embodiment of the Support Body Theory

To construct a long-term building framework adaptable to social functional changes and avoid functional obsolescence, The case design divides the building structure into a fixed support structure and a replaceable infill body, and reserves a unified interface (structure, pipeline functional node) between the two to implement the design strategy of separating the "support body–infill body", thereby enhancing the flexibility, maintainability and life-cycle value of the building in the time dimension. This can enhance the flexibility, maintainability, and life-cycle value of the building over time. This method combines modular assembly, standardized column grids and the concept of circular architecture. As shown in Figure 7, its main contents and implementation key points are as follows:

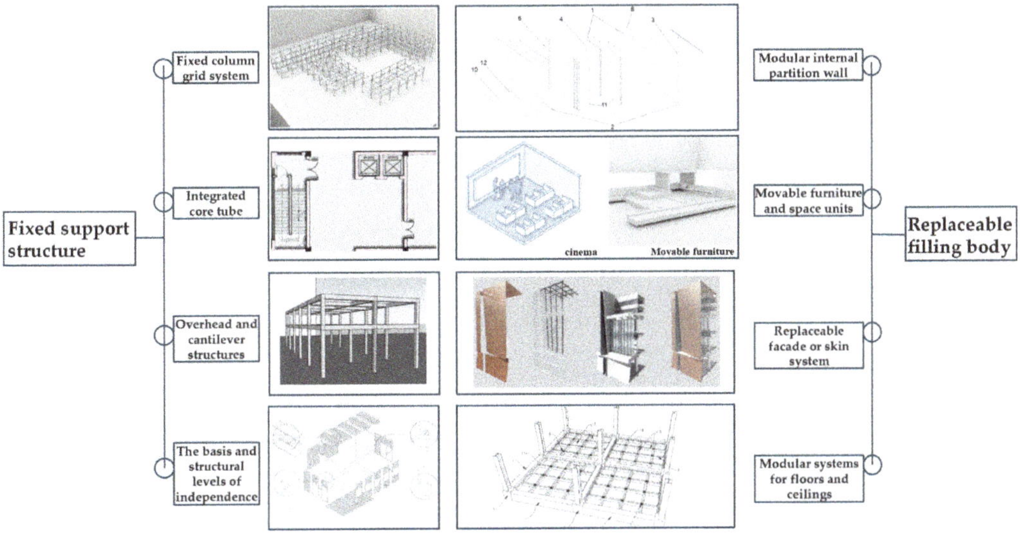

Figure 7. Fixed support structure and replaceable filler.

https://doi.org/10.3390/buildings16010092

(1) Fixed support structure

This design mainly reflects the fixed support structure from the following five aspects: 1. In the fixed column grid system, reinforced concrete is used as the long-term load-bearing system to ensure the safety and durability of the structure. A load-bearing frame is constructed by adopting regular and standardized column spacings (8 m × 8 m), thus providing "grid" support in terms of structure for the internal functional layout, and enabling the space to be reconfigured in the future without being restricted by the structure. 2. Integrated core tube: Elevators, staircases, pipe shafts, and equipment shafts are centrally arranged within the core tube of the building to ensure the stable operation of vertical transportation and electromechanical systems. Meanwhile, the peripheral space features a free layout, enhancing the flexibility of the peripheral infill. 3. Overhead and cantilever structures: Through long-span beams, truss structures, and cantilever beams, the main load-bearing structures are moved to both sides or the outer edges, and the central space is released for functional filling, thereby maximizing the variable space. 4. Foundation and structural levels of independence: Clearly distinguish between load-bearing structures and non-load-bearing components. Adopt a double-layer facade or a demountable curtain wall system, so that the service life of the supporting structure can be set to more than 50 years, while the infill can be replaced periodically (for example, every 5–15 years).

(2) Replaceable Infill System

Similarly, it is reflected in four aspects: 1. Use recyclable wooden structural supports and assemble internal partition walls with detachable panels (such as gypsum boards and wood veneers). Adopt modular assembly for easy decomposition. When the function changes, the partition walls can be quickly removed or reorganized. 2. Movable furniture and space units include furniture walls that can slide, fold, or rotate, sliding rail bookcases, movable storage modules, and folding partition walls. Making the furniture and partitions not only have functions but also play a role in space separation. Users can recombine them according to their usage needs without being fixed to the structural system. 3. The replaceable facade or skin system separates the building curtain wall and facade panels from the main structure and adopts the installation methods of slot type and hanging type, enabling the skin modules to be disassembled, replaced and updated during the operation period. 4. The modular system for the floor and ceiling is adopted. The floor uses an overhead floor system, with reserved space below for mechanical and electrical pipelines and data cable troughs. The ceiling is composed of modular ceiling panels (sound-absorbing panels and grille panels). This allows the ceiling modules to be replaced or adjusted according to functional areas, enabling faster overall construction and more convenient maintenance.

3.2. Functional Space Compounding

This study addresses the issues of scattered functions and inefficient space utilization in community service centers through functional three-dimensional superimposition and spatial boundary ablation. As shown in Figure 8, in terms of building function layout, the first floor primarily accommodates dining and entertainment, convenience services, and some office areas; the second floor includes a gym, a multifunctional hall, and additional office spaces; the top floor is divided into a reading area, an exhibition area, and a youth activity zone. The first-floor centers around an atrium, with functional and circulation spaces arranged around it, such as dining and entertainment areas and convenience services. The west square serves as the main entrance, while the east and north sides act as secondary entrances, and a dedicated entrance is provided for the lecture hall on the south side. Office areas are located on the east side, separated from the main pedestrian flow to ensure a quiet and efficient working environment. Dining areas are situated on the north side to

 https://doi.org/10.3390/buildings16010092

facilitate logistical organization while avoiding conflicts with pedestrian flow. Access to the second floor of the building is diverse, including internal vertical circulation facilities such as elevators, enclosed stairwells, and single-run stairs, as well as direct access via large external steps. The lecture hall in the southeast corner can be accessed directly from the outdoor steps. On the top floor, considering the need for a quiet environment, the north side is designated as a reading area, while the east side is planned for the youth activity center.

Figure 8. Floor plan analysis.

The functional spaces are arranged using the technique of "vertical stratification + horizontal staggering" to enhance the functional adaptability, spatial permeability, and richness of experience of the building. As shown in Figure 9, the specific practices are as follows:

(1) Vertical stratification: in this design, spaces with different functions, natures, or user groups are distributed along the vertical axis to achieve hierarchical configuration of functions, separation of noise and privacy, and optimization of traffic flow diversion. For example, the basement floor is arranged for parking and equipment rooms: High noise, low lighting function; The first floor houses a coffee area, a lecture hall, and a community service center: Highly open and strongly interactive with the external streets; The second floor houses a fitness area, an exhibition hall and activity classrooms: semispan; The west side of the third floor is for reading: silence, Offices and meeting rooms on the north side of the third floor—Highly private. Such a "bottom-up" sequence, progressing from open to

https://doi.org/10.3390/buildings16010092

semi-open and then to private, and from high noise to low noise and then to quiet, forms a typical vertical hierarchical layout.

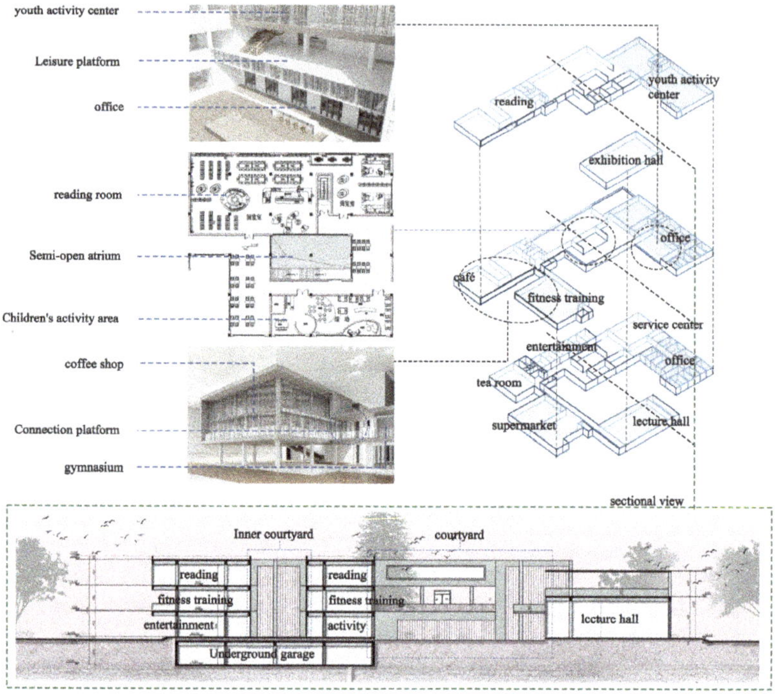

Figure 9. Vertical stratification.

(2) Horizontal staggering: In the planar layout of the building, different functional or spatial units are arranged on the same floor plane through staggering, misalignment, and partial overlapping, creating visual and activity Permeability and interaction. For example, the reading area is connected to the children's activity area through a semispan atrium, which not only achieves functional segmentation but also allows the two spaces to penetrate each other; The office area retreats backward to leave an activity platform. The part of the youth activity center above it is elevated, creating a connection between the upper and lower levels, allowing those on the upper level to view the activities on the lower level; The coffee shop and the fitness space are arranged in an "interlocking" layout through the public corridor to enhance the functional connection, These are horizontal staggering: not simply planar compartments, but a "dialogue between functions".

Combining the two methods, a pattern of "spatial Permeability" is jointly constructed in both the three-dimensional (vertical) and planar (horizontal) dimensions.

3.3. Space Decomposition

After disassembling the large space into small-scale units, on the basis of the compounding of functional spaces, through four strategies of "functional zoning modularization", "hierarchical superposition", "staggered arrangement", and "shared boundaries", the building can maintain overall continuity while achieving higher variability and social interactivity, making the functions more flexible and the space use more comfortable. At the same time, it facilitates later replacement (which is consistent with the supporting structure theory).

https://doi.org/10.3390/buildings16010092

(1) Functional zoning modularization: Divide different functions such as exhibitions, reading, fitness, and rest into standardized modular units, so that each module has relatively independent structural boundaries and functional labels. On the premise of a unified scale, interfaces, and structural nodes, these modules can be freely combined, replaced, or reconfigured. As shown in Figure 10, in the fitness space on the second floor, several "functional boxes" can be preset, including the aerobic area, yoga area, leisure area, communication area, etc. Each box can be individually removed, replaced or repositioned within the support structure.

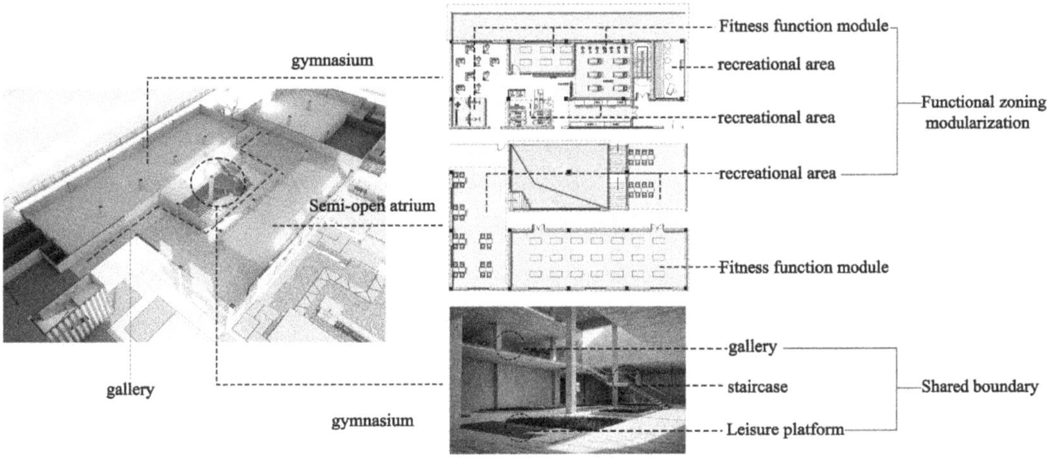

Figure 10. Horizontal staggering.

(2) Shared boundary: Similarly, if modules are completely separated by solid walls, it will instead weaken the overall coherence and interactivity. Therefore, in this case, the boundaries are shared or overlapped through means such as atriums, corridors, semi-open platforms, glass partitions, and vegetation walls, so that the modules are not completely isolated at the boundaries. As shown in Figure 9, after splitting the modules in the reading hall, a shared atrium and a corridor are left in the middle. This corridor not only serves each module but also acts as an intermediary for the sightlines, activities, and circulation routes of multiple modules.

(3) Hierarchical superposition: In addition to modularization, embedding a "box-in-box" structure and setting mezzanines within a largescale volume can further enrich the spatial experience and usage hierarchy. Create multilevel and multifunctional secondary spaces within the main space through nesting. As shown in Figure 11, by embedding the office area into the exhibition space, a closed block is independently embedded inside the large space of the original building, creating a space within a space. This not only enables the creation of good functional zoning but also enriches the spatial hierarchy and experience, making the functions of the large space more diverse.

(4) Spatial Staggering: After disassembling the large space, to enhance the structural sense and dynamic experience of the functional modules, in this case, horizontal misalignment and vertical overlapping arrangement methods are used. Stagger the functional modules on the plane and connect them with "skip floors" or "voids" in height. This makes the circulation route changeable and the line of sight penetrable when moving from one module to another. As shown in Figure 9, the exhibition hall on the second floor extends forward, while the lecture hall on the first floor is set back. The communication area and

https://doi.org/10.3390/buildings16010092

the large staircase are placed under the overhead floor of the exhibition hall and connected by stairs.

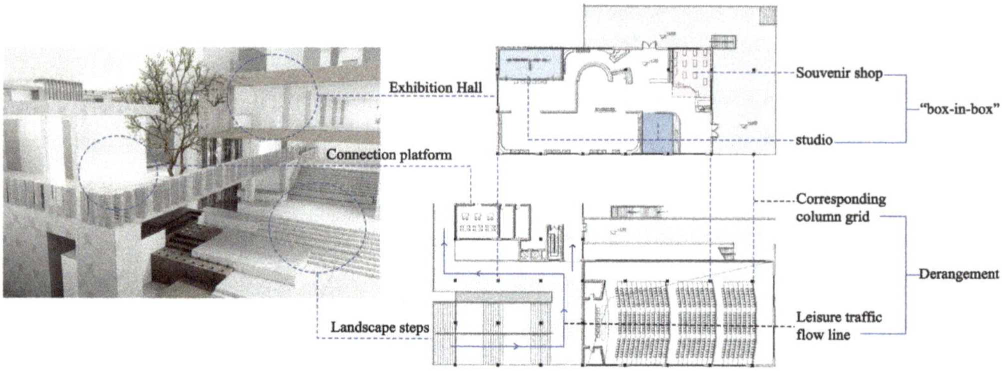

Figure 11. Hierarchical superposition; Staggering.

3.4. Porous Interface

In the building interfaces (walls, facades, partitions), permeable structures are introduced to form spaces that allow light, air, sightlines, or human activities to "penetrate", as shown in Figure 12. This is basically achieved through the following four methods:

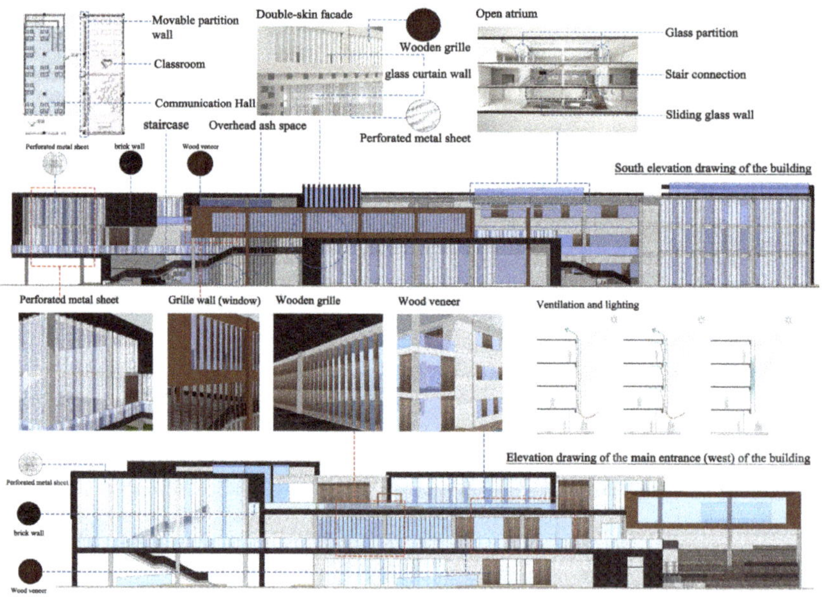

Figure 12. Porous interface.

(1) Hollowed-out wall, Grille wall: In this case, by using materials and structural forms such as perforated metal panels, wooden grilles, and brick-holed walls, the wall surface visually presents a " semi-transparent " state. For example, using wooden grille screens on the facade of the exhibition space allows natural outdoor light and air to penetrate while maintaining a certain degree of privacy. At the same time, adjust the size, density, and

 https://doi.org/10.3390/buildings16010092

arrangement rhythm of the grilles according to functional requirements to regulate the degree of line-of-sight Permeability, light Permeability, and air circulation.

(2) Ventilation corridors, open atriums: Set up shared atriums, corridors, and void walkways, etc., so that the space as an "interface" is not only a partition but also a passage for "transition" and "Permeability". For example, both sides of the atrium are separated from the functional rooms by grilles or glass partitions, but the connection of sightlines and light is retained; Set up a landscape staircase and platform that passes through the second floor from the first floor to the roof, allowing the line of sight to penetrate between floors and the airflow to rise, These spaces together form nodes for air circulation, natural lighting, and interpersonal interaction.

(3) Double-skin facade: A double-layer structure is adopted in the facade system. The outer layer consists of sun shading grilles, perforated panels, and wooden grilles; the inner layer is a glass curtain wall or high-performance glass. For example, perforated metal panels are used on the exterior facade of the lecture hall for shading and decoration. At the same time, they cooperate with the glass wall behind to form a structure of "transparent outside—blocked inside—air layer", thereby promoting ventilation. An air buffer cavity is formed between the two layers, which can play a role in sun shading, ventilation, lighting, thermal control, etc. When the direct sunlight is strong, the curved perforated metal plate can control the direct light and reduce glare. At the same time, the inner-layer glass and the buffer cavity dissipate heat through natural ventilation.

(4) Movable partition: In indoor or semi-outdoor spaces, through components such as sliding doors, folding screens, glass partitions, movable furniture, and walls, an elastic layout with "interfaces that can be both separated and permeable" is established. For example, the youth activity hall is divided into two classrooms and a communication hall by sliding glass walls. When large-scale activities need to be held, the sliding doors are opened, making the space transparent with the line of sight and circulation path connected. Usually, the doors are closed to achieve sound insulation, privacy, and functional zoning.

4. Discussion

4.1. Comparison Between Plot Ratio and Demolition and Renovation Losses

In this case, the support structure theory is used as the main framework, dividing the building's fixed supporting structure and replaceable infill into two parts. The core commonality between the Supports Theory and modular construction lies in their shared design logic: "permanent structure, variable function". Therefore, through existing research on modular buildings, the modular structure can be compared with the traditional structural system. Starting from the two aspects of floor area ratio and demolition and renovation losses, the advantages of the supporting structure can be illustrated:

(1) Although the supporting structure and the modular structure have different origins, they are highly consistent in terms of architectural adaptability and sustainable design concepts. In this case, the advantages of modularity included in the "support structure–Infill System" structure are: 1. Divide the building into the long-term stable "main structure" (support structure–main module) and the replaceable and reconfigurable "filling part" (functional module–internal unit) to achieve flexible adjustment of the space. 2. A replaceable and updatable component system can prevent the entire building from being scrapped due to functional changes, enabling the building to continuously adapt to social and functional updates. 3. By separating the structure from the function, when the building is renovated or updated, only partial replacement is required, thus reducing resource waste and downtime and lowering the "demolition and renovation loss rate".

Therefore, the advantages of the supporting structure can be illustrated by comparing the demolition and renovation losses between modular buildings and traditional buildings.

https://doi.org/10.3390/buildings16010092

The following will conduct a comparative analysis from three aspects. As shown in Table 3, from the perspective of "demolition and renovation losses", the supporting structure building theoretically has more advantages: that is, at the time of demolition and renovation or at the end of the life cycle, its resource waste and replacement cost are lower than those of traditional buildings: 1. It can effectively reduce construction waste and demolition waste. The basic concept of modular construction in supporting-body buildings is to manufacture building modules off-site and then assemble them on-site. This approach greatly reduces the amount of construction waste generated. According to a study by the Waste and Resources Action Programme (WRAP), off-site construction can reduce waste to 1.8%. In the context of traditional demolition, as pointed out by the U.S. Environmental Protection Agency (EPA), the waste generated from demolition accounts for more than 90% of the total construction and development waste [33]. The comparison of these two sets of data implies that the "support structure–Infill System" structure has a potentially lower loss ratio in terms of "demolition waste". 2. Detachable and highly reusable: After modularizing the components of the fixed support structure and the replaceable functional space, the structural components can often be prefabricated in the factory, featuring strong standardization and Demount ability. Moreover, at the end of the life cycle, "module reuse" or "relocation" can be considered instead of just complete demolition. For example, a study on modular timber structures points out that in the "End-of-life" phase (C1–C4), modular structures have the design advantage of being disassembled, while in the corresponding phase of traditional buildings, it is often just demolition + waste disposal [34]. This approach helps to reduce the losses caused by the "structural demolition requirement". In other words, it reduces the resource losses, waste, and replacement costs associated with "demolition and renovation". 3. Lower waste of materials and components and more precise manufacturing: In the "support structure–Infill System" building structure, building components are basically prefabricated in the factory. While standardizing the dimensions of building components, the assembly is also more precise, thus reducing the possibility of on-site adjustment, waste, and changes. For example, some studies have pointed out that modular buildings perform better than traditional methods in terms of material waste, waste transportation, on-site transportation, etc. [35]. Although this study is not entirely focused on the renovation and demolition stage, from the perspective of the end of the life cycle, reducing waste in the initial construction stage will also reduce losses in the renovation and demolition stage.

(2) The floor area ratio (total floor area/land area) can serve as a key indicator for measuring development intensity, spatial utilization efficiency, and potential profitability. In the traditional planning and layout, traffic flow lines (elevators, evacuation passages, corridors), building setbacks, and structural spans (beam-column systems) often lead to the occupation of a part of the effective space. For example, if larger traffic spaces, relatively wide corridors, larger beam depths, or column grid spacings are configured to meet requirements such as fire protection, safe evacuation, natural lighting, and ventilation, the available area will decrease, thereby reducing the FAR (when the land area is fixed). Therefore, the effective space stacking efficiency of traditional structures may be relatively low. In contrast, the layered staggered or three-dimensional modular systems (such as modular rooms and stackable modules) in the "support–filler" structure usually can: Shorten the traffic flow lines, optimize the horizontal and vertical movement lines, so as to reduce the proportion of corridors and public transportation spaces; The separation of the "support structure–Infill System" makes the structure more regular and the panels more standardized, thus reducing the loss of clear headroom of beams and columns and the loss of structural depth. Therefore, a hierarchical staggered or three-dimensional modular system can usually make the structure more compact, minimize the waste of traffic flow

 https://doi.org/10.3390/buildings16010092

lines, and thus increase the effective plot ratio. From the perspective of manufacturing and assembly, the modular structure contained in the "support structure–Infill System" structure may also support a larger clear height, less structural interference, and a more flexible layout, thereby further increasing the usable area [36].

Table 3. Comparison of demolition and modification losses.

Indicator	Modular Building	Traditional Architecture
FAR	Case FAR = **0.86**, vs. traditional structure, **+0.15** improvement.	traditional FAR = **0.71**
Component reusability	**Higher:** The modules are detachable, transferable, and reusable: **Low loss rate**.	**Low**: One-time structure, discarded after demolition: **High loss rate**.
Waste generation volume	**Lower:** Factory prefabrication and standardization reduce waste: **Less waste** during the demolition and renovation phase	**Higher**: Frequent on-site changes and large proportion of overall demolition: **A large amount of waste** during the demolition and renovation phase.
Transformation adaptability	**Stronger**: Modular design facilitates modification/expansion /replacement.	**Weak**: The structure and use are fixed, making it difficult to renovate, and the demolition frequency is high.
Resource replacement requirements	**Low**: Partial module reuse, less structural update	**High**: Often requires reconstruction or full replacement: High resource consumption

By comparing the two design schemes, under the condition of the same land area, use the floor area ratio calculation formula to compare the floor area ratios and illustrate their advantages and disadvantages. A hierarchical and staggered layout is adopted: the total building area is 14,500 m^2, and the land area is 16,700.54 m^2: Comparative calculation reveals a significant efficiency gain: the proposed design achieves an FAR of 0.86, whereas the traditional layout yields an FAR of 0.71. This represents an increase of approximately 0.15, validating the spatial efficiency of the vertical stratification strategy. This efficiency improvement not only enhances the space utilization rate but also partially offsets the initial cost of modular construction.

4.2. Economic and Life-Cycle Verification

Life-Cycle Cost Analysis (LCCA) is an important method for evaluating the economic benefits of a building throughout its entire life cycle, from planning, construction, and use to decommissioning. It considers not only the initial investment but also the comprehensive benefits of operation, maintenance, depreciation, and residual value. Compared with traditional buildings, in the "support structure–infill body" structure, although standardization and modularization have potential savings (up to 20%), there may also be a cost premium [36]. It mainly stems from the costs of factory prefabrication, module transportation, and high-precision assembly. However, this part of the cost can be offset by energy savings and maintenance savings in the long-term operation. Research shows that the modular system in the "support structure–Infill System" structure can shorten the construction period by about 20–40%, enable faster capital turnover, and allow for earlier commissioning, thereby reducing financing costs and interest payments [36]. In addition, the mass production of standardized components increases the material utilization rate by 10–12%, reducing waste and rework.

https://doi.org/10.3390/buildings16010092

During the construction process of the "support structure–Infill System" building, investing in higher-quality materials, energy-efficient systems, and modular designs may require a higher initial investment. However, in the long run, it can significantly reduce maintenance costs, repair frequency, and energy consumption. Life cycle cost analysis typically shows that compared with traditional buildings, costs can be reduced by 15% to 25% [37]. Traditional buildings usually require large-scale repairs around the 15th to 20th year. However, the "supporting structure–infill structure" can achieve "functional reconstruction" by replacing modules, thus avoiding overall shutdown.

The other is the environmental impact (recycling and carbon emissions): At the sustainability level, compared with traditional building practices, the greatest advantage of the "support–infill" building lies in reducing the rework rate, minimizing construction waste, and achieving the goal of carbon emission reduction [38]. Due to the high prefabrication rate, construction waste is reduced by approximately 60–70%, and transportation energy consumption is decreased by about 30% [39]. The supporting structure theory separates the fixed structure from the infill, allowing the infill to be replaced without changing the building structure. The demountability of the infill modules enables the components to be recycled after their service life, creating a potential path for "zero building waste".

Due to the short construction period of "support structure–Infill System" buildings, the on-site machinery usage time is reduced, and the indirect carbon emissions also decline significantly. For example, multiple studies have shown that the emissions of the modular building method are lower than those of the traditional building method. Compared with on-site construction, modular buildings can reduce greenhouse gas emissions by 46.9% [40]. In addition, the modular construction in the "support structure–Infill System" structure relies on off-site prefabrication, so the usage of on-site equipment and machinery is also reduced. For example, these modular materials are precisely cut in the factory first. This practice reduces the need for on-site machinery such as saws and grinders, thereby reducing waste. This practice reduces the need for on-site machinery such as saws and grinders, thereby reducing waste. At the same time, it minimizes the energy and emissions required to operate these tools, reduces the transportation of waste materials, and further lowers the greenhouse gas emissions generated during the construction process [41]. In addition, industrialized production enables more precise material sorting and energy consumption monitoring, providing a feasible technical path for the realization of carbon-neutral buildings.

It can be seen from the comparison and summary of the costs and construction periods between the support-structure buildings and traditional buildings in Table 4. The "support structure–Infill System" structure can effectively shorten the construction period and reduce the whole-life-cycle cost of the building. Although there will be a cost premium, in the long run, it can be offset by energy savings and maintenance savings; Moreover, in terms of low carbon: 1. It can reduce greenhouse gas emissions during the construction phase. 2. Due to the characteristics of unified prefabrication in the factory, construction waste can be significantly reduced. 3. Due to the separation of the fixed support and the replaceable filler, the modules can be partially replaced, which reduces the loss of decomposition and modification while achieving "functional reconstruction".

https://doi.org/10.3390/buildings16010092

Table 4. Comparison table of life cycle costs and environmental impacts between modular buildings and traditional buildings.

Indicator	Modular Building (Complying with the Support-Infill Concept)	Traditional Architecture
LCCA	The whole-life cycle cost can be reduced by **15–25%.**	The life cycle cost is high, and there are frequent repairs in the later stage.
Construction waste reduction	With a high prefabrication rate, construction waste is reduced by **60–70%.**	Large amount of waste and low recovery rate.
Carbon emissions (greenhouse gases)	Compared with on-site construction, greenhouse gas emissions can be reduced by approximately **46.9%.**	There is a large amount of on site processing and high emissions.
Recyclable	Modules can be disassembled and reused: With the potential for "zero waste"	The materials are difficult to recycle, and the losses from disassembly and modification are significant.

Finally, it is a summary of the discussion section. Through comparative analysis, this study compares the traditional building model with the building scheme after applying the "support–infill" theoretical framework, and specifically conducts the following analyses supported by statistics and data:

(1) Comparison of building FAR

Research shows that after adopting the "support–filler" structure, the floor area ratio has increased by approximately 0.15. Through the design of vertical stratification and horizontal dislocation, buildings can more effectively integrate more multi-functional spaces within a limited land area, thereby effectively improving the land use efficiency.

(2) Life Cycle Cost Analysis (LCCA)

Statistical data shows that compared with traditional buildings, buildings adopting modular design and removable fillers can reduce the total cost by about 15–25% over their life cycle. This is mainly reflected in the shortening of the construction period, the reduction in material waste, and the significant decrease in later maintenance costs.

(3) Waste reduction and carbon emission control

Through the adoption of modular design, research shows that construction waste can be reduced by 60–70%. In addition, due to the high-precision production of modular components, the waste in the on-site construction process is significantly reduced, thus helping to reduce carbon emissions throughout the building's life cycle.

5. Conclusions

This paper takes the Supports Theory as the core framework. On this basis, by combining "space Permeability" and the "composite space system", a comprehensive design path that adapts to the changes in social functions and the requirements of life-cycle management is proposed from three aspects: structural variability, spatial complexity, and architectural resilience. A sustainable architectural design method system for community service centers is constructed. Research shows the following:

(1) The separation mechanism of support structure and infill structure can effectively extend the building's life cycle. Through the demarcation of the fixed support structure and the replaceable Infill System, the building achieves flexible adaptability in the time dimension, avoids functional obsolescence, and supports future functional replacement and space reuse. Compared with traditional building structures, it can save 15% to 25% of the

https://doi.org/10.3390/buildings16010092

life-cycle cost and effectively reduce carbon emissions. For example, in on-site construction, it can reduce greenhouse gas emissions by 46.9%.

(2) The vertically stratified and horizontally interlaced spatial composite layout significantly enhances the building's spatial utilization efficiency and social interaction potential. Compared with the traditional building model, the plot ratio can be increased by 0.15. The overlapping and staggering of multi-dimensional spaces not only optimize the functional organization and save the land area but also promote communication among different groups within the community, enhancing the social cohesion and usage vitality of the building.

(3) Spatial Permeability and porous interface design demonstrate advantages in both environmental performance and social behavior. For example, by introducing semi-transparent facades, shared atriums, and movable partitions, buildings achieve multi-dimensional connectivity of light, air flow, and sightlines, enhancing natural ventilation and lighting performance, and at the same time creating an open and shared community spatial atmosphere.

Through the application of this method framework, the mixed-function requirements of community service centers have been addressed, and the policies for low-carbon buildings have been responded to. It provides a new theoretical framework and practical path for dealing with the challenges of functional diversification and sustainable development in high-density urban environments. Meanwhile, this study also has some limitations.

Limitations and Future Work:

While this study establishes a robust theoretical framework for low-carbon functional integration, we acknowledge limitations regarding quantitative validation. Current findings rely on theoretical modeling and comparative calculations (e.g., FAR and waste reduction estimates).

Although the "support-core and infill" design framework proposed in this paper provides a theoretical basis and innovative approach for low-carbon design and functional integration in community service centers, we acknowledge the study's limitations in data validation. Engineering research, especially when proposing new methods and frameworks, often faces significant challenges in verification. Due to the complexity of architectural design and the specific environmental conditions of each project, validation in practical applications usually requires longer timeframes and broader data support. Therefore, the contribution of this paper lies primarily in the proposal of the methodology and the exploration of design concepts. Future research will validate the effectiveness of these design frameworks through more extensive case studies, simulation analyses, and feedback from actual projects, further refining their application strategies. This limitation provides room for future work and lays the groundwork for more empirical research.

Although this study has made certain contributions, there are still some limitations. Although this study proposes a sustainable design framework based on the support theory and composite space organization, the analysis still lacks quantitative verification supported by Building Information Modeling (BIM). Future research will prioritize BIM-driven simulations—specifically Computational Fluid Dynamics (CFD) for ventilation and daylighting analysis—to rigorously quantify the environmental performance benefits proposed in this framework. In the section evaluating spatial flexibility, the ergonomics or human behavior analysis of the system is not included. Without users' physiological and behavioral data, such as movement patterns, activity comfort, crowd density tolerance, and accessibility indicators, the adaptability and usability of the composite spatial system cannot be comprehensively verified. Integrating ergonomic simulation tools, agent-based behavioral models, and post-occupancy evaluation (POE) in subsequent research will contribute to optimizing the functional adaptability of community service centers designed under the support–infill framework.

 https://doi.org/10.3390/buildings16010092

Author Contributions: Conceptualization, M.Z. (Mingrui Zhang); methodology, M.Z. (Mingrui Zhang) and Y.Y.; software, C.Y. and M.J.; validation, M.Z. (Mingrui Zhang) and Y.Y.; writing—original draft preparation, M.Z. (Mingrui Zhang).; writing—review and editing, M.Z. (Menglong Zhang) and Q.Y.; supervision, Q.Y.; funding acquisition, M.Z. (Menglong Zhang) and Q.Y. All authors have read and agreed to the published version of the manuscript.

Funding: This research was funded by Southwest Minzu University Graduate Innovative Research Project. Sponsor: Menglong Zhang.

Data Availability Statement: Data will be available on request.

Acknowledgments: The authors would like to extend their sincere appreciation to the Southwest Minzu University.

Conflicts of Interest: The authors declare no conflicts of interest. The funders had no role in the design of the study; in the collection, analyses, or interpretation of data; in the writing of the manuscript; or in the decision to publish the results.

References

1. Statistical Bulletin on the National Economic and Social Development of the People's Republic of China in 2023. Available online: https://www.stats.gov.cn/xxgk/sjfb/tjgb2020/202402/t20240229_1947923.html (accessed on 29 February 2024). (In Chinese)
2. Zhan, D.; Zhang, W.; Chen, L.; Yu, X.; Dang, Y. Research progress and its enlightenment of urban public service facilities allocation. *Prog. Geogr.* **2019**, *38*, 506–519.
3. Zhang, Q.; Zhao, Z.; Wang, Y.; Ma, Y.; Xi, K. Difficulties in the Construction and Operation Mechanisms of Community Embedded Service Facilities from the Perspective of Urban Renewal. *J. Urban Plan.* **2025**, 71–77. (In Chinese) [CrossRef]
4. Shen, Y.; Lai, Y. Leveraging Resources, Building Consensus, and Forming Norms: A Study on the Mechanism of Digital Empowerment of Social Organizations to Improve Community Governance. *Chin. Public Adm.* **2023**, *39*, 22–29. (In Chinese)
5. Shi, Y. Promote Community Development and Governance Through the Construction of Community Service Complexes. 2022. Available online: https://www.cssn.cn/shx/zlyzx/202211/t20221102_5558268.shtml (accessed on 18 December 2025).
6. Institute of National Opening-Up Strategy; University of International Business and Economics. Annual Report on Construction And Development of China Urban Community. 2024. Available online: https://www.pishu.com.cn/skwx_ps/bookdetail?SiteID=14&ID=15707095 (accessed on 18 December 2025).
7. Lv, Y.; Bi, J.; Yan, J. State-of-the-art in low carbon community. *Int. J. Energy A Clean Environ.* **2018**, *19*, 175–200. [CrossRef]
8. He, M.; Yuan, X.; Chen, X.; Sun, C.; Tian, Q.; Han, H. Exploration of Low-Carbon Community Space Planning and Design Based on the Concept of Carbon Neutrality. *J. Low Carbon Econ.* **2021**, *10*, 70. (In Chinese) [CrossRef]
9. Zhu, K.; Tang, J.; Jiang, N.; Zhang, W. Application of Zero Carbon Building Technique in the Community Service Center. *Build. Energy Effic.* **2019**, *47*, 11–16. (In Chinese)
10. Pitre, K. Community Hubs in Ontario: A Strategic Framework and Action Plan. 2015. Available online: https://dr6j45jk9xcmk.cloudfront.net/documents/4815/community-hubs-a-strategic-framework-and-action.pdf (accessed on 18 December 2025).
11. McShane, I.; Coffey, B. Rethinking community hubs: Community facilities as critical infrastructure. *Curr. Opin. Environ. Sustain.* **2022**, *54*, 101149. [CrossRef]
12. Jacobs, J. *Dark Age Ahead: Author of the Death and Life of Great American Cities*; Vintage Canada: Toronto, ON, Canada, 2010.
13. Khan, F.M.; Pafka, E.; Dovey, K. Extremes of mixed-use architecture: A spatial analysis of vertical functional mix in Dhaka. *City Territ. Archit.* **2022**, *9*, 31. [CrossRef]
14. UNEMORI ARCHITECTS. Sukagawa Community Center in Fukushima by UNEMORI ARCHITECTS. 2021. Available online: https://www.gooood.cn/sukagawa-community-center-in-fukushima-unemori-architects.htm (accessed on 18 December 2025).
15. Chen, L.; Cheng, Y.; Zhou, Z.; Wen, Y. The Impact of Differences in Renovation Models of Abandoned Boiler Rooms on Community Vitality—A Case Study of Shenyang, China. *Buildings* **2025**, *15*, 1807. [CrossRef]
16. Habraken, N.J. *Supports: An Alternative to Mass Housing*; Routledge: London, UK, 1999.
17. UNEP. *Here's How Buildings Contribute to Climate Change—And What Can Be Done About It*; UNEP: Nairobi, Kenya, 2025.
18. Walsh, S.J.; Shotton, E. Integrating design for adaptability, disassembly, and reuse into architectural design practice. *Sustainability* **2024**, *16*, 7771. [CrossRef]
19. UN. Transforming Our World: The 2030 Agenda for Sustainable Development. 2015. Available online: http://bit.ly/TransformAgendaSDG-pdf (accessed on 11 January 2025).
20. Mazur-Wierzbicka, E. Circular economy: Advancement of European Union countries. *Environ. Sci. Eur.* **2021**, *33*, 111. [CrossRef]

 https://doi.org/10.3390/buildings16010092

21. Butková, M. Scale of Community Centre: Clarification of the Relation Between Scale and Multifunctionality of Community Buildings. 2021. Available online: https://reference-global.com/2/v2/download/pdf/10.2478/alfa-2021-0016 (accessed on 18 December 2025).
22. Althaus, E.; Christensen, L. Community centres in increasingly diverse neighbourhoods. Policies and practices of community building in post-war housing estates in Switzerland. *Cid. Comunidades Territ.* **2022**. Available online: https://journals.openedition.org/cidades/5160 (accessed on 18 December 2025).
23. Iqbal, A.; Nazir, H.; Qazi, A.W. Exploring the 15-Minutes City Concept: Global Challenges and Opportunities in Diverse Urban Contexts. *Urban Sci.* **2025**, *9*, 252. [CrossRef]
24. Brzezicki, M. The architectural design of light-permeable facades—A summary of recent trends and observations. *Tech. Trans.* **2019**, *12*, 5–30. [CrossRef]
25. Oyedeji, T. From Porous Cities to Living Skins: Reimagining Public Space and Facades as Ecological Interfaces. 2022. Available online: https://www.academia.edu/136854976/FROM_POROUS_CITIES_TO_LIVING_SKINS_REIMAGINING_PUBLIC_SPACE_AND_FACADES_AS_ECOLOGICAL_INTERFACES (accessed on 18 December 2025).
26. Shen, G.; Bao, X.; Yao, J.; Yu, S. Experimental investigation on wind loads of planar porous façades. *J. Build. Eng.* **2025**, *103*, 112199. [CrossRef]
27. Leupen, B.; Heijne, R.; van Zwol, J. *Time-Based Architecture*; 010 Publishers: Rotterdam, The Netherlands, 2005.
28. Kendall, S.H.; Teicher, J. *Residential Open Building*; Spon Press: London, UK, 2010.
29. Wang, Z.; Wang, Y. Research on Vertical Space System of Mixed-Use Complex. *Int. J. High-Rise Build* **2015**, *4*, 153–160.
30. Xie, X.; Ding, W. An interactive approach for generating spatial architecture layout based on graph theory. *Front. Archit. Res.* **2023**, *12*, 630–650. [CrossRef]
31. Yamu, C.; Van Nes, A.; Garau, C. Bill Hillier's legacy: Space syntax—A synopsis of basic concepts, measures, and empirical application. *Sustainability* **2021**, *13*, 3394. [CrossRef]
32. Maheswari, N.; Arvanda, E.; Kusuma, N. Permeable interior: Unfolding threshold space within transit corridor. *Proc. IOP Conf. Ser. Mater. Sci. Eng.* **2019**, *523*, 012059. [CrossRef]
33. MBI. The Environmental Impact of Traditional vs. Modular Construction. 2021. Available online: https://www.modular.org/2021/07/14/the-environmental-impact-of-traditonal-vs-modular-construction (accessed on 18 December 2025).
34. Krajewska, A.; Siewczyńska, M. Modular Timber Structures As a Proposal to Reduce the Carbon Footprint in Civil Engineering. *Drewno* **2025**, *68*, 216. [CrossRef]
35. Wang, Y.; Lam, S.-K.; Wu, Z.; Gong, L.; Li, H.; Jiang, M. Does the Modular Construction Project Outperform the Traditional One? A Comparative Life Cycle Analysis Study in Hong Kong. *Buildings* **2025**, *15*, 2811. [CrossRef]
36. Bertram, N.; Fuchs, S.; Mischke, J.; Palter, R.; Strube, G.; Woetzel, J. *Modular Construction: From Projects to Products*; McKinsey & Company: Capital Projects & Infrastructure: Chicago, IL, USA, 2019; Volume 1, pp. 1–34.
37. Velent. Beyond Speed: How Modular Construction Impacts Building Performance and Lifecycle Costs. 2025. Available online: https://velent.com/beyond-speed-how-modular-construction-impacts-building-performance-and-lifecycle-costs (accessed on 18 December 2025).
38. Zohourian, M.; Pamidimukkala, A.; Kermanshachi, S.; Almaskati, D. Modular Construction: A Comprehensive Review. *Buildings* **2025**, *15*, 2020. [CrossRef]
39. Kamali, M.; Hewage, K.; Sadiq, R. Conventional versus modular construction methods: A comparative cradle-to-gate LCA for residential buildings. *Energy Build.* **2019**, *204*, 109479. [CrossRef]
40. Pervez, H.; Ali, Y.; Petrillo, A. A quantitative assessment of greenhouse gas (GHG) emissions from conventional and modular construction: A case of developing country. *J. Clean. Prod.* **2021**, *294*, 126210. [CrossRef]
41. Aghasizadeh, S.; Tabadkani, A.; Hajirasouli, A.; Banihashemi, S. Environmental and economic performance of prefabricated construction: A review. *Environ. Impact Assess. Rev.* **2022**, *97*, 106897. [CrossRef]

https://doi.org/10.3390/buildings16010092

buildings

MDPI

Article

The Effects of Spatial Layout on Efficiency of Safe Evacuation After Conversion of an Exhibition Building to a Fangcang Shelter Hospital

Zhanzhi Wan [1,*], Fangming Yang [2], Tiejun Zhou [2] and Chao Li [2]

[1] School of Art and Design, Chongqing Jiaotong University, Chongqing 400074, China
[2] Faculty of Architecture and Urban Planning, Chongqing University, Chongqing 400045, China; 202315131110@stu.cqu.edu.cn (F.Y.); arch_ztj@cqu.edu.cn (T.Z.); 20231501009@stu.cqu.edu.cn (C.L.)
* Correspondence: wanzz@cqjtu.edu.cn

Abstract: When a building normally used for exhibitions is converted into a Fangcang shelter hospital in emergency situations, its original space combination, functional flow line, and safety exits are significantly changed. When the building becomes densely populated, if an accident such as a fire, explosion, or earthquake occurs, then safe evacuation will be a serious challenge. This study systematically considers the characteristics of the building space and functional flow line after the conversion of an exhibition building to a Fangcang shelter hospital. Pathfinder software was used to simulate representative scenarios of a Fangcang shelter hospital and to analyze the main spatial factors affecting evacuation efficiency in terms of evacuation time, spatial congestion characteristics, and the exits used by personnel. Then, a targeted design optimization strategy was proposed based on the accessibility of safety exits and the internal space layout of the building. Finally, a simulation was used to verify the effectiveness of the design strategy. The results of this study provide solid theoretical support and methodological guidance for the spatial arrangement of exhibition buildings converted into Fangcang shelter hospitals so as to effectively improve the efficiency of safe evacuation and promote the resilience and safety of exhibition buildings.

Keywords: Fangcang shelter hospital; exhibition building; spatial layout; emergency evacuation; simulation

1. Introduction

In recent years, major infectious public health emergencies, such as SARS (2003), H1N1 (2009), and COVID-19 (2019), occurred worldwide, posing a danger to public health [1]. During the COVID-19 pandemic, the large number of infected patients quickly taxed the resources of cities around the globe. With the continued increase in infections, patients urgently needed to be isolated or treated, resulting in a serious shortage of beds in healthcare facilities [2]. Therefore, when facing a public health emergency caused by an infectious disease, cities must be able to establish sufficient medical space to ensure that mildly or asymptomatically infected people can be isolated and treated so as to effectively control the spread of the virus and reduce the number of larger-scale infections and severe cases.

The concept of Fangcang shelter hospitals was proposed and implemented by China in response to the emergence of the COVID-19 pandemic [3]. During an outbreak, large space public buildings in cities, such as exhibition buildings, stadiums, and large warehouses, are converted into Fangcang shelter hospitals through reconstruction or expansion [4]. Fangcang shelter hospitals enable the temporary isolation of infected people on a large scale, thereby strengthening the function of urban healthcare systems and enhancing public safety.

As large space public buildings, exhibition buildings are characterized by high ceilings and a large space that can be flexibly arranged according to different uses. In the case of a public health and safety emergency, an exhibition building can be quickly converted into a Fangcang shelter hospital for patient isolation and treatment. The China Optics Valley Convention & Exhibition Center (COVCEC) in 2020 [5], Institución Ferial de Madrid (IFEMA) in 2020 [6], Baltimore Convention Center (BCCFH) in 2020 [7], and Shanghai National Exhibition and Convention Center (NECC) in 2022 [8] were all transformed into Fangcang shelter hospitals for the centralized treatment of mildly ill and asymptomatically infected patients, playing a key role in curbing the spread of infectious disease outbreaks [9].

Whether an exhibition building is used for exhibitions or medical functions, many people gather in the building's space. In the event of an emergency such as a fire, earthquake, or terrorist attack, people must be quickly evacuated from indoors to safe outdoor areas. However, as the evacuation process is prone to crowding, jamming, and stampeding, which may cause casualties, the most important task is to ensure safety.

China's codes, such as the Design Code for Exhibition Building (JGJ 218-2010) [10] and the Code for Fire Protection Design of Buildings (GB 50016-2014) [11], make detailed provisions for the safe evacuation and spatial layout of exhibition buildings. As for the conversion of exhibition buildings into Fangcang shelter hospitals, specifications such as Technical Specification for Large Space Building Retrofitted Fangcang Shelter Hospitals (T/CECS 1206-2022) [12] and the Design Standards of Infectious Disease Emergency Medical Facilities for Novel Coronavirus (2019-nCoV) Infected Pneumonia (T/CECS 661-2020) [13] have been issued. However, as Fangcang shelter hospitals are an emerging concept, their specifications have been hurriedly formulated [14], and, thus, cannot always comprehensively and effectively guide design work related to their internal space arrangement and the safe evacuation of people.

Research on the safe evacuation of exhibition buildings has focused on the impact of the booth layout [15–17] and evacuation signage [18,19]. However, there are significant differences in terms of spatial requirements and bed arrangements when an exhibition building is used as a Fangcang shelter hospital. Studies have shown that the arrangement of rooms and facilities as well as changes in paths and their widths [20,21] have an impact on the evacuation paths and efficiency. A good design can effectively improve evacuation efficiency [22]. Current studies on the conversion of exhibition buildings into Fangcang shelter hospitals focus on indoor environmental quality [23,24], thermal comfort and energy consumption [25,26], and the application of intelligent management and control systems [4,5]. No research has been conducted on the influence of spatial layout changes on safe evacuation. To summarize, compared with relevant research on exhibition buildings, that on safe evacuation after conversion to Fangcang shelter hospitals is significantly lacking. Existing standards, studies, and practical applications of safety evacuation have been insufficiently targeted, and the impact of the conversion of exhibition buildings to Fangcang shelter hospitals on safe evacuation has not been fully considered. This poses a serious challenge to the efficiency and scientificity of personnel evacuation; thus, there is an urgent need to explore and deepen the relevant research.

In this study, Thunderhead Pathfinder software (2019.3.1217) was used to simulate representative evacuation scenarios, which were used to conduct an in-depth study of factors influencing the spatial layout of exhibition buildings on the safe evacuation of people in unexpected situations after their conversion to Fangcang shelter hospitals, while considering aspects such as evacuation time, spatial congestion characteristics, and the exits used by personnel. Based on this, an optimization design strategy for the space layout of Fangcang shelter hospitals was proposed and verified in terms of two aspects: namely, the accessibility of safety exits and the internal space layout of the building, with the aim of providing reasonable suggestions and guidance.

The remainder of this study is organized as follows: Section 2 introduces the characteristics of the building space layout after its conversion to a Fangcang shelter hospital. In Section 3, a simulation model of safe evacuation is established, and Pathfinder is used to simulate typical evacuation scenarios. In Section 4, the simulation results are collected and analyzed, and the main spatial layout factors affecting safe evacuation are also analyzed. In Section 5, targeted optimization design strategies are proposed, and their reliability and effectiveness are verified. Section 6 presents the conclusions.

2. Architectural Spatial Characteristics of an Exhibition Building Converted into a Fangcang Shelter Hospital

The spatial arrangement of an exhibition building is characterized by versatility and flexibility. Its main function in daily use is to hold exhibitions. The main goal of its architectural space and functional layout is to maximize its use and effect for exhibitions. When converted for use as a Fangcang shelter hospital to isolate and treat patients in the case of a public health emergency, its use mode is changed. Fangcang shelter hospitals are characterized by complex environments and diverse types of medical personnel and admitted patients [27]. In this case, the formation of closed-loop process control must be ensured, with the division of clean and contaminated areas. The nature of the use of an exhibition building in different scenarios leads to different spatial attributes (Table 1).

Table 1. Characteristics of different use functions and people in exhibition buildings.

The Time Period Used	Daily Use	Emergency Use
Building function	Exhibition	Fangcang shelter hospital
Personnel type	Exhibition visitors, exhibitors, and hall managers	Healthcare workers, patients with infectious diseases (patients are mainly asymptomatic or have mild cases of infectious diseases, are less than 60 years old, and are able to take care of themselves), management, and logistic service staff
Personnel activity	Viewing, networking, touring, discussions	Hospitalization, consultation, eating, washing, and work study
Activity time characteristics	Mobility and short term	Fixed and long term

To prevent viruses from spreading externally through the air, according to the specification requirements, the building layout of a Fangcang shelter hospital must meet the medical process design principle of "three zones and two pathways", where the three zones are deemed to be contaminated, semi-contaminated, and clean, and the two pathways are for healthcare personnel and patients [3,4] (Figure 1). Additionally, it is necessary to fully consider the setup of different entrances and exits of the building, such as a patient entrance and exit, healthcare personnel entrance and exit, a logistic material entrance, and a contaminated material exit. There should be corresponding flow lines, such as for healthcare personnel, patients, and contaminated material. These help to ensure the safety of patients and healthcare personnel and to avoid the mixing of different flow lines and functions.

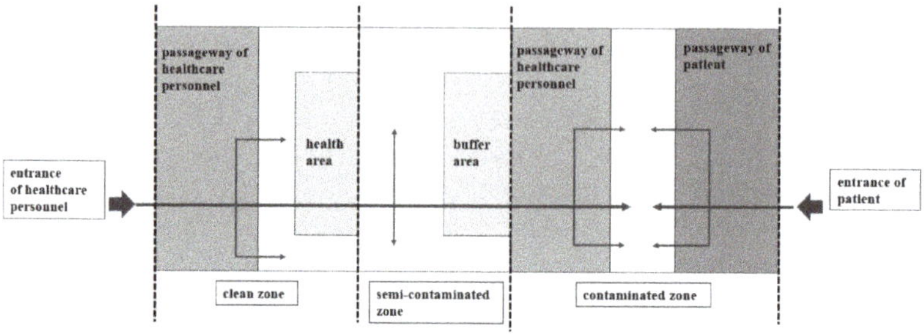

Figure 1. Schematic diagram of "three zones and two pathways".

According to specification requirements, such as those of Technical Specification for Large Space Building Retrofitted Fangcang Shelter Hospitals (T/CECS 1206-2022) as a centralized place for temporary isolation and treatment, a Fangcang shelter hospital has strict requirements on the functional zoning of the building and the flow for hygienic use. The original exhibition space is designated as a contaminated area, meant mainly for the functions of patient admission and treatment. As an exhibition building has multiple exits, new temporary functional areas are set up and connected to different safety exits. Functional areas are designated as contaminated (covering toilets and showers), semi-contaminated (healthcare work), and clean (healthcare living). To minimize virus outflow, a Fangcang shelter hospital retains only one patient entrance, one patient exit, one or two healthcare entrances, one material entrance, and one contaminated material exit. Some entrances are connected to temporary additions such as toilets, showers, or healthcare work areas, while the remaining entrances and exits are sealed off to all people in the building (Table 2). Compared to the exhibition building's normal use, this reduces the number of entrances and exits. However, this makes it impossible for some personnel to evacuate through the nearest entrance or exit, resulting in a longer and more complicated evacuation route.

Table 2. Schematic diagram of architectural spatial function patterns of exhibition buildings.

Use Type	Normal: Exhibition	Emergency: Fangcang Shelter Hospital
Schematic plan	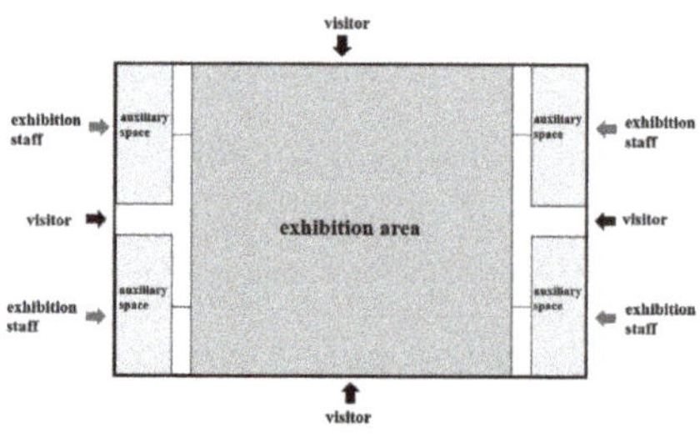	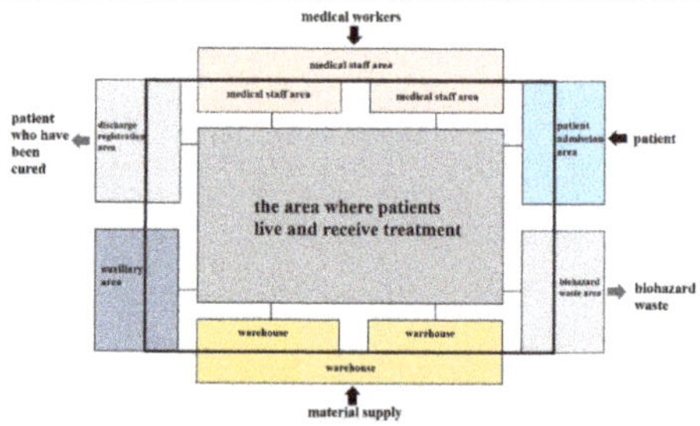
Current status photo		

3. Methodology

3.1. Pathfinder Evacuation Simulation Software

Many researchers have used Pathfinder software to simulate the evacuation of people in public buildings, seeking an optimized solution for evacuation in disaster scenarios [28,29]. Developed by the American Thunderhead Engineering Consultants, Pathfinder is an intuitive and easy-to-use intelligent emergency evacuation evaluation system. It presents a visual display of the evacuation paths and times of people during simulation by systematically defining personnel parameters (e.g., the number of people, walking speed, type of people, and safety exit selection preferences) [30]. It can construct a model according to the actual size of a building, perform a graphical virtual rehearsal of the evacuation behavior of personnel, and visualize the simulation results. Therefore, its application functions could meet the simulation needs in this study.

3.2. Scenario Creation

According to the characteristics of the architectural space arrangement of the Fangcang shelter hospital described in Section 2 of this paper, a typical space scenario was constructed. The most common rectangular plane was selected to carry out a simulation of the safe evacuation of personnel. Pathfinder was used to construct the evacuation scenario model, with the length, width, and height of the building set to 156 m $\times$ 72 m $\times$ 18 m, respectively, and the floor area set to 11,232 m^2. A total of eight safety exits were set up in the building, including one on each short side (numbered 1 and 5) and three on each long side (numbered 2–4 and 6–8). The width of the double doors in the building was 1.8 m and that of the single doors was 1 m. The width of the pathway was set to 2.5–6.5 m. The width and location of the safety exit and the width of the passage in the exhibition hall all met the requirements of the specification.

The exhibition building was converted into a Fangcang shelter hospital for use in emergency situations based on the Technical Specification for Large Space Building Ret-rofitted Fangcang Shelter Hospital (T/CECS 1206-2022), the Design Standard of Infectious Disease Emergency Medical Facilities for Novel Coronavirus (2019-nCoV) Infected Pneumonia (T/CECS 661-2020), etc. The Fangcang shelter hospital could accommodate 880 nursing unit beds laid out in a fishbone shape, with bed areas on both sides and the healthcare area in the middle. There were eight safety exits connected to the restrooms, healthcare entrances and exits, and functional rooms for patient admission and examination (Figure 2).

3.3. Personnel Parameter Settings

The activities of the personnel in the Fangcang shelter hospital are diversified. During the day (08:30–23:00), most personnel are freely distributed in different locations, engaging in social behaviors such as entertainment and communication. At night (23:00–08:30 the next day), they gradually disperse and return to their beds for rest. The gathering status and activities of the personnel vary significantly from daytime to nighttime.

The individual characteristics and initial positions of the personnel to be evacuated were set according to the spatial layout of the Fangcang shelter hospital and the activity characteristics of personnel at different time periods [31]. Therefore, the simulation of personnel evacuation was divided into two scenarios: a day-time scenario, where personnel move freely in the building space, and a nighttime scenario, where personnel mainly return to their beds for rest. When an emergency occurs in the Fangcang shelter hospital, personnel must be evacuated to a safe outdoor area. In the evacuation process, personnel cannot cross obstacles. They first escape to a pathway and then to the different safety exits through the pathway.

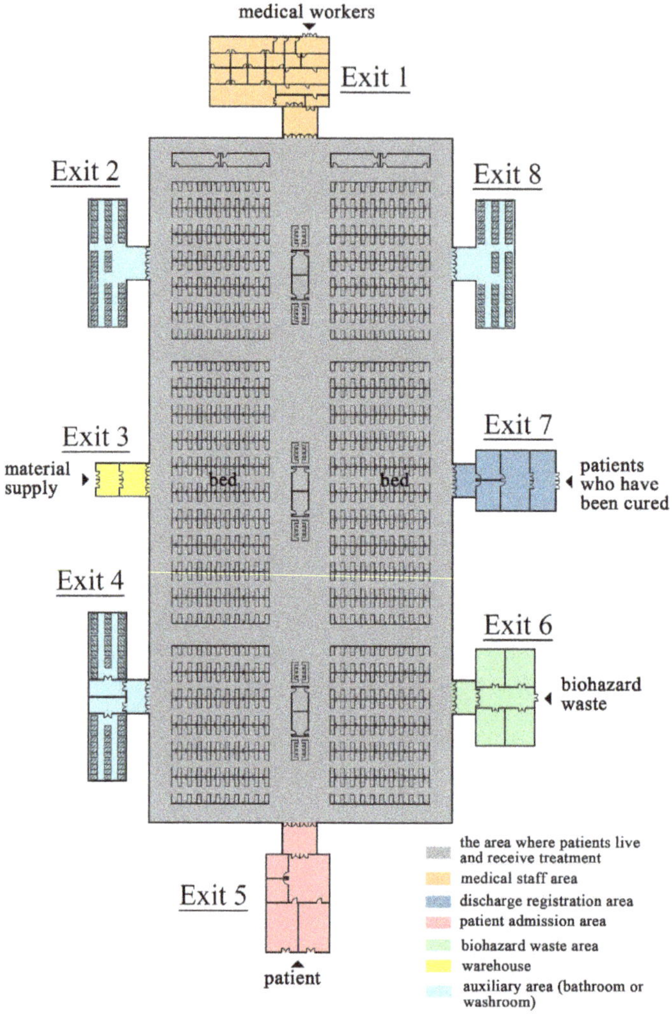

Figure 2. Layout of Fangcang shelter hospital.

As stated above, the hospital could accommodate 880 beds. According to the requirements of relevant design specifications, it is necessary to equip medical personnel according to a bed-to-nurse ratio of 1:0.2 and a physician-to-nurse ratio of 1:5. For every 100 beds, there are also shifts (24 h rotating shifts of 6 h). One police officer, two security guards, and one cleaner are also required. The total number of people in the simulation was finally determined to be 1128.

The main purpose of this simulation was to study the impact of the spatial layout on the evacuation of personnel. Therefore, the type of personnel was simplified, and the male-to-female ratio was set to 1:1. The evacuation speeds of different types of personnel were determined according to relevant studies and standards [32–34], as shown in Table 3. The personnel had an average shoulder width of 42–50 cm, and the buffer zone had a minimum value of 0.48 m. Other relevant parameters were taken as the system default values of the Pathfinder software.

In Pathfinder software, personnel evacuation behavior mode settings are mainly divided into two types: SFPE mode and Steering mode [30]. Compared to SFPE mode, Steering mode focuses on individual interaction behavior and pathway selection in evacuation scenarios. This model emphasizes that individuals are influenced by other pedestrians and environmental factors during the evacuation process, and their choice of safety exits can more accurately reflect the dynamic obstacle avoidance and congestion phenomena observed during the evacuation process. Therefore, Steering mode was selected when running the simulations to more accurately simulate the evacuation of personnel in the environment of a Fangcang shelter hospital [35].

Table 3. Evacuee settings.

Personnel Type	Amount	Speed (m/s)	Notes
Patients	88	Males: 0.45–0.6 Females: 0.35–0.5	10% of the population in Fangcang shelter hospitals are elderly patients (men or women over 60 years of age) [36]
	792		Mildly ill adult patients evacuate at the same rate as healthy adults
Doctors	36	Males: 0.9–1.2, Females: 0.7–1.0	-
Nurses	176		-
Police officers	9		-
Security guards	18		-
Cleaners	9		-

Note: Individuals other than patients (e.g., doctors, nurses, police officers, and security guards) are not infected by a virus and are required to wear medical protective clothing. The effect of protective clothing on evacuation speed is not considered in this study [37].

4. Results

In this study, Pathfinder was used to simulate the evacuation of people after the con-version of an exhibition building to a Fangcang shelter hospital, and it was used to obtain relevant research data. The simulation results were compared and analyzed in terms of evacuation time, spatial congestion characteristics, and exits used by personnel in order to study the main influencing factors of the spatial layout of the hospital on the safe evacuation of people.

In the environmental settings of this simulation, only the evacuation simulation was carried out, without separately considering the influence of fire and smoke on evacuation. This is because, in emergencies other than a fire, such as terrorist attacks and sudden earthquakes, the influence of smoke can be ignored and thus does not need to be given priority attention [38]. Moreover, according to existing research results, it has been confirmed that fire smoke in an exhibition building can be effectively controlled [19]. The ceiling height of an exhibition building is usually in the range of 10–20 m, and enough space can be reserved at the top of the building to store smoke. At the same time, smoke can be quickly exhausted through natural and mechanical ventilation so as to effectively control its adverse effects on personnel safety [39].

4.1. Analysis of Evacuation Time

The evacuation time is the total time taken from the time of sounding the emergency alarm and people responding to it to the completion of safe evacuation [40]; it is expressed as follows:

$$T_{REST} = T_a + T_p + T_m, \quad (1)$$

where T_a is the disaster alarm time, T_p is the people's reaction time, and T_m is the evacuation action time.

In this study, $T_a = 5$ s and $T_p = 10$ s. The initial random distribution of people may lead to differences in the results of each simulation. To improve the accuracy of the simulation data, the same evacuation scenario was simulated five times with the same parameters, and the average evacuation time was determined [41].

The simulation results indicate that the evacuation of a total of 1128 people from an unconverted exhibition building (Scenario 3) is completed within 85.3 s, while the evacuation times after conversion to a Fangcang shelter hospital are 225.3 s and 230.8 s in the daytime (Scenario 1) and nighttime (Scenario 2), respectively (Figure 3). A comparison of Scenarios 1–3 clearly demonstrates that the spatial layout of the building has a significant effect on the efficiency of safe evacuation, which is consistent with the results of previous research [17]. It also demonstrates the importance of studying the spatial arrangement after conversion to ensure safe evacuation.

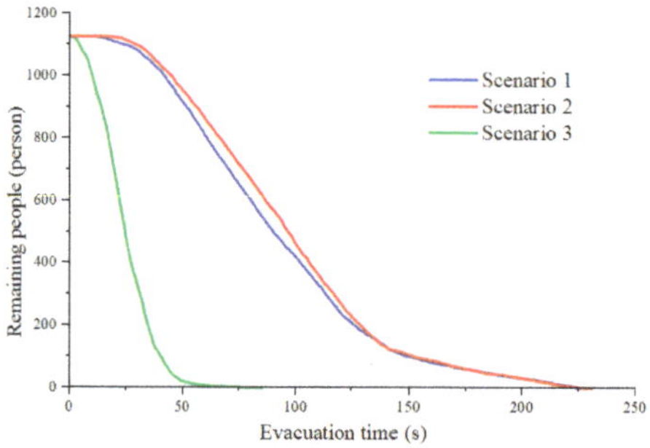

Figure 3. Variation curves of remaining number of people in building and evacuation time under three scenarios.

Further comparison is conducted on the evacuation of people at different times of day and night after conversion. In the daytime (Scenario 1), people are distributed in different areas for social activities. In the nighttime (Scenario 2), people return to their beds for rest, and their distribution within the space is more uniform. Although there are differences in the characteristics of the use of the space in these two scenarios, the variation curves of the number of remaining people and the evacuation time show that the change patterns in Scenarios 1 and 2 are the same, and there is no significant difference in the evacuation time between day and night.

4.2. Analysis of Spatial Characterization of Congestion

Congestion during evacuation is closely related to the spatial arrangement. Crowd density is the best indicator of whether an area is congested. In this study, 2 people/m^2 was taken as the critical value for identifying a congested area; i.e., when this value was reached or exceeded, it was determined that there was congestion or a risk of it [42].

Under normal circumstances, people are free to move around or rest in a Fangcang shelter hospital. In an emergency, they receive evacuation information and evacuate from their locations toward the nearest safety exit. A comparison of the evacuation densities of Scenarios 1 and 2 (Table 4) shows that, when the evacuation time is 20 s, in both scenarios, people are concentrated in the internal pathway, evacuating toward the nearest safety exit. In this process, the highest density is observed in each pathway because, during evacuation, people rush, become concentrated, and flow in large numbers to the evacuation pathway.

Additionally, the evacuation distance from the safety exit leads to a sharp increase in the density of people in the evacuation pathway, thus creating congestion.

Table 4. Evacuation density maps for different scenarios in Fangcang shelter hospital.

Evacuation Time	20 s	80 s	140 s
Daytime (Scenario 1)	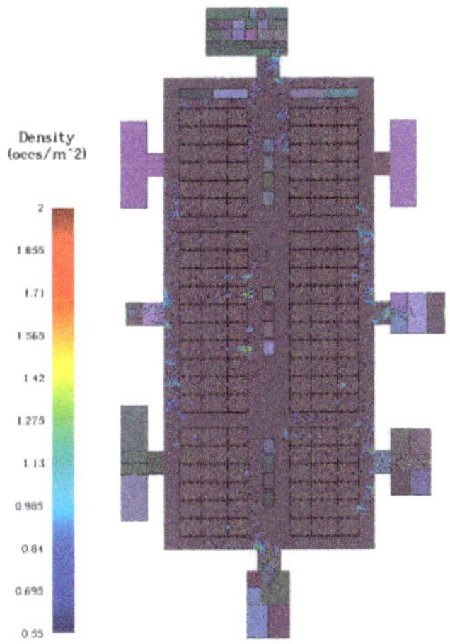	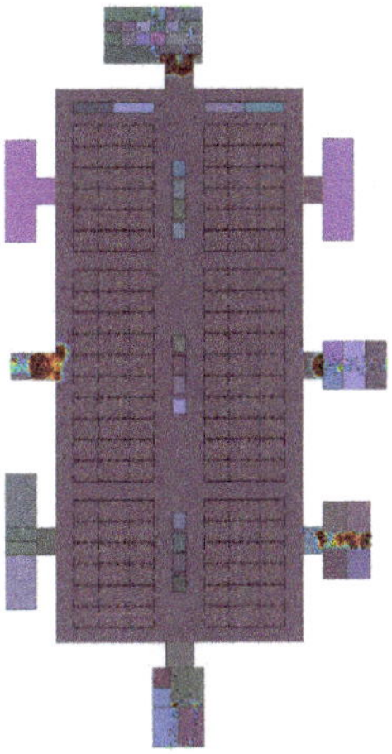	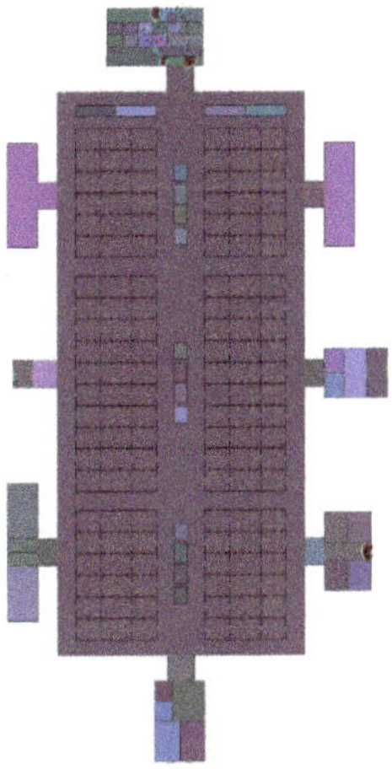
Nighttime (Scenario 2)	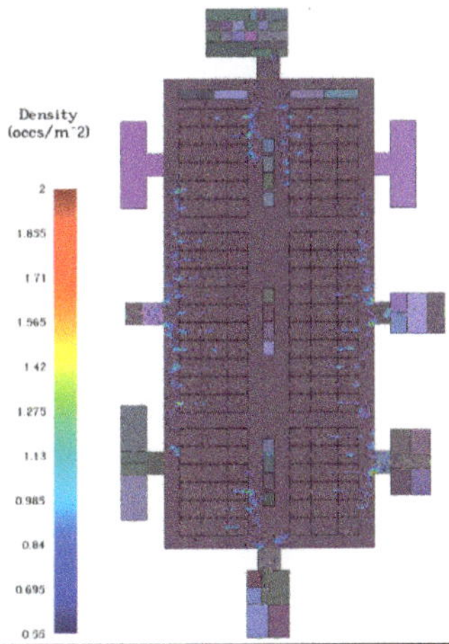	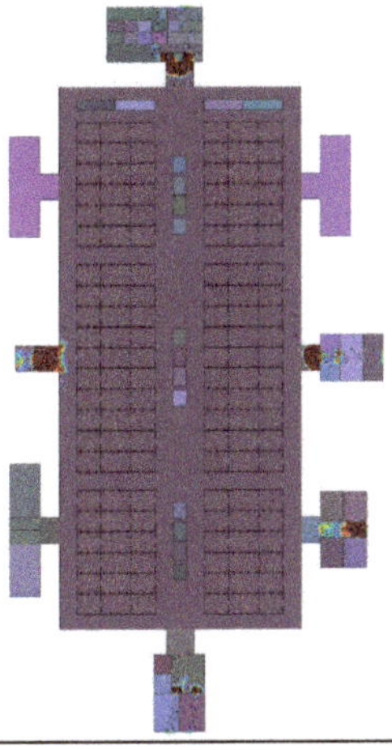	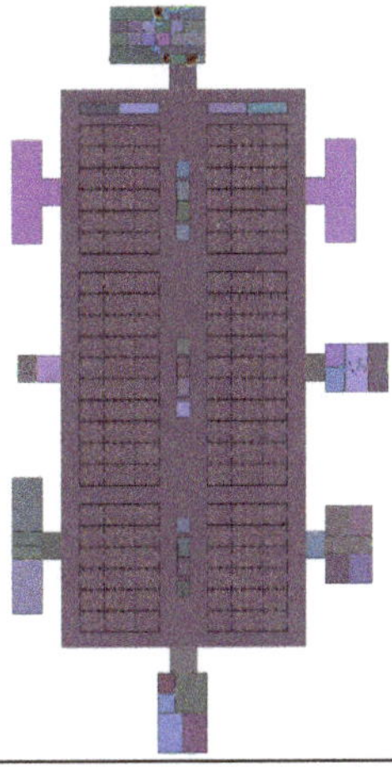

When the evacuation time is 80 s, the greatest density of evacuees is observed at the various safety exits, where congestion occurs; this is mainly due to the change in function and flow setup after conversion. Some safety exits (exits 2, 4, and 8) are used to connect functional rooms such as showers and toilets, which do not have direct access to outside, resulting in a reduction in the number of safety exits. Moreover, some exits are used as sterilization areas for healthcare access, resulting in narrower evacuation pathways and complicated routes. People are evacuated at the Fangcang shelter hospital's remaining safety exits (exits 1, 3, 5, 6, and 7). When the evacuation time is 140 s, most people have been evacuated, but a few are stranded in the newly added functional rooms, such as the healthcare living area. These rooms are more complex because of the health and safety requirements of the clean and contaminated flow lines and the division of internal rooms; as a result, people are not familiar with the environment and are hence unable to evacuate quickly.

The simulation results show that some key areas in the Fangcang shelter hospital, such as pathway intersections, safety exits, and the connections between safety exits and new facilities, are prone to evacuation congestion, which has an important impact on safe evacuation. Therefore, the rational planning of the spatial layout of Fangcang shelter hos-pitals is important, as is the setting of safety exits so as to reduce congestion, improve the efficiency of safe evacuation, and minimize the risk of casualties.

4.3. Analysis of Evacuation Flow Rate at Safety Exits

As can be seen in Figure 4, people mainly evacuate through safety exits 1, 3, 5, 6, and 7. The width of the safety exit is the sum of the widths of all doors. The flow rate of evacuees passing through these exits fluctuates over time; eventually, it levels off and returns to a steady state, indicating that evacuation has been completed. The higher flow rate through exits 3 and 7 is due to their location being in the central area of the hospital, where they can facilitate evacuation on both sides of the structure. Exit 1, however, has a higher flow rate in the middle and at the end of the evacuation because people are unfamiliar with the evacuation routes due to the complexity of partitioning within the additional functional rooms to which the exit is connected. As for safety exits 2, 4, and 8, after the conversion of the exhibition building, some safety exits are closed due to their connection to ancillary functional rooms. Thus, people cannot pass through these exits, which, in turn, exacerbates the congestion of other safety exits. This highlights the key role of the rational setting of the use or closure of safety exits in the evacuation process.

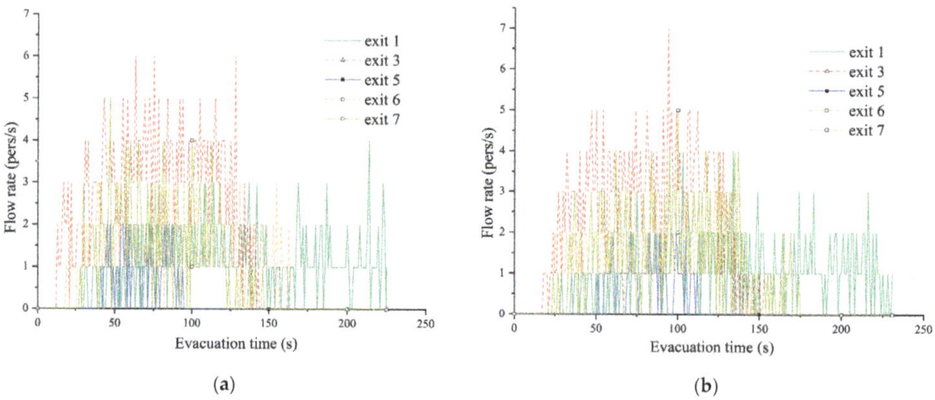

Figure 4. Variation curves of flow rate of people at different exits in Fangcang shelter hospital. (**a**) Daytime (Scenario 1); (**b**) Nighttime (Scenario 2).

5. Discussion

Based on the simulation results presented in Section 4, the congestion situation and its causes in the space area of the Fangcang shelter hospital were analyzed. It was found that the safety exits and spatial layout are not reasonable, significantly reducing evacuation efficiency. To address this problem, targeted optimization design strategies were proposed for the aspects of safety exit design and space layout optimization in Fangcang shelter hospitals. Their effectiveness was verified through empirical evidence in a bid to enhance the efficiency of safe evacuation.

5.1. Optimization Suggestions

(1) Improving of evacuation efficiency through accessibility of safety exits

The conversion of some safety exits in the Fangcang shelter hospital causes congestion during evacuation, resulting in the inability to realize timely and rapid evacuation. After conversion, it is necessary to connect the safety exits to the new temporary functional rooms, leading to the closure or repurposing of some of them; this reduces the number of safety exits directly connected to the outside, which seriously affects their accessibility. This is similar to the situation found by Brzezińska et al. [43]; i.e., upon the conversion of a public building into a Fangcang shelter hospital, failure to adjust the location and number of safety exits will have a significant impact on the safe evacuation time.

Therefore, it is necessary to flexibly adjust the connection relationship between the safety exits and additional functional rooms according to the actual setup of the safety exits. If it is not possible to increase the number of safety exits, and existing exits must be occupied, the provision of safety exits can be optimized through micro-modification. The design should realize the free combination of building safety exits and new functional rooms in a "plug-and-play" manner while improving the accessibility of safety exits so as to ensure quick evacuation. An effective solution is to install temporary safety exits in the pathways connecting safety exits to the new functional rooms (Table 5). As can be seen in Table 5, compared with Scenario a, Scenarios b–d are optimized in different ways for the connection of existing safety exits. Additionally, intelligent control means can be used to ensure that these temporary exits are closed in the normal state; only in emergency situations can they be opened and quickly put into use so as to enhance the efficiency of evacuation.

On this basis, simulations were carried out for the four scenarios, where the number of people passing through the safety exits in the area was set to 200 so as to compare and analyze evacuation under the combination of the four safety evacuation pathways. To obtain accurate results, a simulation with the same parameters was carried out five times for each group, and the average evacuation time was determined.

Figure 5 shows that the evacuation time of Scenario a is the longest, indicating that, in this case, the evacuation of added rooms through safety exits is not conducive to reasonable evacuation. In addition, in the comparison of Scenarios b–d, it is more favorable to choose different directions for evacuation when safety exits are added, thus avoiding possible congestion caused by same-direction evacuation. Scenario b has the shortest evacuation time, reflecting a more favorable design scheme for modification.

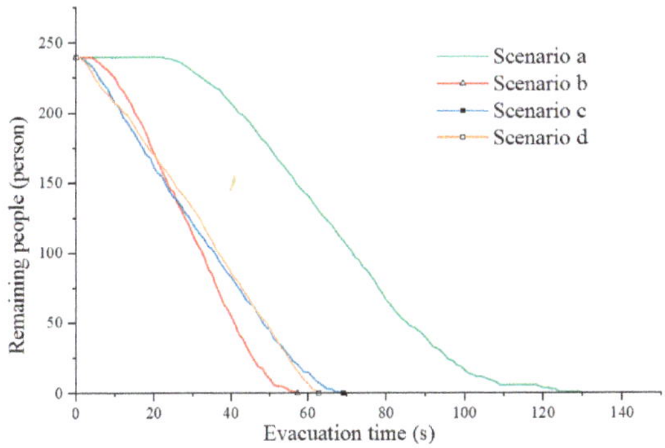

Figure 5. Variation curves of remaining number of people versus evacuation time after safety optimization.

Table 5. Optimized design measures for provision of safety exits in Fangcang shelter hospitals.

Scenario		Schematic Diagram of Safety Exit Setup	Design Notes
Before optimized design	a	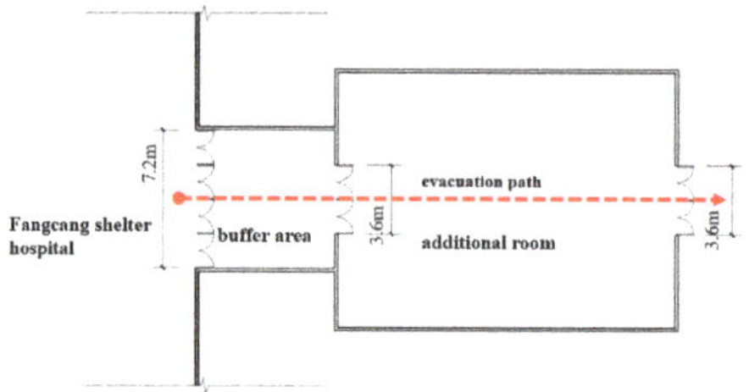	When Fangcang shelter hospital is in use, existing measures connect all safety exits to additional temporary functional rooms.
	b	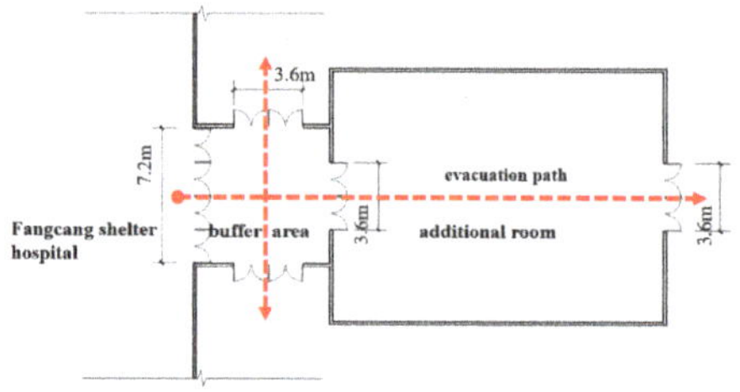	Installation of additional safety exits on both sides of transition space to avoid need to evacuate personnel through additional temporary functional rooms. Control opening and closing of safety exits as required.
After optimized design	c	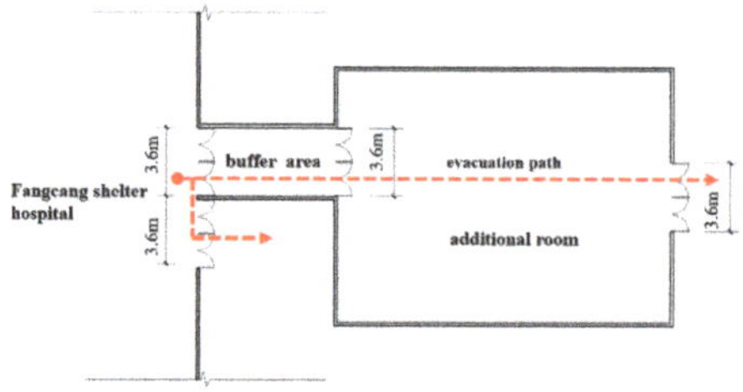	Occupancy of security exits is reduced, consistent with need for new temporary functional rooms. For security exits not connected to transition space, access control is installed so that security exits can be opened and closed as required via remote intelligent control.
	d	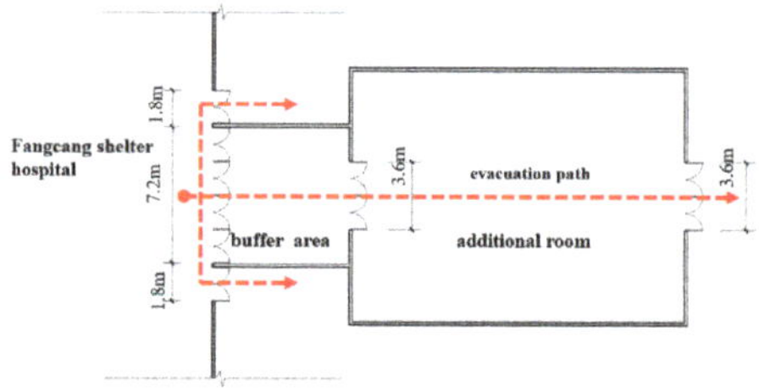	Where conditions permit, wall settings are conditionally altered, new safety exits can be added around the original safety exits, and these can be opened and closed as needed via remote intelligent control.

(2) Improving of evacuation efficiency through space layout

To reduce evacuation congestion, it is necessary to adjust the spatial arrangement of the Fangcang shelter hospital. The architectural space is mainly made up of bed areas and pathways. Wang et al. [44] found that evacuees are more inclined to choose straight paths over turning paths. Therefore, assuming a constant number of beds, the numbers and configurations of beds in different areas can be reasonably adjusted to ensure that safety exits and evacuation pathways are connected in a straight line as much as possible so as to reduce spatial obstructions in bottleneck areas and enhance the efficiency of space utilization (Figure 6). This will increase the throughput of the bottleneck nodes and improve accessibility to the safety zone.

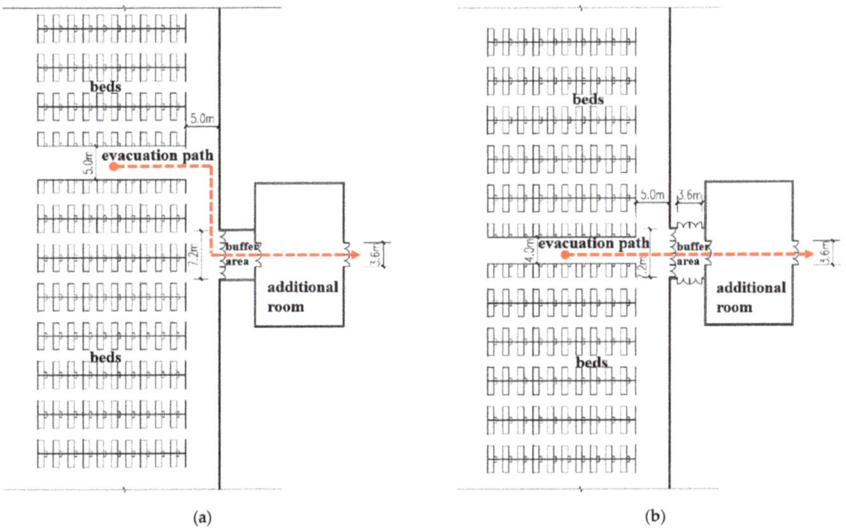

Figure 6. Optimization design measures for evacuation routes and bed layout in Fangcang shelter hospital. (**a**) Before optimized design; (**b**) after optimized design.

Given a fixed total floor area, too great a width of the evacuation pathway will result in a waste of space inside the building and a reduction in the number of beds. The width of the evacuation pathway can be adjusted to enhance the overall smoothness of evacuation. The width of an evacuation pathway in a Fangcang shelter hospital is generally set in the range of 2.5–6.5 m. To determine the optimal width, in this simulation test, the length of the evacuation pathway was set to 20 m, and the width was incremented in 0.5 m intervals. The number of people in the simulations was set to 150, 200, 250, 300, 350, and 400. To obtain accurate results, the simulation was carried out with the same parameters five times for each group, and the average evacuation time was determined.

The simulation results show that when the width exceeds 3.5 m, the variation curve of the trend of the evacuation time with the evacuation pathway width becomes flatter, indicating that the increase in the width of the evacuation pathway has less impact on the efficiency of evacuation (Figure 7). Therefore, without special requirements, it is more appropriate to set the width of the evacuation pathway between 3.5 and 4.5 m. In addition, when arranging the medical equipment and facilities, it is necessary to ensure that they do not affect the width of the evacuation pathway, so as to avoid unnecessary obstructions in emergency situations.

5.2. Simulation Verification

Based on the optimization design strategy proposed in Section 5.1, the bed layout and accessibility of safety exits were adjusted. With the number of beds maintained at 880, the overall bed layout was reorganized to ensure that the evacuation routes were connected to corresponding safety exits in a straight line as much as possible. In addition, new direct routes were added to safety exits 1–8, which were external to the new facilities, so as to enhance the accessibility of safety exits and enable quicker evacuation. The width of the pathway was adjusted to 3.5–4.5 m (Figure 8). To verify the effectiveness of the optimization strategy, the same personnel parameters were used to simulate the evacuation of the optimized design.

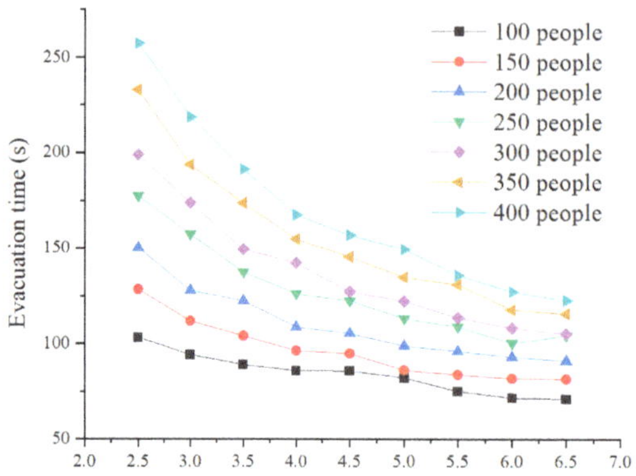

Figure 7. Variation curve of evacuation time with width of evacuation pathway for different numbers of evacuees.

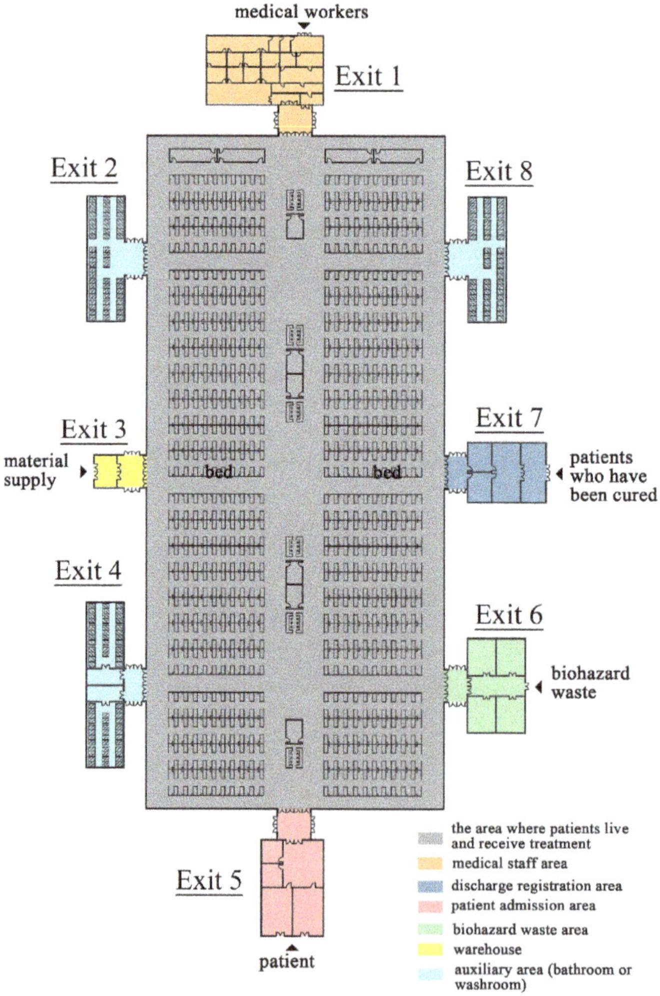

Figure 8. Layout of Fangcang shelter hospital after optimized design.

The simulation results show that the optimized design of the space layout significantly reduces the evacuation time in both the daytime and nighttime scenarios. In the daytime scenario, the evacuation time is reduced to 98.5 s, which is 126.8 s less than before optimization; in the nighttime scenario, the evacuation time is 93.5 s, which is 137.3 s less than before. The evacuation efficiencies are significantly improved, with increases of 56.3% and 59.5% in the daytime and nighttime scenarios, respectively (Figure 9). This change clearly demonstrates that the optimized design strategy effectively alleviates congestion during evacuation, ensures that people can be evacuated in a safer and more orderly manner in emergencies, and gives full play to the important role of optimized design measures in safe evacuation.

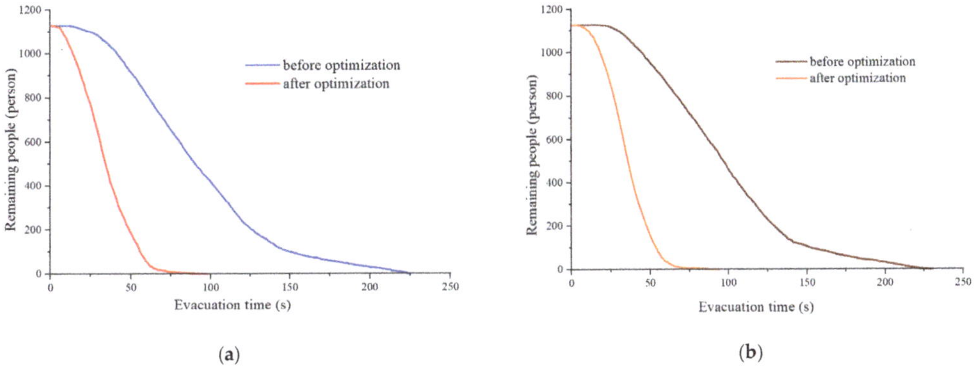

Figure 9. Comparison of number of evacuees in Fangcang shelter hospital before and after optimized design. (**a**) Daytime; (**b**) nighttime.

6. Conclusions

The conversion of an exhibition building to a Fangcang shelter hospital is an important method of response to urban public health emergencies and safety incidents, and it improves the resilience and security of cities. The most important goal is ensuring the safety of the people in the hospital. In this study, the spatial characteristics of a Fangcang shelter hospital were investigated, typical evacuation scenarios were constructed for simulation, and an in-depth analysis was performed to determine the main spatial factors leading to congestion. Based on this, an optimization design strategy of the spatial layout was proposed in terms of safety exits and the design of the internal spatial layout of the building. Evacuation simulations were carried out to verify the effectiveness and feasibility of the design optimization strategy. The main conclusions are as follows.

(1) After the conversion of an exhibition building to a Fangcang shelter hospital, the spatial layout is changed, which impacts the safe evacuation of people. Thus, it is necessary to reorganize the space, function, and flow line as well as to ensure that different functional zones correspond to reasonable functional flow lines and safety exits.

(2) Congestion in a Fangcang shelter hospital is concentrated in pathways and at safety exits. The safety exit design for a Fangcang shelter hospital is a factor that must be considered. The opening and setup of the safety exits must fully consider evacuation needs. Safety exits cannot be completely closed. Instead, micro-modifications should be made to flexibly set up safety exits and pathways so as to improve the efficiency of safe evacuation.

(3) The combination of beds and pathways in the spatial layout of Fangcang shelter hospital is the key to improving the efficiency of personnel evacuation. A simulation revealed that the efficiency of emergency evacuation is optimal when the width of the evacuation pathway is between 3.5 and 4.5 m, with further increases in the width showing no obvious improvements. Therefore, the combination layout of beds and pathways can be designed based on this conclusion.

This study discusses the impact of spatial layout changes on the safe evacuation of people after the conversion of an exhibition building into a Fangcang shelter hospital, filling a research gap. It also provides guidance for arranging the spatial layout, optimizing the bed layout and evacuation pathway factors, and improving evacuation efficiency and safety so as to promote the sustainable development of the safety of large-scale public buildings such as exhibition buildings.

In future research, more simulation scenarios will be set up according to different disaster types so as to provide a more comprehensive study of the influence of the spatial layout change of a Fangcang shelter hospital on the safe evacuation of personnel.

Author Contributions: Conceptualization, Z.W.; methodology, Z.W.; validation, C.L.; writing—original draft, Z.W. and T.Z.; writing—review and editing, F.Y.; supervision, T.Z.; simulation experiment, F.Y.; data curation, C.L.; date analysis, F.Y.; project administration, T.Z. All authors have read and agreed to the published version of the manuscript.

Funding: This research was funded by the Chongqing Social Science Planning Project (Grant No. 2024BS118), the Humanities and Social Sciences Research Program of Chongqing Municipal Education Commission (Grant Nos. 25SKGH102 and 24SKGH116), the Science and Technology Research Program of Chongqing Municipal Education Commission (Grant No. KJQN202400725), the National Natural Science Foundation of China (Grant Nos. 52278005 and 52108032), and the Research Initiation Fund Project of Chongqing Jiaotong University (Grant No. F1230063).

Data Availability Statement: The data that support the findings of this research are available on request from the corresponding author.

Acknowledgments: We would like to thank the editors and anonymous reviewers for their comments and helpful suggestions, which greatly improved the manuscript.

Conflicts of Interest: The authors declare no conflicts of interest.

References

1. Lee, J.M.; Jansen, R.; E Sanderson, K.; Guerra, F.; Keller-Olaman, S.; Murti, M.; O'sUllivan, T.L.; Law, M.P.; Schwartz, B.; E Bourns, L.; et al. Public health emergency preparedness for infectious disease emergencies: A scoping review of recent evidence. *BMC Public Health* **2023**, *23*, 420. [CrossRef]
2. Pan, A.; Liu, L.; Wang, C.; Guo, H.; Hao, X.; Wang, Q.; Huang, J.; He, N.; Yu, H.; Lin, X.; et al. Association of Public Health Interventions with the Epidemiology of the COVID-19 Outbreak in Wuhan, China. *JAMA* **2020**, *323*, 1915. [CrossRef]
3. Chen, S.; Zhang, Z.; Yang, J.; Wang, J.; Zhai, X.; Bärnighausen, T.; Wang, C. Fangcang shelter hospitals: A novel concept for responding to public health emergencies. *Lancet* **2020**, *395*, 1305–1314. [CrossRef] [PubMed]
4. Fang, D.; Pan, S.; Li, Z.; Yuan, T.; Jiang, B.; Gan, D.; Sheng, B.; Han, J.; Wang, T.; Liu, Z. Large-scale public venues as medical emergency sites in disasters: Lessons from COVID-19 and the use of Fangcang shelter hospitals in Wuhan, China. *BMJ Glob. Health* **2020**, *5*, e2815. [CrossRef] [PubMed]
5. Zhou, Y.; Wang, L.; Xu, Y.; Ding, L.; Tang, Z. Intelligent Fangcang Shelter Hospital Systems for Major Public Health Emergencies: The Case of the Optics Valley Fangcang Shelter Hospital. *J. Manag. Eng.* **2022**, *38*, 05021010. [CrossRef]
6. Valdenebro, J.; Gimena, F.N.; López, J.J. The transformation of a trade fair and exhibition centre into a field hospital for COVID-19 patients via multi-utility tunnels. *Tunn. Undergr. Space Technol.* **2021**, *113*, 103951. [CrossRef]
7. Jones, J.A.; Siddiqui, Z.K.; Callahan, C.; Leekha, S.; Smyth, S.; Preas, M.A.; Ficke, J.R.; Cabunoc, M.K.F.; Kantsiper, M.E.; the CONQUER COVID Consortium. Infection Prevention Considerations for a Multi-Mission Convention Center Field Hospital in Baltimore, Maryland, During the COVID-19 Pandemic. *Disaster Med. Public Health Prep.* **2021**, *16*, 2643–2650. [CrossRef]

8. Jiang, J.; Yin, Z.; Li, J.; Jia, L.; He, R.; Yang, W.; Yang, J.; Fan, H.; Zhang, S.; Wang, Y.; et al. Temporal and Spatial Distribution of SARS-CoV-2 Aerosols in a Large-Scale Fangcang Shelter Hospital in Shanghai, China. *Engineering* **2023**, *28*, 222–233. [CrossRef]
9. He, Q.; Xiao, H.; Li, H.-M.; Zhang, B.-B.; Li, C.-W.; Yuan, F.-J.; Yu, S.-S.; Zhang, F.; Kong, P. Practice in Information Technology Support for Fangcang Shelter Hospital during COVID-19 Epidemic in Wuhan, China. *J. Med. Syst.* **2021**, *45*, 42. [CrossRef]
10. *JGJ 218-2010*; Design Code for Exhibition Building. China Architecture & Building Press: Beijing, China, 2010.
11. *GB 50016-2014*; Code for Fire Protection Design of Buildings. China Planning Press: Beijing, China, 2014.
12. *T/CECS 1206-2022*; Technical Specification for Large Space Building Retrofitted Fangcang Shelter Hospitals. China Architecture & Building Press: Beijing, China, 2022.
13. *T/CECS 661-2020*; The Design Standards of Infectious Disease Emergency Medical Facilities for Novel Coronavirus (2019-nCoV) Infected Pneumonia. China Architecture & Building Press: Beijing, China, 2020.
14. Ji, H.; Huang, L.; Li, H.; Zhang, J.; Jing, Q.; Deng, J.-Y. A Comparative Study on Technical Standards for the Design of Emergency Medical Facilities in China in the Context of COVID-19. *Buildings* **2022**, *12*, 1502. [CrossRef]
15. Kamoun, F.; El Barachi, M.; Belqasmi, F.; Hachani, A. A smart dynamic crowd evacuation system for exhibition centers. *Procedia Comput. Sci.* **2021**, *184*, 218–225. [CrossRef]
16. Li, Z.; Xu, W.A. Pedestrian evacuation within limited-space buildings based on different exit design schemes. *Saf. Sci.* **2020**, *124*, 104575. [CrossRef]
17. Wan, Z.; Zhou, T.; Luo, N. Effect of Layout of Exhibition Hall on the Evacuation Efficiency in Exhibition and Convention Center. *Saf. Environ. Eng.* **2020**, *27*, 106–115.
18. Wan, Z.; Zhou, T.; Tang, Z.; Pan, Y.; Zhang, L. Smart Design for Evacuation Signage Layout for Exhibition Halls in Exhibition Buildings Based on Visibility. *ISPRS Int. J. Geo-Inf.* **2021**, *10*, 806. [CrossRef]
19. Wan, Z.; Zhou, T.; Xiong, J.; Pan, G. Smart Safety Design for Evacuation Signs in Large Space Buildings Based on Height Setting and Visual Range of Evacuation Signs. *Buildings* **2024**, *14*, 2875. [CrossRef]
20. Vermuyten, H.; Beliën, J.; De Boeck, L.; Reniers, G.; Wauters, T. A review of optimisation models for pedestrian evacuation and design problems. *Saf. Sci.* **2016**, *87*, 167–178. [CrossRef]
21. Zhang, Z.; Ling, W.; Yang, Z.; Wei, X.; Wang, H. A congestion prediction model for optimizing emergency evacuation design of university libraries in China. *J. Build. Eng.* **2025**, *99*, 111537. [CrossRef]
22. Shen, N.Z.; Cui, J.W.; Peng, M.; Xu, H.; Wu, Z.Q. An Optimization Method of Large Building Space Based on Evacuation Simulation. *Int. J. Simul. Model.* **2024**, *23*, 435–446. [CrossRef]
23. Guo, W.; Wang, J.; Tang, H. Investigation on construction characteristics and indoor environmental quality of Fangcang shelter hospitals during the COVID-19 pandemic in Shanghai, China. *Build. Environ.* **2024**, *262*, 111796. [CrossRef]
24. Ma, L.; Zheng, X.; Guo, Y.; Zhong, W.; Zhang, Y.; Qian, H. A multi-zone spatial flow impact factor model for evaluating and layout optimization of infection risk in a Fangcang shelter hospital. *Build. Environ.* **2022**, *214*, 108931. [CrossRef]
25. Yang, Y.; Yang, H.; Li, Q.; Zhang, L.; Dong, Z. Simulation Study on Airflow Organization and Environment in Reconstructed Fangcang Shelter Hospital Based on CFD. *Buildings* **2023**, *13*, 1269. [CrossRef]
26. Miletić, M.J.; Stamenković, M.G.; Djordjevic, M.; Komatina, D.F. Emergency conversion of sports halls into temporary hospitals caused by COVID-19 pandemic: Case studies of thermal comfort and energy consumption analysis. *Therm. Sci.* **2023**, *27*, 797–809. [CrossRef]
27. Li, Z.; Liang, T.; Zhang, J.; Song, G.; Leng, Y.; Qiao, L.; Zhan, Y.; Zhang, W. Characteristics and Occurrence Scenarios of Staff Exposed to SARS-CoV-2 Omicron in Shanghai Lingang Fangcang Shelter Hospital, CHINA. *Infect. Drug Resist.* **2023**, *16*, 10. [CrossRef]
28. Abir, I.M.; Ibrahim, A.M.; Toha, S.F.; Shafie, A.A. A review on the hospital evacuation simulation models. *Int. J. Disaster Risk Reduct.* **2022**, *77*, 103083. [CrossRef]
29. Xu, H.; Wei, Y.; Tan, Y. Optimization of emergency evacuation in complex rail transit station. *J. Build. Eng.* **2024**, *95*, 110321. [CrossRef]
30. Thunderhead Engineering. *Pathfinder*, version 2019.3.1217; Thunderhead Engineering: Manhattan, KS, USA, 2019.
31. Li, J.; Shu, Y.; Chen, N.; Wang, F.; Li, H. 'Re-socialisation' in isolated spaces: A case study on the social organisation of Fangcang shelter hospital patients under extreme spatial conditions. *Indoor Built Environ.* **2022**, *31*, 1210–1223. [CrossRef]
32. Ding, N.; Chen, T.; Zhu, Y.; Lu, Y. State-of-the-art high-rise building emergency evacuation behavior. *Phys. A Stat. Mech. Its Appl.* **2021**, *561*, 125168. [CrossRef]
33. Frohnwieser, A.; Hopf, R.; Oberzaucher, E. Human Walking Behavior—The Effect of Pedestrian Flow and Personal Space Invasion on Walking Speed and Direction. *Hum. Ethol.* **2013**, *28*, 20–28. [CrossRef]
34. Li, J.; Wang, J.; Jin, B.; Wang, Y.; Zhi, Y.; Wang, Z. Evacuation of Nursing Home Based on Massmotion: Effect of the Distribution of Dependent Elderly. *KSCE J. Civ. Eng.* **2020**, *24*, 1330–1337. [CrossRef]

35. Snohová, A.; Kučera, P. Comparative Study of SFPE and Steering Modes in Pathfinder to Optimise Evacuation Routes. *Buildings* **2024**, *14*, 2361. [CrossRef]
36. Zhang, F.Q.; Liu, H.C.; Qu, L.; Yang, Y.W.; Weng, Y.M.; Fang, Y. Clinical features of COVID-19 Omicron variant cases in makeshift hospital of National Exhibition and Convention Center (Shanghai). *China Trop. Med.* **2023**, *23*, 53–57.
37. Liu, Y.; Xia, Z.; Wang, L.; Fan, J.; Zeng, Q.; Liu, Y. Development status and trend of antivirus medical protective clothing. *J. Text. Res.* **2021**, *42*, 195–202.
38. Chen, C.; Li, Q.; Kaneko, S.; Chen, J.; Cui, X. Location optimization algorithm for emergency signs in public facilities and its application to a single-floor supermarket. *Fire Saf. J.* **2009**, *44*, 113–120. [CrossRef]
39. Huang, Y.; Wang, E.; Bie, Y. Simulation investigation on the smoke spread process in the large-space building with various height. *Case Stud. Therm. Eng.* **2020**, *18*, 100594. [CrossRef]
40. Cooper, L.Y. A concept for estimating available safe egress time in fires. *Fire Saf. J.* **1983**, *5*, 135–144. [CrossRef]
41. Hao, Y. Numerical Simulation Study on the Influence of Gate on Dormitory Evacuation in Colleges and Universities. *Saf. Secur.* **2022**, *43*, 13–17.
42. Zhong, G.; Zhai, G.; Chen, W. Optimization on Spatial Distribution of Shelter through Dynamic Evacuation Simulation of High Density Urban Area-Xinjiekou Case. *KSCE J. Civ. Eng.* **2022**, *26*, 4760–4776. [CrossRef]
43. Brzezińska, D.; Barański, M.; Bryant, P.; Haznar-Barańska, A. The safe evacuation of persons from a building operating within COVID-19 restrictions. *Build. Serv. Eng. Res. Technol.* **2022**, *43*, 461–472. [CrossRef]
44. Wang, D.; Li, N.; Wu, S.; Zhou, T. The Impact of Corridor Directional Configuration on Wayfinding Behavior during Fire Evacuation in Underground Spaces: An Empirical Study Based on Virtual Reality. *Fire* **2024**, *7*, 294. [CrossRef]

buildings

MDPI

Article

Research on the Construction and Sustainable Development of Cave Dwellings in Mogou Village, Mengzhou City

Wenhao Feng and Ying Meng *

School of Architecture, Southwest Minzu University, Chengdu 610225, China; 240833002012@stu.swun.edu.cn
* Correspondence: mying@swun.edu.cn; Tel.: +86-136-1809-2860

Abstract: Cave dwellings in the Loess Plateau and western Henan region are ancient architectural forms that offer significant potential for rural revitalization and cultural heritage preservation. This study examines their role in Mogou Village through field surveys and mathematical analysis, highlighting their contributions to urban–rural integration, rural tourism, and cultural sustainability. Key findings include the mean radiant temperature (MRT) in cave dwellings remaining stable between 22.7–25.9 °C, facilitating lower indoor temperatures in summer. These dwellings maintain an APMV within the Class I thermal environment ($-0.5 \leq$ APMV ≤ 0.5), achieving a 90% satisfaction rate for thermal comfort. Despite excellent thermal insulation, improvements in ventilation design are necessary. Moreover, cave dwellings attract urban residents, fostering urban–rural integration and rural tourism development. They also preserve cultural heritage through traditional construction techniques and philosophical thought. Preserving local characteristics while protecting traditions is essential for social harmony, economic development, and cultural inheritance, supporting sustainable urban–rural integration.

Keywords: cave dwellings; indoor environment; sustainable urbanization; Mogou Village

1. Introduction

Cultural heritage is an important carrier of China's excellent traditional culture. General Secretary Xi Jinping pointed out that "Cultural relics and cultural heritage carry the genes and bloodline of the Chinese nation and are non-renewable and irreplaceable resources of excellent Chinese civilization [1]". Outstanding cultural heritage enhances national pride and identity. It helps preserve cultural independence and uniqueness in a world of diverse and intermingling cultures [2,3]. Traditional dwellings, as key parts of historical and cultural heritage, hold villagers' historical memories, production and life wisdom, artistic achievements, and regional characteristics [4]. Often built with local materials, these dwellings adapt well to their environment. They are energy-efficient, environmentally friendly, and land-saving. In particular, the materials used have good insulation and heat resistance, effectively reducing building energy consumption. They are of great value to sustainable urbanization [5,6].

Cave dwellings are a unique type of residential architecture found worldwide. They have attracted extensive research from both domestic and international scholars. For example, Dr. Mark from the United States conducted field investigations in places like Shaanxi and Shanxi, examining the structure, function, and decoration of cave dwellings and writing 'Chinese Cave Dwelling Culture'. This work has greatly promoted cultural exchanges between China and other countries. Similarly, Professors Aoki Shiro, Miyano Akihiko, and

Chagawa Masayoshi from the Tokyo Institute of Technology in Japan have visited China multiple times to investigate the loess cave dwellings in the Yellow River basin. Their comparative studies, covering architectural technology, living culture, and environmental adaptability, have broadened the research horizons for traditional dwellings in East Asia and provided diverse ideas for the protection and inheritance of cave dwellings [7–9].

Domestic scholars have also been continuously researching traditional dwellings and cave dwellings. Associate Researcher Wang Tianyi from Sichuan University has conducted an in-depth study on the classification, spatial and temporal distribution, and technological evolution of pre-Qin cave dwellings in the Loess Plateau [10]. Professor Hou Jiyao has discussed the value of cave dwellings in the field of cultural folklore. This includes aspects such as village morphology and decorative characteristics. He has proposed strategies for protection and reuse, exploring sustainable development paths [11].

In modern society, cave dwellings have enhanced their value and vitality. They have achieved this by improving the environment, upgrading facilities, and integrating with tourism development. This allows their traditional characteristics to blend organically with modern life [12,13]. This study conducts experiments on the indoor environment of Mogou Village. It aims to reveal the value of cave dwellings in thermal insulation and in providing a comfortable thermal environment for residents. It also explores the important role of the rational development of cave dwellings in increasing the income of local residents, promoting rural revitalization, and advancing sustainable urbanization.

2. Materials and Methods

2.1. Experimental Principles and Framework

This study employed a combination of on-site monitoring and questionnaire surveys to evaluate the energy efficiency and indoor environmental quality of the cave dwellings in Mogou Village. To collect indoor environmental data, temperature and humidity sensors (Beijing Huaxia Risheng Technology Co., Ltd., Beijing, China), air quality monitors (Hangzhou Polytech Co., Ltd., Hangzhou, China), and light sensors (Chengdu Xinxin Electronics Technology Co., Ltd., Chengdu, China) were installed within the cave dwellings. One traditional cave dwelling and one conventional brick-and-concrete residence were selected as research subjects, with the sample size determined based on the results of a preliminary experiment and statistical requirements to ensure the reliability and generalizability of the study findings. The monitoring period lasted for one week in mid-July 2024, during which a continuous 24 h period covering both daytime and nighttime was chosen to fully evaluate the thermal performance of the cave dwellings under different temperature conditions. To ensure the accuracy of the sensor data, all sensors were subjected to a simple calibration before installation: temperature sensors were compared with a standard thermometer and adjusted to match the readings at room temperature; humidity sensors were compared with a standard hygrometer and adjusted in the indoor environment; air quality sensors were tested with standard gases and adjusted to match the standard values, with each sensor being calibrated three times and the average value being taken. To ensure the reliability of the data, the following straightforward and feasible data validation methods were employed in this study: eliminating obviously abnormal data points (such as values exceeding the sensor range or significantly differing from surrounding data) to ensure data rationality and conducting multiple measurements (at least three times) for each sensor to ensure that the readings remained consistent within a reasonable error range. The mean radiant temperature (MRT) calculation formula, as shown in Equation (1), was used to process the data, with the results presented in graphical form to illustrate the differences in indoor and outdoor temperature variations between cave dwellings and conventional houses.

$$MRT = \sum_{i=1}^{n} A_i T_i / \sum_{i=1}^{n} A_i \quad (1)$$

where A_i is the visible area of each surface (m^2) and T_i is the temperature of that surface (°C).

This study employs the Predicted Adaptation Mean Vote (APMV) to assess the indoor thermal environment of rural residences on the Loess Plateau during the summer season, as indicated in Equation (2).

$$APMV = PMV/(1 + \lambda PMV) \quad (2)$$

where $APMV$ is the Predicted Adaptation Mean Vote and λ is the adaptive coefficient. For residential buildings in cold regions, when $PMV \geq 0$, $\lambda = 0.24$; when $PMV < 0$, $\lambda = -0.5$.

PMV is the Predicted Mean Vote, which is determined based on the actual clothing and living habits surveyed. The metabolic rate is taken as 1.0 met, the clothing thermal resistance in summer is taken as 0.5 clo, and the indoor air temperature, relative humidity, air velocity, and mean radiant temperature (MRT) are obtained from the tests.

Additionally, to gather comprehensive feedback on living comfort, a structured questionnaire was designed and administered to residents of Mogou Village. The questionnaire included both closed-ended and open-ended questions to capture detailed insights into residents' experiences and perceptions of their living conditions. The questionnaire was divided into several sections: Demographic Information, Living Comfort, Facilities and Services, and Tourism Impact. A stratified random sampling approach was employed to ensure representation across different types of cave dwellings (traditional, homestays, and libraries) and various demographic groups. A total of 150 households were selected for the survey, with a response rate of 85%, resulting in 128 completed questionnaires. In addition to the questionnaire surveys, field visits were conducted to observe local tourism development projects such as the commercial street and children's amusement park. Interviews with local officials and business owners provided reliable data on the income of the local original residents. These qualitative insights were triangulated with the quantitative data from the questionnaires to provide a comprehensive understanding of the impact of tourism on the local community.

2.2. Study Area

Mogou Village is located in the northwest of Henan Province (Figure 1).

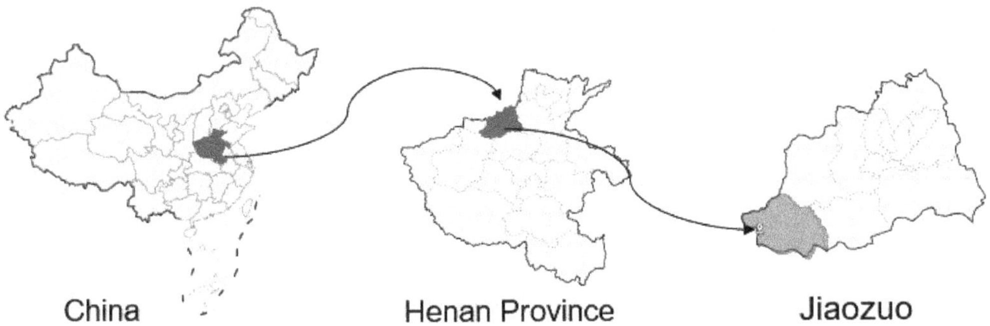

Figure 1. Location analysis of Mogou Village (drawn by the author).

The figure indicates that the village is situated in the transitional zone between the Taihang Mountains' foothills and the Yellow River Plain. It is bordered by the Taihang Mountains to the north and the Yellow River to the south. The village's terrain is higher in the north and lower in the south, surrounded by gullies on three sides, with the Shunjian Reservoir to the rear and the Wen River flowing through the west (Figure 2a). It is a typical concentration of the southern Taihang cave dwelling culture [14]. DEM stands for Digital Elevation Model, which is used to illustrate the topographical conditions of the area. It shows that the local terrain is suitable for the construction of cave dwellings by the original residents (Figure 2b). The village is situated in a steep loess hilly area with deep soil layers, primarily composed of loess parent material. Below this, there are uneven layers of sand, which typically exhibit a columnar structure and are loose in texture, with weak resistance to erosion. This leads to severe soil erosion in the area, resulting in varying degrees of residual loess tablelands, crisscrossing gullies, and depths reaching over 30 m [15]. The climate is classified as a warm temperate continental monsoon climate, with distinct seasons: warm springs, hot and rainy summers, cool autumns, and cold, dry winters.

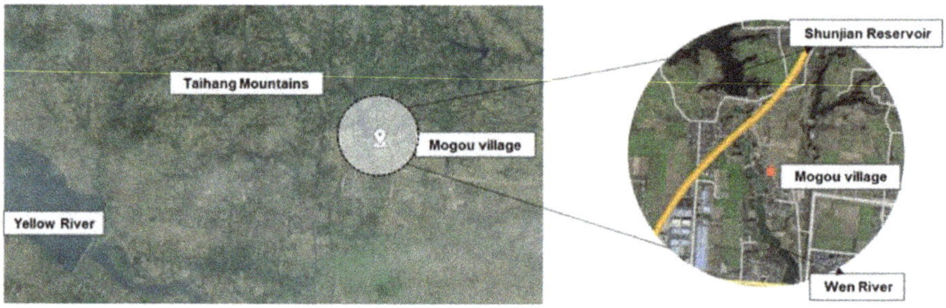

(**a**)

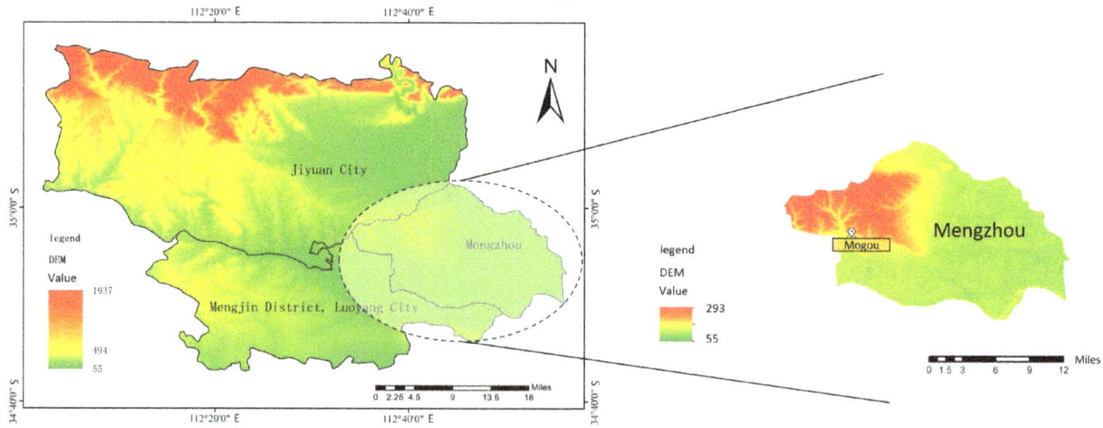

(**b**)

Figure 2. (**a**) Relationship between Mogou Village and mountains, water systems, and reservoirs. (**b**) Elevation analysis of the city (drawn by the author; elevation data sourced from the National Fundamental Geographic Information Database of China).

Cave dwellings are ingeniously excavated along the sides of mountains and gullies according to the terrain and topography. The typical dimensions of these dwellings are a depth of 7–8 m, a height of over 3 m, and a width of about 3 m (Figure 3). The cave dwellings are closely arranged, stacked in layers, and arranged in a harmonious and orderly manner (Figure 4b). To date, Mogou Village has a total of 183 cave dwellings, of which 131 can be traced back to the Ming and Qing dynasties (Figure 4a). These dwellings from the Ming and Qing dynasties were primarily used as residences by local inhabitants, as well as for storing goods and raising livestock. With a history of several hundred years, they hold significant value for studying the living and housing conditions of people at that time. As historical heritage, their rational development can also add much vitality to the tourism industry.

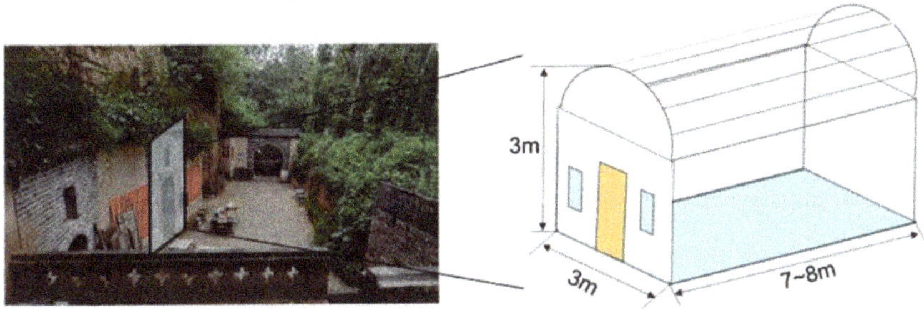

Figure 3. Cave dwellings in Mogou Village (drawn by the author).

(**a**) (**b**)

Figure 4. Qing Dynasty cave dwellings. (**a**) Existing Qing Dynasty cave dwellings in the village (photograph by the author). (**b**) Layout and internal dimensions of Qing Dynasty cave dwellings (drawn by the author).

The village was once a typical traditional agricultural planting village. With the implementation of the rural revitalization strategy, Mogou Village has developed ecological agriculture and tourism by establishing the "Mogou Village Ecological Agriculture Development Co., Ltd.". This has led to the development of a tourism industry based on the village's rich historical and cultural heritage and well-preserved cave dwellings, focusing on "ecological sightseeing, leisure vacations, and health and wellness". As a result, the villagers' production and living standards have been improved, and the village's economic development and cultural heritage have been strengthened.

3. Results

Energy efficiency in buildings is a primary objective of energy policies at the regional, national, and international levels [16]. Cave dwellings are a unique form of architecture that embodies the profound cultural heritage and ecological wisdom of the Loess Plateau region [17]. They serve not only as living spaces but also as models of harmonious coexistence with the natural environment, highlighting energy conservation and environmental protection.

Cave dwellings are characterized by durability, economic efficiency, energy conservation, and environmental protection, as well as land conservation [18]. A local folk song vividly depicts the living experience of cave dwellings: "When esteemed guests visit my home, do not mock the absence of tile-roofed houses, for the earthen caves are like divine caves, warm in winter and cool in summer". This not only reflects the comfort of living in cave dwellings but also the local residents' affection and praise, as well as their deep emotional attachment to cave dwelling culture.

3.1. Energy-Saving and Environmental Protection Characteristics

Firstly, the design of the cave dwelling's shape is also an important reflection of its energy-saving and environmentally friendly characteristics. The arched structure is not only aesthetically pleasing but also provides excellent insulation. The arched structure effectively reduces the flow of warm air, slowing down the internal air circulation of the cave dwelling, thereby reducing heat loss [19]. Additionally, the arched structure can resist external pressure to a certain extent, enhancing the stability of the building (Figure 5). The arched structure depicted in the figure is a unique architectural hallmark of the Loess Plateau region, closely related to the local loess soil. The characteristics of loess enable the excavation of arched cave dwellings to better withstand the pressure from above, ensuring the stability of the dwellings.

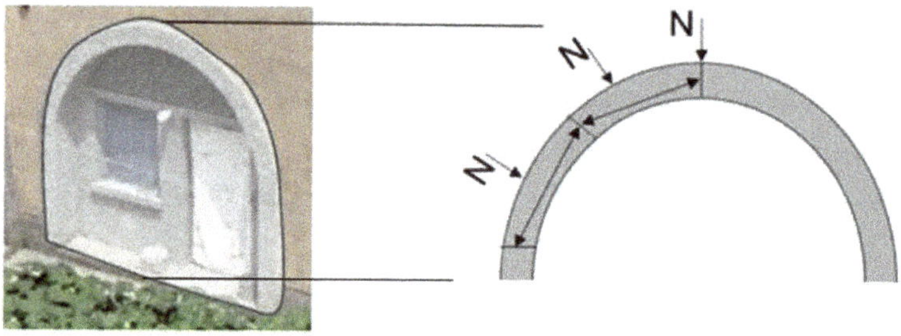

Figure 5. Schematic diagram of the arched structure of cave dwellings (drawn by the author).

Secondly, cave dwellings fully utilize locally available materials, especially loess soil, for construction. The particle structure of loess soil provides excellent thermal insulation, effectively reducing heat transfer through the walls and thus offering superior insulation performance [20,21]. In winter, the insulation properties of loess help maintain indoor warmth, reducing the need for heating; in summer, they slow down the rise of indoor temperature, reducing the use of air conditioning [22]. This natural insulation is due to the high porosity and moderate pore size of loess, which allows the interior of cave dwellings to form an effective thermal buffer layer. And during the construction process, the interior of cave dwellings is filled with materials such as loess soil and straw, which provide excellent insulation (Figure 6). The straw and other fillers illustrated in the figure form an insulating layer [23], reducing direct heat conduction and convection, thereby further enhancing the thermal insulation performance of the cave dwellings.

Figure 6. Loess and straw serve as building materials (drawn by the author).

Radiative heat transfer between room surfaces and the human body has a significant impact on the sensation of warmth or cold. In Mogou Village, the mean radiant temperature (MRT) of cliff-side cave dwellings ranges from 22.7 °C to 25.9 °C throughout the day. The narrow variation range of MRT values within the cave dwellings contributes to maintaining a stable indoor temperature. Specifically, the average MRT of the cave dwellings is 24.3 °C, with a standard deviation of 0.94 °C. Furthermore, the results indicate that the MRT of cave dwellings is consistently lower than the indoor air temperature throughout the day, with a difference of no more than 0.4 °C. This characteristic plays a significant role in reducing indoor temperatures during the summer months. In contrast, the MRT of traditional dwellings ranges from 25.8 °C to 28.5 °C throughout the day (Figure 7) [24], with an average of 27.2 °C and a standard deviation of 0.85 °C. Although the variation range is also small, the MRT is not always lower than the indoor air temperature. During the daytime when temperatures are higher, the MRT exceeds the indoor air temperature, leading to an increase in indoor temperature. The stability of cave dwellings is attributed to their thick outer walls, which significantly diminish the impact of outdoor temperatures. As mentioned above, cave dwellings utilize loess and straw as building materials, which possess excellent heat storage properties and a high volumetric heat capacity. When there are drastic changes in outdoor temperature, the heat transfer between them and the covering structure is slowed. By storing heat themselves, they reduce the amount of heat transferred to the interior during the day, achieving indoor cooling.

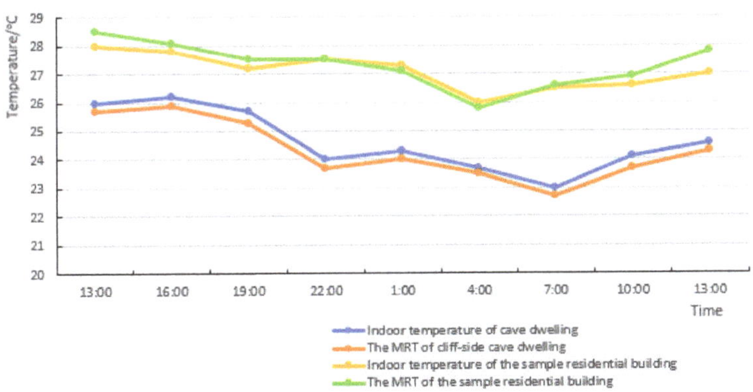

Figure 7. The mean radiant temperature (MRT) of cave dwellings and sample residential buildings (drawn by the author).

For the cliff-side cave dwellings and the sample residential buildings in Mogou Village, the variation curves were derived through Equation (2) (Figure 8). A value of 0 signifies the optimal thermal comfort condition indoors. The findings indicate that the majority of cliff-side cave dwellings in the village are situated within a Class I thermal environment ($-0.5 \leq$ APMV ≤ 0.5) [25], with an average value of -0.08 and a standard deviation of 0.16, corresponding to a satisfaction rate of 90%, thereby essentially satisfying the requirements for thermal comfort. In contrast, the sample residential buildings exhibit values within the -0.5 to 0.5 range only between 7:00 and 13:00, adhering to the Class I thermal and humidity environment standard. During other periods, the values fall outside this range, into the Class II standard (greater than 0.25 or less than -0.5), resulting in a slightly warm thermal sensation and suboptimal indoor thermal comfort in the absence of auxiliary cooling measures. The average for the sample residential buildings was found to be 0.49, with a standard deviation of 0.16, indicating higher variability and less consistent thermal comfort compared to cave dwellings. Compared to brick-and-concrete rural houses, cave dwellings demonstrate smaller fluctuations, enabling a more consistent human thermal sensation throughout the day. These results highlight the thermal efficiency and comfort provided by cave dwellings compared to traditional residential buildings, emphasizing the potential for integrating traditional architectural wisdom into modern sustainable building practices.

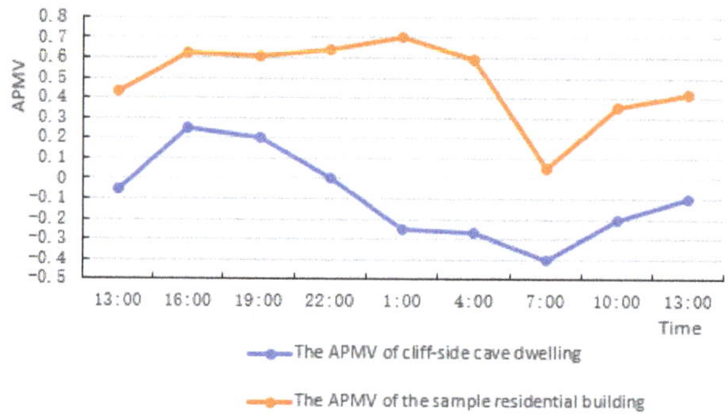

Figure 8. The APMV of cave dwellings (drawn by the author).

It is important to acknowledge the limitations of this study. The study's findings are based on a specific geographical location, Mogou Village, which may limit the generalizability of the results. Additionally, the environmental monitoring was conducted over a short period, which may not capture long-term trends or seasonal variations. The potential impact of these factors on the thermal performance should be considered when interpreting the results. Despite these limitations, the study provides valuable insights into the thermal efficiency of cave dwellings compared to traditional residential buildings, emphasizing the potential for integrating traditional architectural wisdom into modern sustainable building practices.

The study involved on-site monitoring of the indoor environmental quality of cave dwellings in Mogou Village. The results indicate that cave dwellings maintain a more stable indoor temperature compared to ordinary brick-concrete residences. The indoor thermal environment essentially meets the comfort requirements of residents, and the human thermal sensation can be kept at a consistent standard. Additionally, on non-rainy days, the indoor illuminance can reach 300–500 lux during the day, eliminating the need for additional lighting. However, radon concentration monitoring results show that some cave dwellings have slight radon exceedances.

Overall, cave dwellings maintain their energy-saving and environmentally friendly characteristics through various aspects such as building materials and structural design. These features enable cave dwellings to adapt to the climate of the Loess Plateau region and also provide valuable concepts for modern architectural design. With the increasing severity of global climate change and energy crises, the advantages of cave dwellings are expected to gain wider recognition and application, setting an example for energy conservation and environmental protection efforts. Compared to modern building standards, the ventilation system design of cave dwellings is relatively simple and cannot meet the high requirements of modern buildings for air quality. Therefore, it is recommended to introduce modern ventilation technologies, such as fresh air systems, in the renovation of cave dwellings to improve indoor air quality.

3.2. The Efficient Utilization of Land Resources

Integrating locally available building materials and bioclimatic design principles derived from existing vernacular residences into the renovation of modern residential buildings will enhance awareness of strategies for reducing energy consumption [26]. Cave dwellings make full use of the terrain and the characteristics of loess soil, efficiently utilizing land resources while adapting to the climate environment, demonstrating great efficiency and wisdom. This unique form of architecture not only cleverly utilizes the geographical features of the Loess Plateau but also makes significant contributions to the conservation and protection of land resources.

The disturbance to the ground during the construction of cave dwellings and traditional residences is different (Figure 9). As shown in the figure, firstly, the construction process of cave dwellings has minimal disturbance to the land. Traditional above-ground construction involves land leveling and clearing to accommodate foundation requirements. In contrast, cave dwellings, being underground or semi-underground, minimize land occupation. Their construction avoids extensive land disturbance and vegetation destruction. This results in more surface space being left intact [27]. Consequently, cave dwellings conserve land resources and provide conditions conducive to vegetation growth, thereby supporting ecological balance.

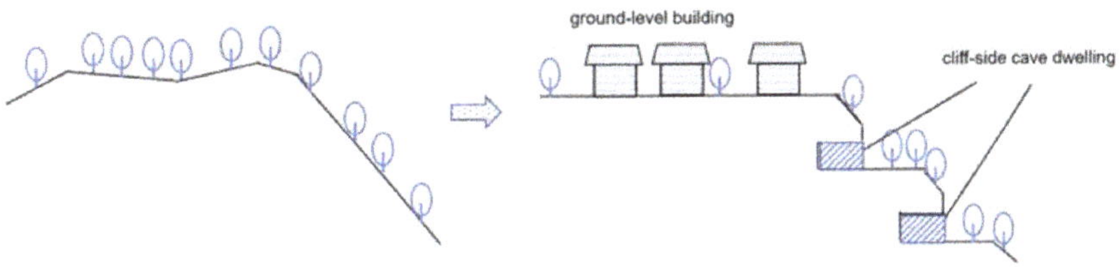

Figure 9. A comparison of the impact on original surface vegetation during the construction of semi-underground cave dwellings and traditional above-ground buildings (drawn by the author).

Secondly, the construction materials for cave dwellings are sourced from locally available loess soil. Loess is the primary building material for cave dwellings. It is not only easily accessible but also cost-effective. Using loess as a construction material reduces reliance on external building materials such as timber and stone, as well as lowering transportation costs and energy consumption. This approach of sourcing materials locally not only stimulates local economic development but also minimizes environmental impact.

Thirdly, when constructing cave dwellings, certain ecological principles are typically followed, such as minimizing impact on underground water sources, avoiding disruption of the water table, and protecting the surrounding environment. This respect and protection of land resources make cave dwellings not just a living space but also an ecological architecture that coexists harmoniously with the land, ensuring its sustainable use.

In summary, cave dwellings not only conserve land and reduce dependence on external resources, but also protect the ecological environment, embodying the concept of harmony between humans and nature. This sustainable approach to land use holds significant practical meaning and exemplary value for the green development philosophy advocated in today's society.

3.3. Cultural Characteristics of Cave Dwelling Architecture

Designated as a provincial intangible cultural heritage in 2006 and a national intangible cultural heritage in 2008, cave dwellings are not merely simple living spaces but also historical witnesses that carry profound cultural connotations [28]. Originating from the long-term interaction between the local people and the natural environment, this unique architectural method reflects the ancient humans' reverence for nature and their deep affection for the land. The construction and living practices of cave dwellings are rich in cultural content, including architectural techniques, lifestyles, and folk customs, which together form the unique cultural landscape of the Loess Plateau region [29].

Firstly, the spatial organization of cave dwellings profoundly reflects their unique and rich cultural characteristics. From the perspective of family culture, cave dwellings often take the form of courtyards (Figure 10a), with multiple caves surrounding the courtyard, allowing family members to live closely together. This layout strictly adheres to the principle of seniority, reinforcing the family's hierarchical concepts and emotional bonds. It embodies the traditional family culture's respect for family order and gathers deep family emotions.

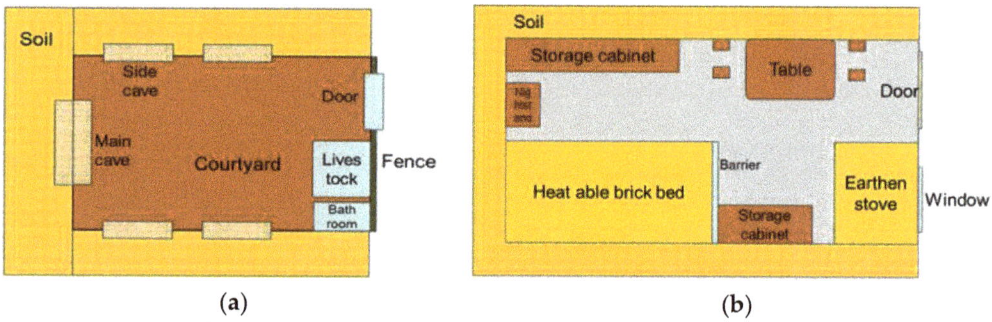

Figure 10. Schematic diagram of a traditional cave dwelling. (**a**) Traditional cave dwelling courtyard. (**b**) Internal functional layout of a traditional cave dwelling (drawn by the author).

Secondly, in terms of functional organization, cave dwellings are practical and economical. Their functional layout is compact and efficient, skillfully integrating basic living functions such as sleeping, living, and dining. A single cave dwelling often serves as a bedroom, living room, and dining area. The placement of simple furniture like beds and tables is compact and rational. The kang (a traditional heated brick bed) can be used for sitting and resting during the day and as a warm bed at night, making full use of the limited space (Figure 10b). This reflects the local residents' frugal and simple lifestyle, where everything is used to its fullest extent, meeting practical living needs and positively adapting to a challenging living environment.

Thirdly, cave dwellings have symbolic expressions with positive connotations. Traditional wood carvings on the doors and windows represent the wish for smooth and prosperous lives, while meander patterns symbolize continuity and a long history. These symbols carry the family's memories and the ancestors' aspirations for a better life, ensuring that cave dwelling culture continues through time and reflecting the residents' commitment to preserving and inheriting traditional culture (Figure 11) [30]. In Figure 11a, people arrange wooden strips into the shapes of flowers or orderly patterns. In Figure 11b, the doors and windows of the cave dwellings are carved into double-happiness patterns, symbolizing the auspicious meaning of "double happiness arriving together". These features reflect the aesthetic exploration and pursuit of beauty during the construction of cave dwellings. Many designs of cave dwellings are inspired by nature. The natural yellow color of the cave dwellings blends with the surrounding loess environment, symbolizing nature itself. In the decoration of cave dwellings, there are many symbols of natural elements, such as patterns of flowers, birds, fish, and insects, reflecting their lifestyle of close contact with nature and expressing a cultural concept of adapting to and respecting nature.

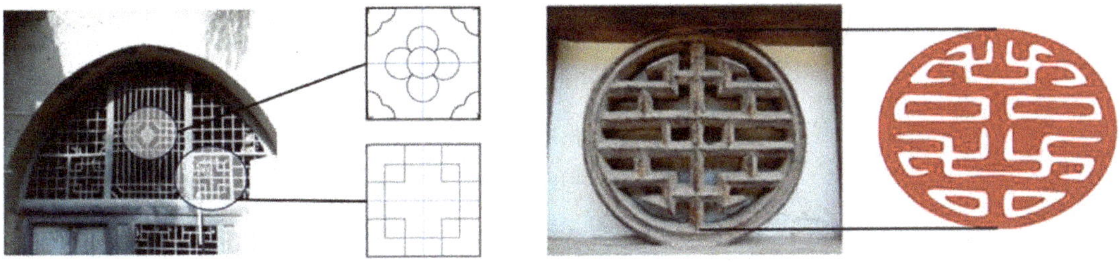

Figure 11. Traditional wood carvings on cave dwelling doors and windows. (**a**) Geometric pattern window grills. (**b**) Text pattern window grills (double happiness character) (drawn by the author).

4. Discussion

In the context of rapid urbanization and the integration of urban and rural development, various traditional cultures face the risk of homogenization. The protection and inheritance of cave dwellings should be regarded as a cultural awareness to resist cultural homogenization and maintain the uniqueness of regional culture. The protection, inheritance, and utilization of cave dwellings not only have significant academic value for the study of traditional Chinese architecture, lifestyles adapted to regional environments, and folk cultures [31], but also provide strong momentum for promoting the integration of culture and tourism, as well as the development of agritourism.

4.1. Promote Further Integration of Urban–Rural Relations

The cave dwellings in Mogou Village have played a unique role in promoting urban–rural integration and have driven local urbanization and sustainable economic development. In the process of improving the living conditions of villagers, these dwellings have provided a more energy-efficient and comfortable living environment. The harmonious coexistence of cave dwellings with the natural environment has further enhanced the villagers' sense of belonging and happiness. In recent years, Mogou Village has developed the "Old Home Mogou Scenic Area", integrating traditional cave dwellings with modern tourism to create a destination for leisure, sightseeing, and experiential activities. With the growth of the local tourism industry, the village's reputation has been enhanced, attracting an increasing number of visitors who come to explore and experience the local culture. Visits to the cave dwellings allow tourists to gain a deeper understanding of the area's history and culture and to experience traditional agricultural life. This, in turn, has stimulated the development of related industries such as catering, accommodation, and transportation, further fueling the local economic boom.

Cave dwellings have become a significant factor attracting urban populations to rural areas. Over the past three years, Mogou Village has received an average of over 550,000 visitors annually. Additionally, with the rise of the tourism industry, more than ten urban residents have opened shops in the area. This indicates that the development of tourism has had a positive impact on local population movement and economic activities. As living standards improve, some urban residents may choose to relocate to more picturesque rural areas to experience the charm of nature. This has led to a concentrated development model, promoting the flow of urban–rural elements, achieving integration between urban and rural areas, and driving rural revitalization.

4.2. Promote the Rapid Development of Rural Tourism

Rural tourism is an experiential activity that provides opportunities for sightseeing, leisure, vacationing, experiencing, entertainment, and fitness, based on the natural environment and pastoral landscapes of rural areas. Cave dwellings are hailed as the "fossils of agrarian loess culture" and "breathing architecture", making them an important attraction for tourists. Their warm winters and cool summers offer a unique living experience for visitors. Before the development of local cave dwellings and the promotion of rural characteristic tourism in Mogou Village, the village relied on traditional arable land cultivation for subsistence, which generated meager income. Young people left their hometown to seek employment elsewhere, leaving behind mostly the elderly and children of school age. This led to severe population loss and an accelerated aging population within the village. However, since the village seized the local feature of "cave dwellings" to develop rural tourism, the situation has gradually improved: since 2023, the "Old Home Mogou" scenic area has received a cumulative total of 528,300 visitors, with a comprehensive tourism revenue of 7.5054 million CNY, and the average dividend per villager has reached 1700 CNY (Figure 12). The income of local residents has significantly increased, and the village has also attracted young people to work here, promoting the revitalization of the original agricultural village.

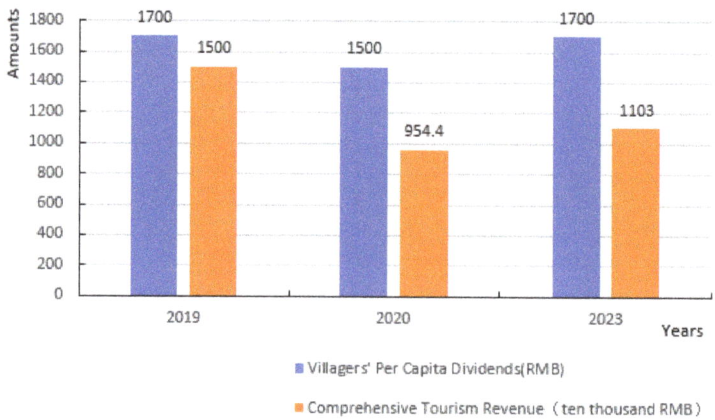

Figure 12. Collective income and per capita dividends of Mogou Village from 2019 to 2023 (the scenic area was closed during the pandemic in 2021 and 2022) (drawn by the author).

Today, the cave dwellings in Mogou Village have been transformed into several uniquely styled boutique cave hotels, such as Wangshu Homestay, Hejia Courtyard, Tinghu

Shangyuan, and Yanan. These cave hotels retain their traditional advantages of being warm in winter and cool in summer, while undergoing modernization and upgrades. They have installed new air circulation systems and other equipment to ensure fresh air, providing a comfortable living experience for tourists. Additionally, they offer a comprehensive experience by utilizing the local unique cuisine and ethnic culture, which in turn encourages surrounding villages to participate and stimulates economic development.

Mogou Village also actively develops cultural experience projects using cave dwellings. A deserted cave dwelling was renovated into the Laomiao Cave Library, which includes various functional areas such as Han Yu's Study, an adult reading room, a children's reading room, and an electronic reading room. The library has a rich collection of books covering various fields, including party building, science and technology, education, literature, children's books, planting, and breeding. It serves not only as a spiritual home for the villagers but also as a "must-visit" spot for rural tourism, attracting numerous visitors for reading and sightseeing.

It has become an important vehicle for protecting traditional dwellings and inheriting local culture, providing valuable insights for the development of rural tourism in China.

4.3. Promote the Protection and Inheritance of Outstanding Culture

The inheritance of construction techniques. The construction techniques of cave dwellings, characterized by their unique adaptability to local conditions and eco-friendly nature, involve distinct processes and methods [31]. Figure 13 illustrates the construction of different types of cave dwellings, such as cliff-side, sunken, and independent styles, each with its own skills and characteristics. For instance, various stages of construction, including site selection, excavation, reinforcement, drainage, and ventilation, are the outcomes of the long-term practices of our ancestors. Building dwellings into cliffs not only reflects the wisdom and creativity of ancient people but also provides inspiration for modern architectural design. When combined with modern building materials and technologies, it creates architectural works that possess regional characteristics and cultural connotations.

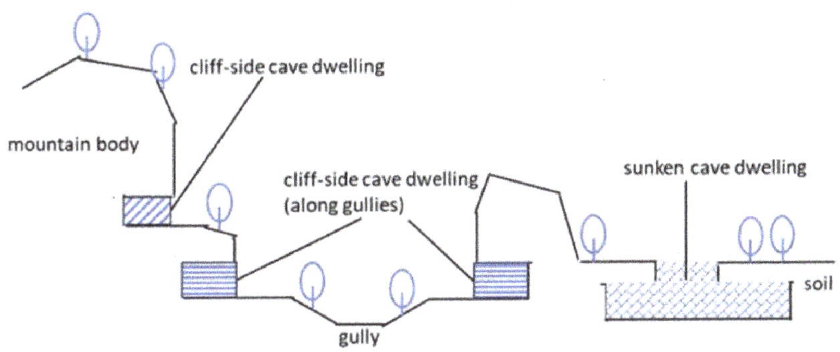

Figure 13. Schematic diagram of cliff-side and sunken cave dwellings (drawn by the author).

Cultural Heritage. Cave dwellings have witnessed the evolution of human habitation from cave dwelling to semi-cave dwelling and then to above-ground structures. The study and preservation of cave dwellings provide insights into the lifestyles, social organization, and economic activities of ancient humans, offering tangible evidence for historical research [32]. Cave dwellings are closely linked to local historical events and revolutionary history, bearing witness to the arduous journey and great victories of the Chinese revolution. These cave dwellings have become significant red cultural heritage sites, carrying people's historical memories and revolutionary spirit [33]. They play an important role in passing on the red gene and conducting patriotism education.

Philosophical Thought. Cave dwellings make full use of local loess resources and terrain, integrating with the natural environment without disrupting the ecological balance, reflecting the "unity of heaven and man" philosophy [34]. This concept of harmonious coexistence between humans and nature offers important lessons for the construction of ecological civilization in today's society, reminding people to respect and protect the environment for sustainable development [35].

4.4. Effective Inheritance of Architectural Functional Organization

Further improvement of cave dwelling lighting and ventilation. Skylights or transparent tiles can be installed on the roof to allow sunlight to pass through. Polycarbonate boards, which are used as transparent roof materials, have excellent light transmission and weather resistance, effectively increasing the amount of light entering the cave dwelling. Additionally, an attached sunroom can be added to the light-receiving side of the cave dwelling (Figure 14a) [36]. The figure illustrates how enhancing the light transmission coefficient of the window structure can facilitate the concentration and reflection of natural light, allowing it to enter the interior more effectively. For ventilation issues, ventilation openings can be set on both side walls of the cave dwelling to promote air circulation, and ventilation openings can also be installed on the roof. By utilizing the principle of hot air rising and cold air descending, natural convection is formed (Figure 14b) [37].

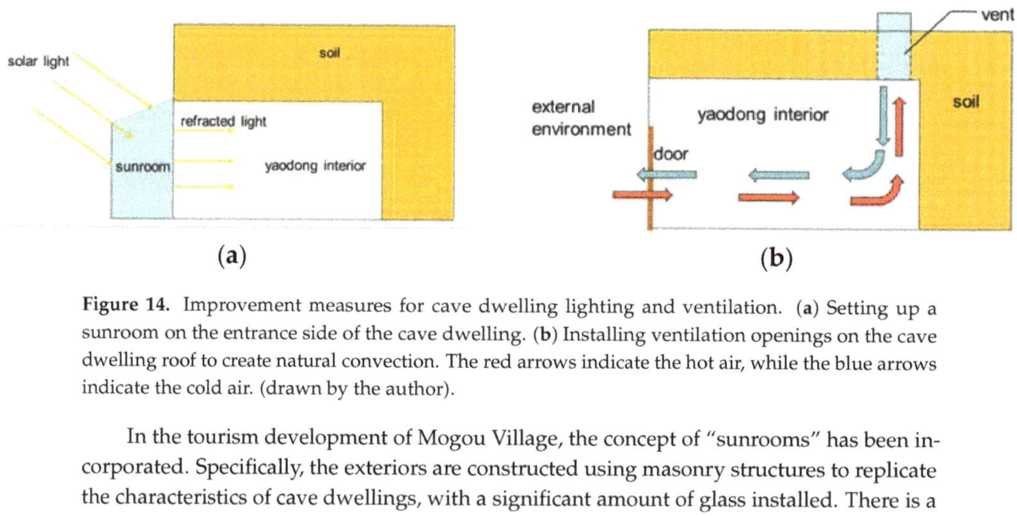

Figure 14. Improvement measures for cave dwelling lighting and ventilation. (**a**) Setting up a sunroom on the entrance side of the cave dwelling. (**b**) Installing ventilation openings on the cave dwelling roof to create natural convection. The red arrows indicate the hot air, while the blue arrows indicate the cold air. (drawn by the author).

In the tourism development of Mogou Village, the concept of "sunrooms" has been incorporated. Specifically, the exteriors are constructed using masonry structures to replicate the characteristics of cave dwellings, with a significant amount of glass installed. There is a certain gap between these additions and the original cave dwellings, which functions as a sunroom. These cave dwellings with added sunrooms are being developed into specialty shops or homestays.

Furthermore, to meet the functional demands of modern life, the layout of the cave dwelling rooms is optimized. While ensuring thermal insulation, the dimensions along the north–south axis are reasonably reduced, and the dimensions along the east–west axis are increased to enlarge the area of south-facing windows, facilitating better lighting in the rear of the cave dwelling. The multifunctional spaces of traditional cave dwellings are divided reasonably, with separate living rooms and dining areas provided for family interactions, entertaining guests, and meals. To accommodate the placement of modern appliances and furniture, the internal spatial dimensions of the cave dwelling can be appropriately adjusted, such as widening the facade and raising the arch, to meet the requirements for large modular furniture.

At the same time, to enhance the indoor environmental quality of cave dwellings, it is recommended to install fresh air systems to improve ventilation conditions and reduce radon concentration. Meanwhile, it is also suggested to use environmentally friendly building materials, such as paints with low volatile organic compounds (VOCs), to reduce indoor pollution [38].

Lastly, the infrastructure has been improved. Complete ventilation and drainage facilities are installed in the bathroom and kitchen areas. Traditional dry toilets are transformed into flush toilets, enhancing the sanitary conditions of the bathrooms. In the kitchen, reasonable ventilation ducts and drainage systems are set up, and modern kitchen appliances are equipped to improve the convenience of kitchen use.

5. Conclusions

Cave dwellings in western Henan and the Loess Plateau region are not only traditional residences but also significant cultural and sustainable development assets. Their design, rooted in energy efficiency, environmental friendliness, and optimal land use, integrates seamlessly with cultural values such as spatial harmony, functional practicality, and symbolic depth. These features position cave dwellings as quintessential examples of sustainable "green architecture". However, to fully realize their potential, future efforts must address radon control and ventilation design, incorporating modern technologies and materials that enhance comfort and health without compromising traditional integrity.

The case of Mogou Village illustrates how the protection and development of cave dwellings can catalyze local economic growth and embody the concept of "rural civilization and ecological livability". This approach not only safeguards historical heritage but also elevates living standards, cultural richness, and overall systemic integrity. It underscores the importance of balanced, sustainable development that respects both natural and cultural environments. This example also highlights the broader potential for regional development through the strategic use of local resources. By transforming traditional dwellings into cultural and economic assets, communities can foster cultural confidence, attract tourism and investment, and drive innovation in traditional crafts. This strategy is essential for rural revitalization and sustainable development, offering a model that balances heritage preservation with modern needs.

Author Contributions: Conceptualization, W.F.; methodology, Y.M.; validation, W.F. and Y.M.; formal analysis, W.F.; investigation, W.F.; data curation, W.F.; writing—original draft preparation, W.F.; writing—review and editing, Y.M.; visualization, W.F.; supervision, Y.M.; project administration, W.F.; funding acquisition, W.F. All authors have read and agreed to the published version of the manuscript.

Funding: This research received no external funding.

Data Availability Statement: All data generated or analyzed during this study are included in this published article.

Acknowledgments: The authors would like to express their sincere gratitude to Meng Ying for her valuable guidance and support throughout this research. We are also grateful to the anonymous reviewers for their constructive comments, which have helped improve the quality of this manuscript.

Conflicts of Interest: The authors declare no conflicts of interest.

References

1. Liu, J.P.; Gong, J. Research on Xi Jinping's Important Expositions on Cultural Heritage. *J. Xiangtan Univ. (Phi-Losophy Soc. Sci.)* **2024**, *48*, 87–93. (In Chinese)
2. Wan, R.; Kong, D.Q.; Zhang, L.X. Study on Bottom Damp-Proof Method of Cave Dwelling. *Adv. Civ. Eng.* **2018**, *2018*, 6. [CrossRef]
3. Zhang, L.; Lu, P.; Lau, R.; Yan, L.J.; Li, X.; Yang, R.X.; Leung, H.H.; Chen, P.P.; Wang, X. Unique traditional villages on the Loess Plateau of China: Historic evolution and challenges to sustainable development of silo-caves. *Herit. Sci.* **2021**, *9*, 19. [CrossRef]
4. Hu, Z.; Josef, S.; Min, Q.W.; Tan, M.; Cheng, F.L. Visualizing the cultural landscape gene of traditional settlements in China: A semiotic perspective. *Herit. Sci.* **2021**, *9*, 19. [CrossRef]
5. Zhang, J.H.; Jiang, L.; Wang, X.Q.; Chen, Z.H.; Xu, S.B. A Study on the Spatiotemporal Aggregation and Corridor Distribution Characteristics of Cultural Heritage: The Case of Fuzhou, China. *Buildings* **2024**, *14*, 121. [CrossRef]
6. Zhou, Z.W.; Liu, Z.; Wang, G.Q. Driving Sustainable Cultural Heritage Tourism in China through Heritage Building Information Modeling. *Buildings* **2024**, *14*, 3120. [CrossRef]
7. Jiménez-Sánchez, M.; Domínguez-Cuesta, M.J.; Aranburu, A.; Martos, E. Quantitative indexes based on geomorphologic features: A tool for evaluating human impact on natural and cultural heritage in caves. *J. Cult. Herit.* **2011**, *12*, 270–278. [CrossRef]
8. Zhu, J.Y.; Tong, L.P. Experimental study on the thermal performance of underground cave dwellings with coupled Yaokang. *Renew. Energy* **2017**, *108*, 156–168. [CrossRef]
9. Serizawa, C. Cave sites in Japan. *World Archaeol.* **1979**, *10*, 340–349. [CrossRef]
10. Shi, J.Y.; Zhang, T.; Bai, L.J.; Fukuda, H.; Gao, W.J. Comparative analysis on the mechanism of climate responsiveness strategies on indoor thermal performance of Yaodong dwellings in the Loess Plateau region of China. *Energy Rep.* **2024**, *11*, 4741–4758. [CrossRef]
11. Xue, J.Y.; Zhao, X.B.; Zhang, F.L.; Xu, D.; Hu, X.F.; Ma, L.L. Shaking table tests on seismic behavior of the underground loess cave of earth building of traditional dwellings. *Eng. Struct.* **2020**, *207*, 110221. [CrossRef]
12. Guo, P.G. Reliability Analysis of Loess Cave Dwellings Based on Fuzzy Failure Criterion. *Math. Probl. Eng.* **2021**, *2021*, 8. [CrossRef]
13. Peng, W.; Huang, Y.; Li, C.; Wang, Y. Exploration of Resident Satisfaction and Willingness in the Renovation of a Typical Old Neighborhood. *Buildings* **2025**, *15*, 293. [CrossRef]
14. Zhao, X.B.; Xue, J.Y.; Zhang, F.L.; Liu, Z.Q.; Du, J.; Hu, X.F. Seismic assessment of loess cave of traditional earth building retrofitted with composite materials. *Structures* **2022**, *40*, 18–35. [CrossRef]
15. Lin, Z.R.; Liang, Y.; Chen, K.Y.; Li, S.S. Ecological Wisdom and Inheritance Thinking of the Traditional Village's Water Resources Management in Taihang Mountains. *J. Asian Archit. Build. Eng.* **2024**, *23*, 424–442. [CrossRef]
16. Asdrubali, F.; Baldinelli, G.; Bianchi, F.; Sambuco, S. A comparison between environmental sustainability rating systems LEED and ITACA for residential buildings. *Build. Environ.* **2015**, *86*, 98–108. [CrossRef]
17. Cheng, L.P.; Liu, W.Z. Long Term Effects of Farming System on Soil Water Content and Dry Soil Layer in Deep Loess Profile of Loess Tableland in China. *J. Integr. Agric.* **2014**, *13*, 1382–1392. [CrossRef]
18. Zhu, J.Y.; Tong, L.P.; Li, R.X.; Yang, J.Z.; Li, H.X. Annual thermal performance analysis of underground cave dwellings based on climate responsive design. *Renew. Energy* **2020**, *145*, 1633–1646. [CrossRef]
19. Sekhon, N.; Novello, V.F.; Cruz, F.W.; Wortham, B.E.; Ribeiro, T.G.R.; Breecker, D.O. Diurnal to seasonal ventilation in Brazilian caves. *Glob. Planet. Change* **2021**, *197*, 14. [CrossRef]
20. Zhao, X.; Nie, P.; Zhu, J.Y.; Tong, L.P.; Liu, Y.F. Evaluation of thermal environments for cliff-side cave dwellings in cold region of China. *Renew. Energy* **2020**, *158*, 154–166. [CrossRef]
21. Shen, N.; Wang, Z.L.; Guo, Q.; Zhang, Q.W.; Wu, B.; Liu, J.N.; Ma, C.Y.; Delang, C.O.; Zhang, F.B. Soil detachment capacity by rill flow for five typical loess soils on the Loess Plateau of China. *Soil Tillage Res.* **2021**, *213*. [CrossRef]
22. Hao, Y.E.; Lan, Y.Q. Study on Mechanical Calculation Model of Arch Ring in Freestanding Stone Cave-Dwelling. *Sustainability* **2023**, *15*, 9545. [CrossRef]
23. Chandel, S.S.; Sharma, V.; Marwah, B.M. Review of energy efficient features in vernacular architecture for improving indoor thermal comfort conditions. *Renew. Sustain. Energy Rev.* **2016**, *65*, 459–477. [CrossRef]
24. Jami, M.; Moufekkir, F.; Mezrhab, A.; Fontaine, J.P.; Bouzidi, M. New thermal MRT lattice Boltzmann method for simulations of convective flows. *Int. J. Therm. Sci.* **2016**, *100*, 98–107. [CrossRef]
25. Yao, R.M.; Li, B.Z.; Liu, J. A theoretical adaptive model of thermal comfort—Adaptive Predicted Mean Vote (aPMV). *Build. Environ.* **2009**, *44*, 2089–2096. [CrossRef]
26. Costa-Carrapiço, I.; González, J.N.; Raslan, R.; Sánchez-Guevara, C. Understanding the challenges of determining thermal comfort in vernacular dwellings: A meta-analysis. *J. Cult. Herit.* **2022**, *58*, 57–73. [CrossRef]

27. Hao, Y.E.; Lan, Y.Q. Force Analysis of Masonry Cave-Dwelling Structure Based on Elastic Center Method. *Appl. Sci.* **2023**, *13*, 4292. [CrossRef]
28. Cardoso, R.C.; Ferreira, R.L.; Souza-Silva, M. Caves' environmental stability shaping subterranean biodiversity in the neotropics. *Acta Oecol.-Int. J. Ecol.* **2024**, *125*, 8. [CrossRef]
29. Zhao, P.H.; Wu, L.Y.; Li, Y.J. Analysis of the form characteristics of traditional dwellings in the Tongchuan area of Shaanxi Province, China. *J. Asian Archit. Build. Eng.* **2025**, *24*, 54–71. [CrossRef]
30. Zhang, S.; Chen, W.W.; Wang, N.Q.; Zhang, J.K. Study on the irrigation-induced damage characteristics of loess cave dwellings. *Bull. Eng. Geol. Environ.* **2025**, *84*, 225. [CrossRef]
31. Zhang, T.; Hu, Q.N.; Ding, Q.; Zhou, D.; Gao, W.J.; Fukuda, H. Towards a Rural Revitalization Strategy for the Courtyard Layout of Vernacular Dwellings Based on Regional Adaptability and Outdoor Thermal Performance in the Gully Regions of the Loess Plateau, China. *Sustainability* **2021**, *13*, 3074. [CrossRef]
32. Xie, X.D.; Huang, X.N.; Yan, J.N. Displacement field and mechanical analysis of a creeping silo-cave and the evaluation about its popularization potentiality. *J. Asian Archit. Build. Eng.* **2024**, *Early Access*, 1–12. [CrossRef]
33. Xu, J.; Lu, Z.L.; Huo, X.P. The evolution and adaptive development of traditional dwelling in Southern Shaanxi, China. *Environ. Sci. Pollut. Res.* **2019**, *26*, 13914–13930. [CrossRef] [PubMed]
34. Ding, C.B.; Zhuo, X.L.; Xiao, D.W. Ethnic differentiation in the internal spatial configuration of vernacular dwellings in the multi-ethnic region in Xiangxi, china from the perspective of cultural diffusion. *Herit. Sci.* **2024**, *12*, 17. [CrossRef]
35. Halilovic, S. Islamic civilization in spain—A magnificient example of interaction and unity of religion and science. *Psychiatr. Danub.* **2017**, *29*, S64–S72.
36. Yang, Y.J.; Wang, K.X.; Zhou, D.; Wang, Y.P.; Zhang, Q.; Xu, D. Improvement of Human Comfort in Rural Cave Dwellings via Sunrooms in Cold Regions of China. *Buildings* **2024**, *14*, 734. [CrossRef]
37. James, E.W.; Banner, J.L.; Hardt, B. A global model for cave ventilation and seasonal bias in speleothem paleoclimate records. *Geochem. Geophys. Geosyst.* **2015**, *16*, 1044–1051. [CrossRef]
38. Moscoso-Garcia, P.; Quesada-Molina, F. Analysis of Passive Strategies in Traditional Vernacular Architecture. *Buildings* **2023**, *13*, 1984. [CrossRef]

buildings

MDPI

Article

A Digital Analysis of the "L"-Shaped Tujia Dwellings in Southeast Chongqing Based on Shape Grammar

Quan Wen *, Yuqi Zhao, Xianwen Huang and Gang Wang

School of Architecture and Urban Planning, Chongqing Jiaotong University, Chongqing 400074, China; 622230190016@mails.cqjtu.edu.cn (Y.Z.); 622240190006@mails.cqjtu.edu.cn (X.H.); 622240190011@mails.cqjtu.edu.cn (G.W.)
* Correspondence: caad2019@cqjtu.edu.cn

Abstract: The Tujia ethnic group is one of the major ethnic groups in China, with a long history and abundant cultural heritage. As a distinctive architectural style, Tujia dwellings have evolved over thousands of years, developing a wealth of construction techniques and embodying the wisdom of local craftsmen. These construction techniques are a valuable asset of Tujia folk dwellings but still rely on the oral tradition among craftsmen. Therefore, it is extremely valuable for enriching the world's architectural system and heritage inheritance to refine these techniques and transform them into regularized digital properties. The "L"-shaped system of Tujia houses is the most common type of Tujia house, featuring both the main house and the wing house, and can distinctly represent the construction technology and style characteristics of Tujia houses. The grammar of "L"-shaped houses is the core part of the grammar of Tujia houses and is also important for analyzing and inheriting the construction technology of Tujia houses. Shape grammar is an analytical method centered on the refinement of rules. This paper takes advantage of its ability to analyze and refine rules, and based on the rich Tujia architectural material library, it summarizes the corpus and refines the grammatical rules of "Generation of the main structure framework", "Roof truss conversion and support", "Side houses and stilted structures", and "Cantilevered elements and corners" into four dimensions, along with many detailed grammars. These rules are transformed into a programming language and parameterized toolkit, providing a detailed summary of the construction logic and techniques. Ultimately, an "L"-shaped construction grammar for Tujia traditional dwellings has been proposed, and with the help of software tools such as Grasshopper, the digital regeneration has been completed.

Keywords: shape grammar; Tujia dwellings; construction technology

1. Introduction

1.1. Tujia Construction Techniques

Chinese traditional houses are residential buildings with regional or national characteristics created and inherited by people of various ethnic groups in the process of long history and development, which have formed a rich variety of architectural forms due to the differences in climate, geographic environment, resources, and culture of different places [1,2], which vividly reflects the relationship between human beings and nature in harmonious coexistence, and the connotation of which can be replaced by "vernacular architecture". Chinese traditional houses are inherited in the context of traditional production and life, usually built by craftsmen and hosts and using local materials and traditional crafts, with strong regional and national characteristics. Traditional houses are directly related to people's households or family residences, and over the past thirty years, academic research on vernacular architecture has been fruitful [3–5].

Compared with different types of residential writings and theses at various stages, it is demonstrated that the research results become constantly enriched, research perspectives become constantly updated, and disciplinary methods become constantly diversified. The research on vernacular architecture has entered into a multi-faceted and multi-disciplinary comprehensive research, and its research methodology covers architecture, urban and rural planning, landscape architecture, and other disciplines; the breadth of the research involves sociology, economics, human geography, design, esthetics, computer science, and other disciplines; and it is also an important part of the contemporary "discipline of human habitat science" and its methodological system.

Tujia folk dwellings are one of the ancient ethnic groups of the Chinese nation, widely distributed in the southeast of Chongqing, northwestern Hubei, and western Hunan, and have an extremely important position in the history of the development of the Chinese nation as well as in the historiography of Chinese architecture [6]. Tujia settlements and architectural form types are influenced by different geographic spaces, production, and lifestyles and interact and penetrate with Han and Miao cultures, forming a building system with strong national characteristics and rich regional characteristics, which is of typical significance for the study of national architectural historiography and the development and evolution of settlements. The construction technology is the most valuable property of the Tujia folk houses and the core of the folk house architectural inheritance, among which, the construction technology of stilted building has been selected into China's national intangible cultural heritage list and the national traditional craft revitalization catalog. In view of the lack of traditional genes and the disconnection between design and construction in Tujia residential architecture, this study will focus on the systematic integration of traditional architectural construction and formal language theory and knowledge system and will start from the excavation of Tujia traditional construction wisdom, the construction of translation theory and method, the integration of grammar, and the digital reproduction, in order to cope with the needs of cultural inheritance and the improvement of the quality of the human environment.

1.2. Conservation Status of Tujia Folk Houses

China's current urban and rural construction systems are vastly different, and compared with the planning, design, and construction of towns and cities in strict accordance with engineering construction standards, which ensures the quality of construction through a high degree of professional division of labor, the rural construction system in the Tujia region has long continued to imitate the self-built, artisan-organized construction model, and the construction techniques of the Tujia dwellings have accordingly relied on the oral tradition of the artisans. Therefore, with the development of rural society and economy, the change in construction system, and the extensive intervention of architectural design, local rural construction is also facing serious challenges: (1) the traditional progressive renewal of the countryside and the inheritance mode of the Tujia folk houses are gradually disintegrating; (2) the local rural buildings are large in volume and wide in scope, and the translation of the national architectural forms to modern farmhouses is lacking, which is difficult to be realized with the general design standard and normative system; (3) the urban architectural design methods and technologies cannot be directly used in the rural construction system, and the cost is high. In order to meet the urgent demand for modern farmhouses with regional characteristics, it is necessary to develop rural architectural design methodology by adapting the design methodology for rural construction.

Early research on Tujia ethnic dwellings primarily focused on the investigation and analysis of their structure, appearance, functionality, and other attributes. The outcomes were documented and organized through text and images, with some simple models established for explanatory purposes, though these lacked informational depth [6–8]. In recent years, with the widespread application of digital technologies and methods such as Building Information Modeling (BIM) in the study of traditional Chinese dwellings, some scholars have introduced these approaches to the conservation of Tujia ethnic dwellings [9,10]. However, while digital technologies like BIM emphasize the comprehensive lifecycle information of buildings, they have not effectively captured or analyzed the underlying construction logic and the craftsmanship wisdom inherent in these dwellings. By leveraging Shape Grammar for rule analysis and extraction, and integrating information to generate adjustable three-dimensional models with visual programming tools, the aforementioned issues can be effectively addressed.

When Shape Grammar was proposed in 1972, it was mainly used as a tool for analyzing design works, and exploring the generative logic of existing design languages through the study of rules based on geometrical shapes [11]; after 2000, Shape Grammar gradually began to be used as an auxiliary tool for the design process, expanding a certain design language or constructing a new design language by applying the rules of shape transformation. For traditional houses, Shape Grammar can be based on the relatively stable plan and elevation paradigm of houses, associated with their fixed geometrical shapes, as well as the rules of "transformation" to adapt to geographic conditions, family size, and functional needs, to explain the rules of this type of building, which is "not designed by architects". It explains the rule-generating method of this "architect-less" building type and elaborates the arithmetic principle of shapes and the hierarchical logic of rules. Under the strategy of rural revitalization, the features of "readability", "process", and "changeability" of shape syntax can be organically combined with digital design, which effectively expands the design method of rural residence and provides a new method for the design process of rural residence generation. In this way, it can effectively expand the rural residence design methodology, provide a scientific basis for the design process of rural residence generation, further reveal the mutual influence of the countryside and the natural environment, human characteristics, and architectural design, and thus further develop precise and intelligent rural design analysis tools to promote the progress of the rural design system.

2. Literature Review

2.1. A Brief Overview of Shape Grammar

Shape Grammar is a theoretical framework based on linguistics and derived from design, initially proposed by George Stiny and James Gips [12]. This theory leverages rules and grammars to recreate contextualized corpora, aiming to capture a design concept analogous to "grammar" in linguistics. It transforms abstract design logic into a tangible "grammar", effectively materializing the rule-based logic inherent in design. With advancements in computer algorithms, Shape Grammar has evolved to become more visual and digital, offering higher degrees of freedom and possibilities in its generative outcomes, all supported by algorithmic processes.

Over time, Shape Grammar has diversified into several branches, with varying definitions provided by different researchers. However, its application logic can be succinctly represented by the formula SG = (S, L, R, I), where:

SG denotes Shape Grammar.

S (Shape) represents a finite collection of material shapes derived through scientific sampling and processing.

L (Label) is a finite set of labels used to categorize shapes.

R (Rule) consists of a finite set of shape transformation rules.

I (Initial Shape) refers to the starting point of the design, which could be a coordinate point, a complex shape derived from existing conditions, or the simplest basic shape.

The rules are typically expressed in the form a→b, where "a" is referred to as the Left-hand Side (LHS) shape and "b" as the Right-hand Side (RHS) shape. According to an open textbook on computer-aided design from the Department of Architecture at the Massachusetts Institute of Technology (MIT), the basic design process for applying Shape Grammar involves the following steps: determining the basic shape, defining spatial relationships, establishing rules, specifying the shape language, and applying these to the design [13]. To apply Shape Grammar in specific architectural design contexts, it is essential to first abstract the plan layout, spatial form, and structural configuration of a building into two-dimensional or three-dimensional geometric shapes. Subsequently, the proportions, scales, and functional relationships within the object of study are distilled into shape rules, thereby converting architectural styles into a design language that facilitates clear and concise descriptions [14] (Figure 1).

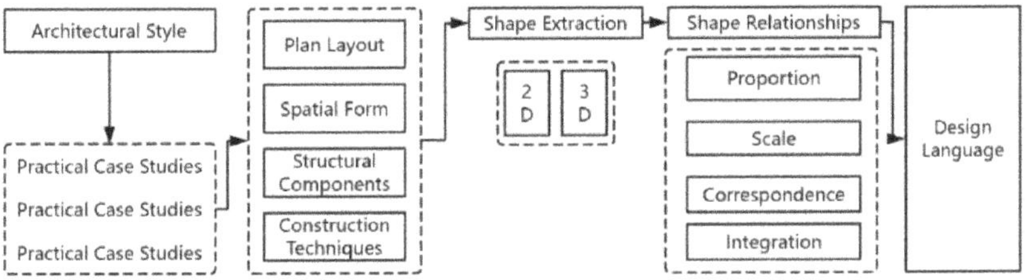

Figure 1. The Process of translating architectural styles into design language using Shape Grammar.

In the field of architecture, Shape Grammar has two primary application directions [11]:

Generative Design: With the advancement of computer science and the expansion of Shape Grammar theory, "generative Shape Grammar" has emerged. This branch uniquely bridges the gap between architects' graphical visual representations and computer symbolic coding. Notable examples include the 2008 application of built-in CGA Shape Grammar by City Engine for rapid city model visualization [15], Veloso's 2018 development of an apartment building floor plan design and residential customization system based on Shape Grammar [16], and Yavuz and Sahika's 2019 exploration of store floor plan generation and evaluation using Shape Grammar [17]. Wang Jiang et al. also developed a Shape Grammar tailored to traditional Chinese rural houses to meet residents' needs [18–20].

Analytical Shape Grammar: This branch is used to analyze classical architecture, traditional architectural forms, or buildings with uniform and well-defined appearances. Initially, Stiny posited that while graphical computation relies on visual perception, Shape Grammar provides a visual design process that transcends graphical computation's predictability [20]. Early research focused on analyzing the generation of complex geometric shapes at the two-dimensional level. For instance, Stiny's 1977 analysis of traditional Chinese residential window patterns, particularly ice-cracked windows, illustrated shape evolution rules through numerous diagrams [21]. Subsequently, Shape Grammar was applied to architectural analysis, with Stiny and Mitchell's 1978 study of the Palladian villa marking the first architectural design application [22] (Figure 2). This pioneering work inspired further analyses of traditional Japanese tea rooms, Buffalo houses in the United States, Frigellio apartments in Trani, and additional Palladian villas [23–26].

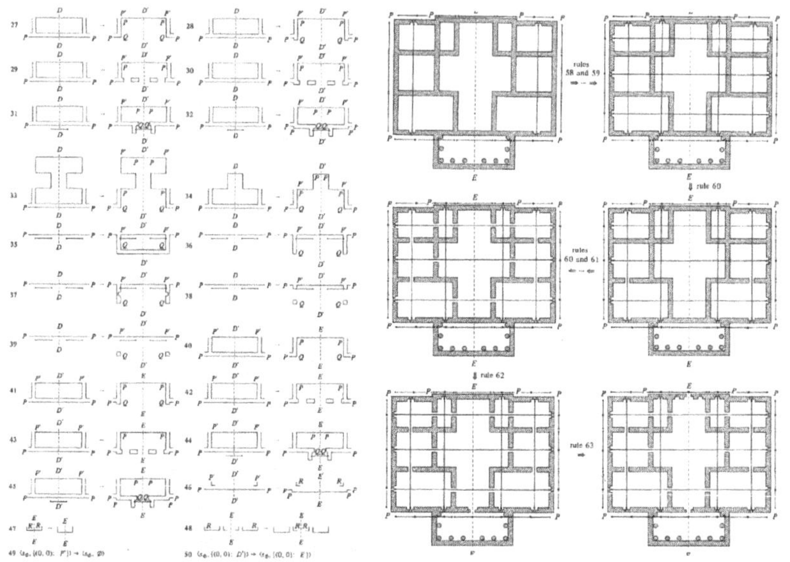

Figure 2. Shape Grammar analyzing Classic architecture—Palladian villa (Source: Stiny).

To date, Shape Grammar remains a widely used tool for parsing Classical architecture [27,28]. Its application underscores the theoretical and methodological strengths of Shape Grammar. As research objects expand and techniques are refined, the scope of Shape Grammar research continues to deepen.

2.2. Shape Grammar Analysis of Traditional Dwellings

Shape Grammar was first introduced to China by Sun Jiaguang in 1986. His article "Shape Grammar and Shape Rules" provided a clear introduction to the definition, composition, and operation of Shape Grammar. He developed a two-dimensional Shape Grammar interactive processing system for generating various architectural floor plan designs [29]. This work gradually introduced Shape Grammar to China. In 1996, Chiou defined Shape Grammar rules based on Taiwan's feng shui and traditional courtyard spatial organization: first the main house, then the courtyard, and finally, the side house and reversely set house [30]. This approach offered a new method for studying traditional Chinese architecture. In 2001, Li Yikang used Shape Grammar to analyze the traditional architectural language of the Chinese Song Dynasty, as described in the "Treatise on Architectural Methods" [31]. He used planar shape rules to express spatial relationships among openings, depths, and columns, and sectional shape rules to describe relationships among beams, columns, and lifting folds. Comparing Song Dynasty buildings with those of other dynasties provided a rational perspective on the evolution of ancient Chinese architectural forms. In 2002, Huang Ruisong explored the design language of arch forms using Shape Grammar, establishing the preliminary architecture of traditional arches [32].

In the field of traditional dwellings, research has expanded significantly. In 2012, Xiong Lu established syntax rules and reproduced parametric models using Guangdong bamboo buildings and Dong drum towers as examples (Figure 3). He later deduced parametric models for Jiangnan waterfront buildings. In 2015, Yang Guoquan used Shape Grammar to analyze the typical rural residence plan in Jiaxing City, providing data support and a theoretical basis for energy simulation [33]. In 2017, Zhang Yuhang used Shape Grammar to analyze Huizhou architecture, proposing an architectural grammar for Huizhou [34] (Figure 4). After Shape Grammar was introduced to China, research expanded to cover multiple dimensions, including architectural monoliths, traditional architectural language, and traditional architectural groups.

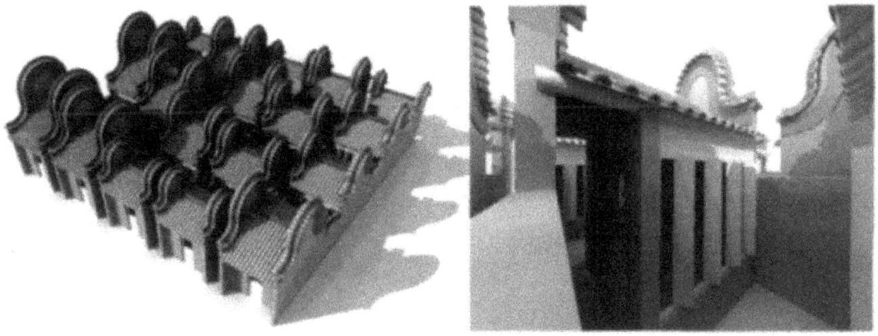

Figure 3. Shape Grammar analysis of traditional Chinese houses—bamboo houses in Guangzhou (Source: Xiong Lu).

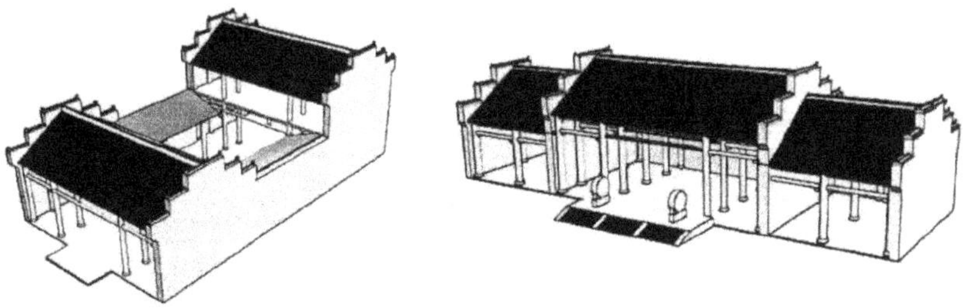

Figure 4. Shape Grammar analysis of traditional Chinese houses—Huizhou architecture (Source: Zhang Yuhang).

In recent years, research on Shape Grammar for vernacular buildings has flourished. Zihao Gu used Shape Grammar for the interpretation and generation of traditional courtyard houses, establishing grammatical rules from site to volume [35]. Kehan Zhang applied it to analyze the typical planes of Gulangyu Island's recent foreign buildings, extracting landscape elements through parameterization to generate digital three-dimensional models [36]. These studies not only tailor reproducible grammatical rules for traditional dwellings but also employ emerging digital technologies to optimize research outcomes [37], infusing new vitality into traditional dwellings. Overall, Shape Grammar shows strong research potential and a broad field for the conservation of traditional houses.

In summary, previous research on Shape Grammar in the study of rural dwellings has primarily been divided into two categories. The first involves using Shape Grammar to summarize relatively simple constraints and material foundations, such as planar layout materials, in pursuit of generating new digital rural dwelling design schemes, as exemplified by scholars like Wang Jiang. The second category employs rule analysis and related tools from Shape Grammar to achieve a more straightforward parameterization of architecture. Scholars like Xiong Lu and Yuhang Zhang have conducted such research.

This study utilizes Shape Grammar to analyze Tujia ethnic dwellings. Initially, it introduces Shape Grammar into the study of the complex traditional wooden framework system of Tujia dwellings. Furthermore, in terms of research depth, this study begins with the basic unit of traditional dwelling construction—the framework—and introduces specific rules into three-dimensional geometric transformations. This approach transcends the previous limitations of Shape Grammar, which were confined to planar combinations or single planes and attempts to explore modern considerations based on traditional dwelling construction patterns.

3. Research Subjects and Methodology Results

3.1. Traditional Folk Houses of the Tujia People of Southeast Chongqing

The Tujia ethnic group, one of China's major ethnic groups with a long history, is primarily distributed in the Wuling Mountains bordering Hunan, Hubei, Chongqing, and Guizhou. In Chongqing, the Tujia people mainly reside in the southeast, including Qianjiang District and the minority autonomous counties of Shizhu, Pengshui, Youyang, and Xiushan (Figures 5 and 6). This region is characterized by a high concentration of ethnic minorities and distinct ethnic features (Table 1), hosting numerous traditional ethnic villages where the Tujia population can constitute over one-third of the local populace [6]. Consequently, studying Tujia traditional dwellings in southeast Chongqing is a vital component of research on traditional dwellings in the region.

Figure 5. Location of southeastern Tujia within the national distribution of Tujia (author's own illustration).

The primary structural form of Tujia dwellings in southeast Chongqing is the column-and-tie construction, one of China's two main types of wooden frameworks. This southern (southwestern) form contrasts with the northern and official raised-beam frames [38]. It shares the superior characteristics of traditional Chinese wooden structures, such as the use of mortise-and-tenon joinery to connect components without nails or iron, relying solely on the wood's shape and size for precise fitting. This reflects a high level of craftsmanship and architectural esthetics [39]. Over a long historical period, the Tujia ancestors perfected this architectural form, developing a unique construction logic and building system.

Given the regional and social context of Tujia dwellings and the demand for local materials, a distinctive "falling-and-hanging pillars" system evolved on the column-and-tie frame, becoming a hallmark of Tujia folk houses. Specifically, in a roof frame, not all columns need to be grounded; some can be inter-columnar or even multi-columnar, suspended in the air. This design, leveraging hanging columns, conserves wood and enhances spatial flexibility, embodying the traditional wisdom of Tujia craftsmen.

Figure 6. Location of southeastern Tujia in Chongqing municipality (author's own illustration).

Living in mountainous areas, the Tujia people in southeast Chongqing face an uneven terrain, making flat building sites scarce. To adapt, they developed a unique structural practice: stilted construction. Historically, the prototype of Tujia stilt houses traces back to primitive nesting houses, embodying the oldest human wisdom of "site adaptation" [40]. Typically built on slopes, these structures utilize the terrain's height differences and employ a hanging design, where part of the building is supported on the ground while the other part is suspended. This reduces terrain damage and protects against mountain torrents and humidity, reflecting the Tujia people's respect for nature and their building wisdom [39].

The stilted sections of Tujia houses are mainly used as compartments, and correspondingly, the floor plans of Tujia houses exhibit various forms and changes. In terms of spatial composition, Tujia houses in southeast Chongqing consist of a main house (also called the "seat house") and compartments. In some cases, only the main house is present, forming a "—"-shaped plan, the simplest form. More commonly, the main house and side rooms combine to form an "L"-shaped plan, the most widely used and characteristic form of Tujia folk houses [40]. The main house and two side rooms can also form a triple or quadrangle house. In practice, the Tujia floor plan system is highly adaptable, often combining flexibly based on actual needs, resulting in more complex variations like the "comb"-shaped plan.

Additionally, the construction wisdom of Tujia folk houses is evident in structural and construction details, such as the flexible arrangement of square elements in different sizes and positions, various roof intersections and combinations, and different frames embedded in joints and overlaps. In summary, the characteristic through-double structure of Tujia folk houses is rational in stress distribution and skillful in construction, adaptable to natural terrain changes and diverse functional needs, and superior in the folk building process in the BaShu region [41]. This study focuses on the most representative "L"-shaped Tujia folk houses for grammatical research, using programming tools to summarize the construction logic of the characteristic column-and-tie-beam structure and highlight the features of Tujia folk houses.

Table 1. Tujia houses in southeast Chongqing.

Village landscape and regional environment	
Architectural form and style	
Falling and hanging pillars, framing features	
Stilted structure and site adaptation	

3.2. Application of Shape Syntax

The purpose of constructing a Shape Grammar is to express a "design language" by extracting or constructing a corresponding design grammar and organizing the different design elements in the shape library with specific rules to generate a language. The basic logic of grammar must contain a "shape" and a corresponding "rule". The application of the rule requires determining the transformation relationship between shapes and the label anchor point, with an initial starting point, i.e., the initial shape. Therefore, creating a Shape Grammar set can consist of the following stages (Figure 7):

(1) Build a shape set and corpus.
(2) Determine spatial relations.

(3) Determine initial shapes.
(4) Determine label positions and shape rules.
(5) The initial shape and shape rules work together to build a Shape Grammar.

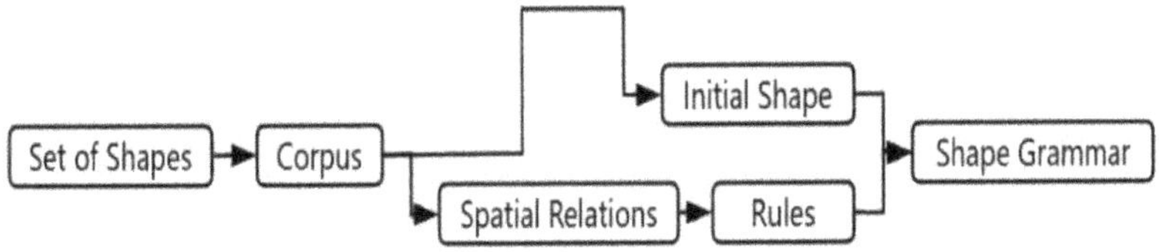

Figure 7. Shape Grammar construction flow (author's own illustration).

For the Tujia ethnic dwellings, the Shape Grammar corpus must originate from a multitude of actual residential cases, with a substantial amount of realistic materials serving as evidence and foundation for this research. Our research team conducted extensive field surveys in the southeast region of Chongqing, employing various research methods including mapping, photography, interviews, and sketching. We extensively collected and translated cases of traditional dwellings from 32 villages across five districts and counties, forming a rich Shape Grammar corpus (Figure 8). Concurrently, this study involved substantial literature review and organization of related findings, drawing upon established summaries of construction techniques for Tujia dwellings to supplement the comparative analysis of the corpus.

Regarding the determination of the "spatial relationship of the shape" and the "initial shape", these can be directly analyzed from the traditional building process of Tujia folk houses. First, the construction process of Tujia traditional dwellings is a process of "building from scratch", evolving from the initial assembly of timber frames to erecting upright frame assemblies and connecting them into a cohesive structure using through-beams and other components, and finally setting up supplementary details, such as connecting beams, cantilevered beams, and plank walls. The construction of Tujia dwellings progresses in layers, gradually enriching and following a clear logic. According to the characteristics of Shape Grammar theory, the initial shape of the grammar is highly inclusive. Besides adopting a basic geometric shape, it can also start directly from a specific coordinate point [12]. This aligns well with the Tujia construction feature of "building from scratch".

Second, as previously mentioned, the column-and-tie-beam structure of Tujia houses still clearly reflects the characteristics of traditional Chinese mortise-and-tenon joinery, where the entire frame relies on the nesting and overlapping relationships between different elements. This means that the relationships between "shapes" can be interpreted through conventional geometric transformations, such as "movement", "symmetry", "rotation", etc. This feature provides a basis for setting label positions and further determines the choice of parsing strategy. Common parsing strategies in Shape Grammar include "grid", "subdivision", "addition", etc. [11]. Through the superposition of the addition strategies, the process of building and geometric transformations between constructions can be more clearly reproduced. In conclusion, in this study, a coordinate point (in this case, the geometric center point of the column on the ground in the first bay of the roof frame of the main house) is taken as the initial shape, and the additive parsing strategy of Shape Grammar is used to conduct a grammatical study on Tujia dwellings in southeast Chongqing.

The analysis and expression of grammatical rules are the most important part of this study, derived from two sources. One is summarizing and analyzing the corresponding shapes of the corpus to extract commonalities. The other is collecting and organizing relevant information about house-building customs and comparing them with corpus examples for verification. In short, the extraction of grammatical rules can be synthesized from literature, interviews, research, and analysis but must ultimately resonate with the actual building corpus to ensure the scientific validity of the method. The presentation of grammatical rules was initially expressed as text and formula transformations, supplemented by two-dimensional diagrams. With technological advancements, the rules are also presented in the form of digitized three-dimensional results. This research will use visual programming tools to translate grammar rules into computer language and algorithm toolkits, synchronizing digital presentation. The research results include rule descriptions, rule application illustrations, grammar toolkits, and digital models (Figure 9).

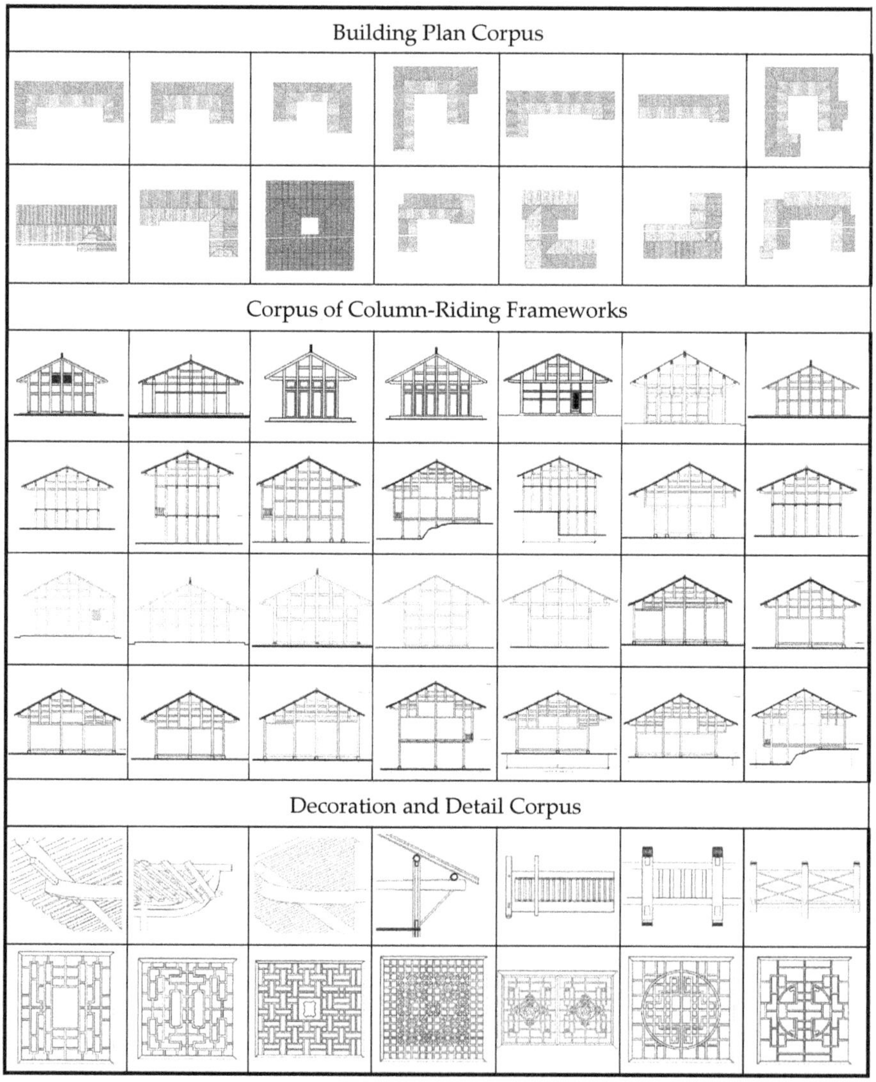

Figure 8. Corpus collection of column-riding frameworks.

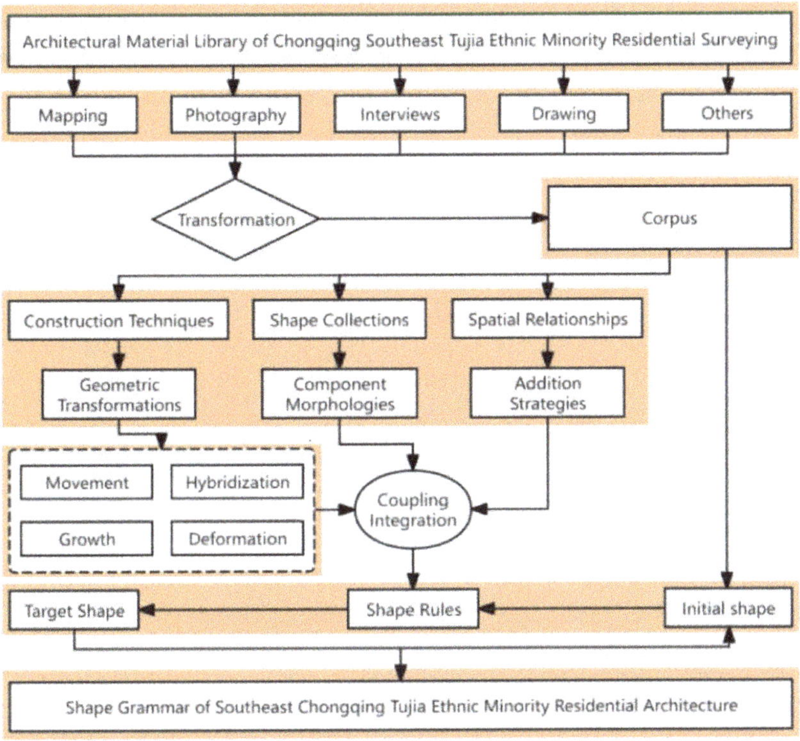

Figure 9. Application strategies of Shape Grammar for Tujia dwellings in southeast Chongqing (author's own illustration).

3.3. Grammar Construction of "L"-Shaped Tujia Folk Houses

The "L"-shaped configuration of Tujia residences, comprising a main house and a side room, embodies the construction characteristics of both the "main house" and the "hanging-legged room". This design distinctly represents the construction technology and stylistic features of Tujia dwellings. The grammar of "L"-shaped houses is central to the overall grammar of Tujia houses and is crucial for analyzing and preserving Tujia construction techniques. In this study, following the traditional construction logic of the Tujia people, four primary grammatical sets were successively extracted: "Generation of the Main Structural Framework", "Roof Truss Conversion and Support", "Side Houses and Stilted Structures", and "Cantilevered Elements and Corners". Additionally, supplementary grammatical details were identified. Through the additive strategy of Shape Grammar, these elements were incrementally combined to complete the overall grammatical construction [42].

(1) Subject Construct Generation Grammar:

The house frame is the most important part of Tujia houses and the entire structural system. The main frame of Tujia houses includes the column network system determined by the "falling and hanging pillars" as well as the beams connecting the roof frame in the longitudinal and transverse directions. Only after the main structural framework is completed can the construction grammar of characteristic elements, such as eave corridors and stilted structures, be further superimposed on the main framework, along with supplementary grammatical elements like cantilevered beams and corner connections.

In the traditional Tujia house-building process, the construction of the main structural framework begins with a single roof frame. The roof frame is first set up on the ground, followed by the erection of the rows of fans, and the entire roof frame is connected by the beams and joined squares. Thus, at the grammatical level, analyzing the grammatical rules of the one-bay roof frame leads to the grammatical rules of the entire main frame. A roof frame consists of a series of ground-touching or suspended columns and transverse beams, which not only represent the establishment of the roof frame at the structural level but also determine the depth of the house in terms of functionality and plan scale. Additionally, the main frame of the main house and its dimensions can be further determined by arranging multiple bays of roof frames at a distance from each other according to the modulus associated with the openings of a Tujia house.

Based on the corpus of Tujia houses in southeast Chongqing, a representative "four falling with five hanging pillars" configuration was selected as a generative case for this study, and a grammatical synthesis was carried out.

In the southeast region of Chongqing, Tujia dwellings utilize local fast-growing tree species, such as fir and pine, as the primary construction materials for the structure of the houses. Locally sourced slate and small blue tiles are selected for flooring and roofing. After simple processing, these building materials inherently possess a high degree of geometric uniformity. Consequently, the grammatical rules of a single house frame are specifically reflected in the dimensional constraints and geometric limitations of components such as columns and through-beams. Some of the geometric and positional relationships of these components can be directly determined by comparing preliminary literature and the corpus library. The specific values and rules of this part of the data are presented in the following table (Table 2):

Table 2. Values and descriptions of some of the main frame components.

	Value	Basis	Explanation
Main Hall Width	5.0 m (14.8 feet)	'Zhang-Ba-Ba System', Traditional Construction Custom	The width of the main hall is set at 5.0 m. The width of the secondary bays decreases by 0.33 m successively.
Height of Central Column	6.3 m (18.8 feet)	Literature, "Yabai" Custom	The height of the central column is 6.3 m. The height of the building is related to the depth of the roof truss. For a "five-column-four-chess" structure, the building height is typically 18.8 feet.
Height of First Purlin	1.9 m (5.78 feet)	Corpus, Literature	The height of the first purlin is 1.9 m. This height is determined to meet the requirement for human passage.
Height of Top Purlin	27 mm below the top of the first hypostyle column	Literature, "Yabai" Custom	The height of the top purlin is measured 27 mm below the top of the first golden column.
Crossbeam Section Ratio	Height to Width Ratio: 1:3	Corpus, Traditional Construction Custom	The cross-sectional ratio of the crossbeam is 1:3. The specific dimensions of the crossbeam are flexible, depending on the materials used and the requirements, but the general height-to-width ratio is 1:3.

The remaining part requires continued analysis of the geometric transformation rules based on the established values and reserved variables. Based on the above analysis and the corpus analysis, the basic variables for the generation grammar of a single roof truss can be determined as follows: Bu Chi, Bu Shui, column height, column diameter, through-beam height (excluding the head-through), and through-beam cross-sectional width. These variables serve as the foundation for further analysis of the grammatical rules. First, since the roofs in the southeastern Chongqing region are mostly straight slopes with bamboo-pole water, there is a formulaic relationship between Bu Chi, Bu Shui, and the difference in column heights:

$$BuShui = \tan\theta = \frac{step\ length\ difference\ in\ adjacent\ columu\ heights}{Bu\ Chi}$$

Therefore, through the constraints of Bu Chi and Bu Shui, grammatical rules can be established to qualify the height difference in the columns. Given that the height of the central column has a common value, the heights of the hypostyle columns and eave columns can be correspondingly determined, thereby establishing the Bu Shui relationships and forming of the roof surface (Figure 10).

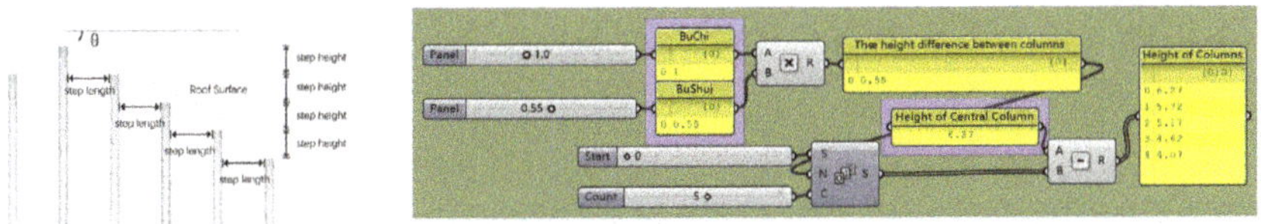

Figure 10. Determine the heights of each column based on the Bu Shui/Bu Chi method.

Except for the first through-beam, which has a relatively fixed position, the position of the top through-beam is related to the height of the hypostyle columns (Table 2). The placement of the remaining through-beams is relatively flexible, reflecting the wisdom of the Tujia people. Due to common limitations in materials and supplies, it is often impossible to meet the demand for evenly spaced beams. The specific positions of the through-beams in Tujia dwellings are flexibly adjusted based on their basic connection and load-bearing functions, resulting in a rich variety of forms. Moreover, when the dimensions of the through-beam materials are inappropriate, the uneven arrangement can provide opportunities for secondary material adjustments, greatly reducing material waste [12]. Based on this, the positions of all through-beams are reserved as initial variables and incorporated into the specific layout grammar to better reflect the flexibility of through-beam arrangement.

The logic for generating columns and through-beams has been established, and by superimposing the generation grammar of cantilevered beams, the grammar for a single roof truss is also established. Regarding the bay dimensions of the side houses, the traditional Tujia "Zhang-Ba-Ba System" (eight-based dimension rule) can be referenced. The "Zhang-Ba-Ba System" is a local construction custom that applies numbers ending in "eight" to traditional architectural dimensions [43]. Typically, the main hall bay is set at one zhang four chi eight cun (approximately 4.736 m), and the widths of subsequent bays are reduced by one zhang sequentially, such as one zhang three chi eight cun (approximately 4.416 m), one zhang two chi eight cun (approximately 4.096 m), etc. By overlaying a numerical operation of bay dimensions on a single roof frame and establishing corresponding grammatical rules for longitudinal connecting components, such as the "main gate beam" above the eaves, floor beam, and purlins, the overall structural logic of the building is replicated (Figure 11).

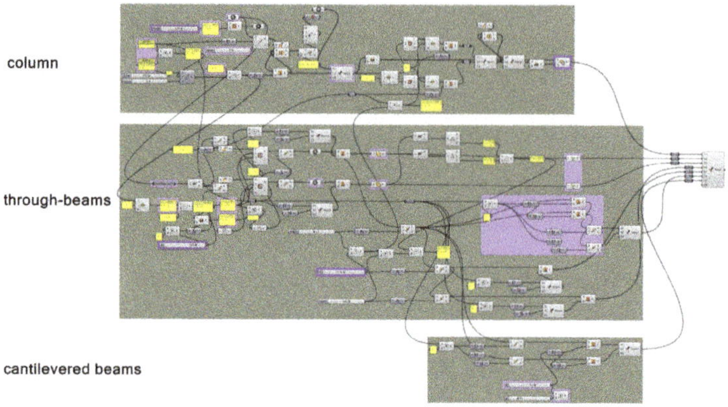

Figure 11. Grammar for the generation of a single roof truss.

At this point, the rule constraints for the main structure of the building have been sorted out. Based on these rules, a "Framework Grammar" toolkit was developed within the Grasshopper platform, using an additive strategy to gradually build the generation grammar of "initial shape—central column—upright column—column rider—through-beam—main framework". This is also the most important part of the grammar for the "L"-shaped Tujia dwellings (Table 3).

Table 3. Main framework construction process from initial coordinate points.

	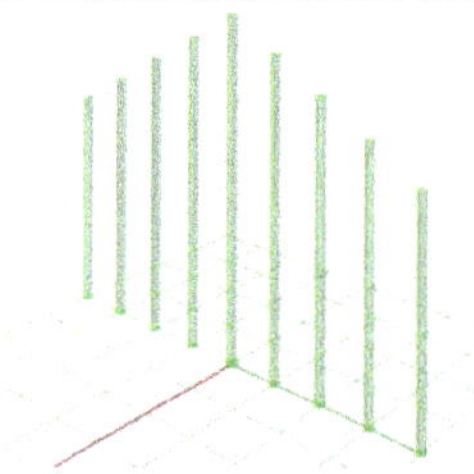	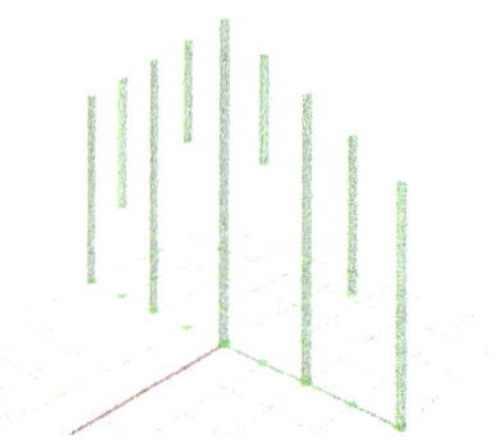
STEP 01	STEP 02	STEP 03
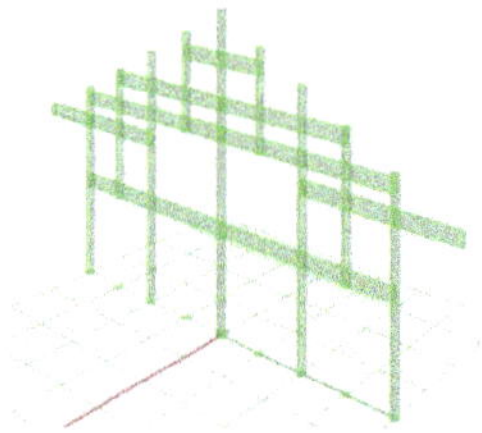	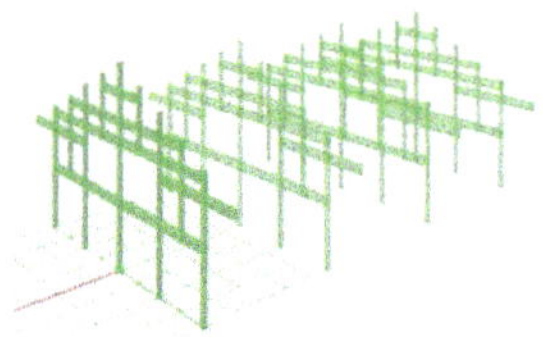	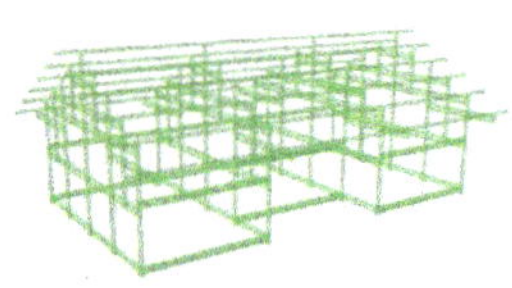
STEP 04	STEP 05	STEP 06

(2) Roof Frame Conversion and Support Grammar

The structural rules of the main frame not only determine the construction of the main house but also significantly influence the generative logic of the roof frame of the side house. For instance, according to the corpus comparison, the intersection line (i.e., the diagonal ridge line) of the roofs of the main house and the side house in Tujia dwellings in Southeast Chongqing is 45° in plan projection. This implies that the purlins of the main house and the side house are evenly distributed in both directions, meaning that the side house has the same step length as the main house frames, and the step length of the main house affects the side house simultaneously. Therefore, after the main frame is established, the next step is to determine the intersection between the frames of the main house and the side house.

The connection methods between the main house and the side house vary, including "L"-shaped intersections and "T"-shaped intersections. There are also cases where the main house and side house do not connect at all. A distinctive practice is to leave a separate transfer space at the intersection and then install an "umbrella handle column" at the intersection part of the ridge of the side house [44]. Simultaneously, a slanting ridge is extended from the top of the column to the corner column to support the purlins in both directions, thereby completing the transfer of the roof frame in both directions. The umbrella handle column can be considered an outstanding achievement in the construction of traditional Tujia houses, and its invention plays a key role in the intersection of the Tujia main house and the side house. Its setup can be divided into two scenarios [45]:

First scenario: When the height of the ridge purlin of the main house and the side house is the same, the umbrella handle column directly falls to the ground to support the weight of the two ridge purlins of the main house and the corner diagonal ridge purlin of the side house. From the plan projection, the umbrella handle column is located at the intersection of the main ridge purlin of the main house, the horizontal ridge purlin of the side house, and the corner diagonal ridge.

Second scenario: When the height of the main house and the ridge purlin of the side house are different, the umbrella handle column is not at the intersection of the main ridge purlin of the main house, the ridge purlin of the side house, or the corner ridge. Instead, it is located at the intersection of the horizontal ridge purlin, the ridge purlin of the main house one or two steps back from the purlin, and the corner ridge. This results in a hollowed-out section at the intersection of the two roofs, forming the unique "Yaquekou" feature on the roofs of the Tujia ethnic group.

According to the author's research, Tujia houses in the southeast of Chongqing typically follow the rule that the height of the rooms is lower than that of the main house, and the height of the inverted seat is lower than that of the rooms, reflecting the traditional architectural hierarchy. Therefore, most roof intersections adopt the "Yaquekou" practice, and the corresponding umbrella-like columns are usually located at the height of the main house's gold columns to support the slanting spine (Figure 12).

Figure 12. "Yaquekou" roof intersection method (different ridge heights between main house and wing rooms).

Based on the previously determined foundation dimensions and main framework, the location of the stilted side rooms and the main house is used as the basis to sequentially determine the placement of the umbrella-like columns. This further defines the diagonal support grammar and connects the tie beams to complete the structural transition between the main house and the side rooms (Figures 13 and 14).

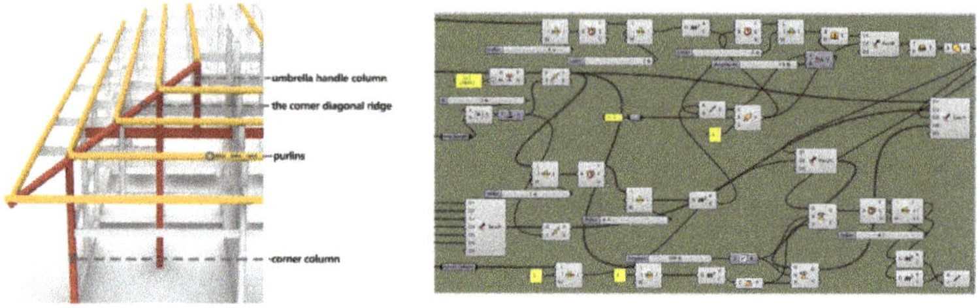

Figure 13. "Umbrella-like Columns" and the grammar of support conversion (author's own illustration).

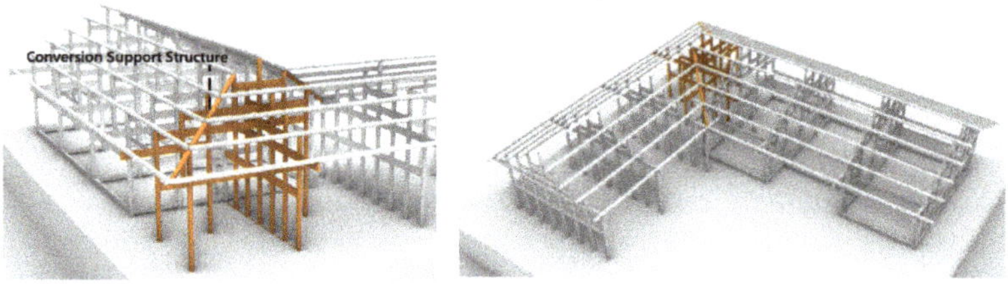

Figure 14. Roof truss and support conversion grammar model (author's own illustration).

(3) The Side Houses and Stilted Structure Grammar:

The stilted structure is the most representative component of Tujia folk houses, characterized by their distinctive national features. Given the mountainous environment, the side houses of Tujia folk houses typically adopt the practice of stilt houses. This structural approach allows for adaptation to varying site slopes and height differences, providing greater flexibility in selecting building sites on mountain slopes.

Depending on the site conditions, the method of elevating the side house varies (Figure 15). If the main house and the side house are relatively flat, the Tujia folk house compartment usually employs the "flat ground lifting" method. In this case, the floor of the side house is raised to match the eaves of the main house, with the lower part of the floor elevated to create overhead space.

Figure 15. Two approaches to wing rooms based on different site conditions.

The wisdom of Tujia folk houses in site adaptation is more evident when the site has a significant height difference. In such cases, the room cannot maintain the same construction plane as the main house. The common practice is to maintain the usable space on the second floor of the room, aligning it with the main house while allowing the ground floor to follow the site's slope downward, supported by high stilts to accommodate the site's changes. Currently, if the space on the ground floor of the side house is small, it is typically used for storage or livestock. If the space is larger, it can be enclosed to serve as an indoor space. These two practices reflect the adaptive treatment of Tujia houses according to actual needs, embodying the so-called "flat sky, uneven ground" approach. Regardless of how the stilted structure and the site change, the eaves are always kept neat when constructing the house. In this study, to restore the characteristics of Tujia houses as much as possible, we selected side houses with a high stilted structure and grounding for grammatical analysis.

The height difference in the site directly determines the height of the stilts. By using a program to set up the house plan and simulate the site's height difference, the syntax rules for the "grounding stilted structure" can be further developed (Figures 16 and 17).

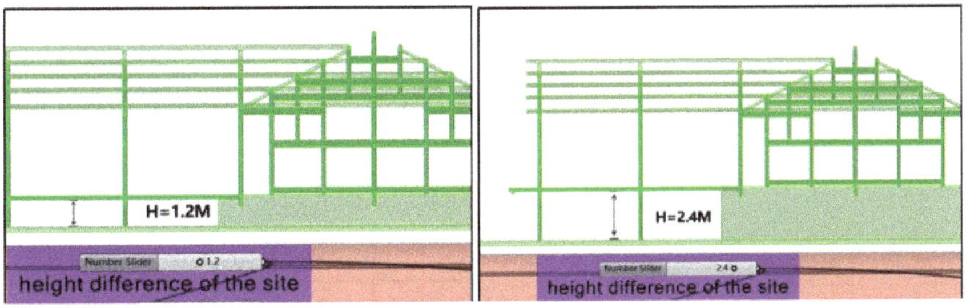

Figure 16. Adapting to different site-level differences through the use of stilt foundations in wing rooms (author's own illustration).

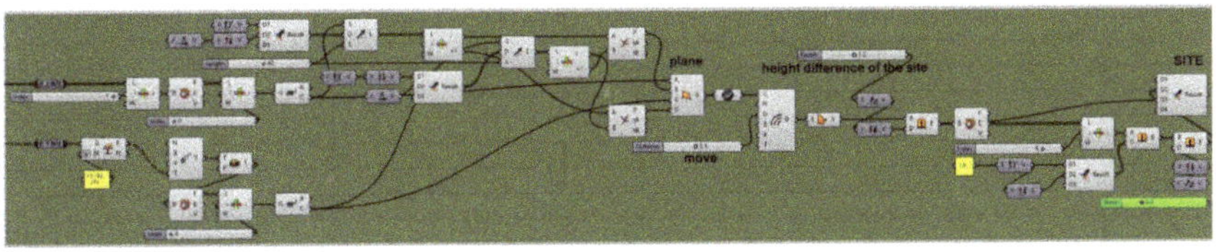

Figure 17. Two approaches to side houses based on different site conditions (author's own illustration).

Given the different main house frames, the side house incorporates an extended corridor and a high stilted structure. To ensure the integrity of the frame, additional beams are added to connect and provide load-bearing support. For example, the extended corridor outside the pick columns is supported by the extended beams and supplemented by the structure of the string beams. The gable end columns are connected to the gable corridor of the "square", and the floor pillow is interspersed in the grounding columns on the pick square to form a complete structural whole. Based on this, the syntax of the construction of side houses and stilted structures was determined by taking the most common double-sided picks in the L-shaped Tujia dwellings in southeast Chongqing as the object of study.

(4) The Cantilevered Elements and Corners Grammar

With a series of grammatical determinations from the main frame to the umbrella handle column transfer to the side house and the stilted structure, the architectural main body of the "L"-shaped residential grammar has been completed. The next step is to further analyze the unique "ox horn picking" component and the corner tower with a prominent shape to complete the construction of the residential grammar.

The Tujia corner building is the most intensive part of the structure and the most prominent part of the house's outline, serving as one of the important symbols for recognizing Tujia houses (Figure 18). The wall of the corner house is usually covered with a hipped gable roof-style roof, supported by a series of purlins, columns, trailing beams, and picking beams in different positions. The structural connection is very complicated, often resulting in positional conflicts where multiple components need to be inserted in the same location. The response to this problem is an innovative Tujia residential construction skill: a more curved pick.

Figure 18. Complex corner structures and cantilevered elements in Tujia architecture.

In Tujia corner buildings, the use of curved wood is most concentrated. If the position of the pick beam were straight wood, conflicts would inevitably arise due to the unchanged height of the eaves, which need to be supported by the pick beam. Using curved wood ensures the same support for the eaves while avoiding conflicts. Additionally, since most of the curved picks used by the Tujia people are sourced from naturally curved timber, this practice allows for flexible arrangement of the curved picks based on the actual availability of materials.

As illustrated in Figure 18, cantilever beam 1 intersects with the corner column at point A. If cantilever beam 2 were also a straight beam, it would inevitably intersect with the corner column at point A, leading to a positional conflict. This conflict would be even more pronounced if there were additional cantilevered elements at the corner. However, when cantilever beam 2 is designed as a curved cantilever, it does not need to intersect with the corner column at point A, thereby avoiding the positional conflict. Moreover, since the curved cantilevers in Tujia architecture are often sourced from naturally curved timber, this practice allows for flexible arrangement of the cantilevers based on the actual availability of materials. This is a concentrated manifestation of the Tujia people's construction wisdom (Figures 19 and 20).

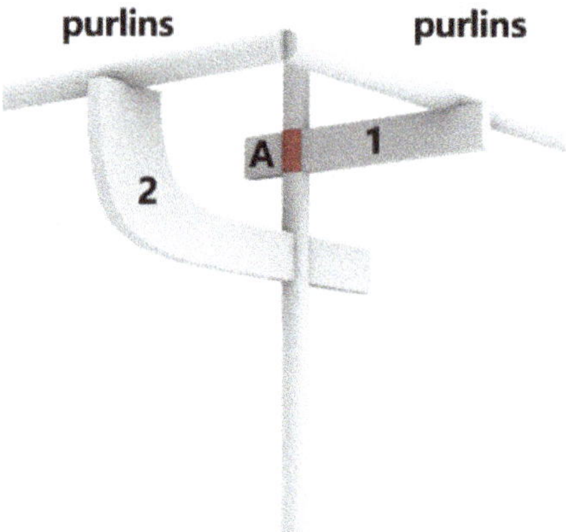

Figure 19. Cantilever curvature resolves spatial conflicts (author's own illustration).

Figure 20. Complex corner structures and cantilevered elements in Tujia architecture (image source: Sun Yan).

Among the various types of cantilevered elements, the most representative is the "ox-horn cantilever". Named for its significant curvature that resembles an ox horn, this structural feature is emblematic of Tujia architecture. To authentically reproduce the Tujia architectural style, this study has developed a morphological generation grammar for the ox-horn cantilever. The curvature of the ox-horn cantilever originates from naturally grown timber, making it extremely difficult to simulate using conventional circular arcs or curves. Additionally, the non-uniform cross-sectional dimensions of the ox-horn cantilever further complicate its morphological analysis. Through extensive comparison of architectural materials, this study has found that the shape of the ox-horn cantilever closely resembles the curves of trigonometric functions, specifically the sine and cosine functions in mathematics. Based on extensive data comparison, this study employs a combination of two cosine functions to construct the morphological grammar of the ox-horn cantilever. This approach not only accurately captures the unique curvature of the ox-horn cantilever but also provides a flexible framework for its representation in architectural design (Figures 21 and 22).

Figure 21. Morphology and simulation of ox-horn cantilevers.

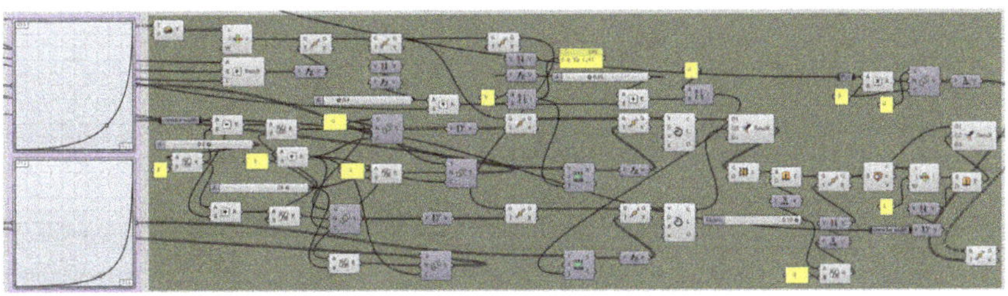

Figure 22. Cosine-curve generation of ox-horn cantilever morphological grammar.

3.4. Generation and Practical Application and Verification of the "L"-Shaped Tujia Ethnic Dwelling's Architectural Grammar

Adhering to the structural characteristics and construction logic of Tujia ethnic dwellings in southeast Chongqing, this study extracts a set of grammatical rules from the form and style features of the Tujia ethnicity, including the generation of the main framework, conversion of the main wing, side houses with stilted structures, and the transformation of corners with cantilevered elements. It further customizes multiple detailed grammatical rules for each framework, ultimately completing the generation of the "L"-shaped dwelling construction grammar (Figure 23). This grammar is comprehensively regulated by variables such as the step size, pitch, component dimensions and positions, site height differences, component morphological curves, and the presence or absence of column riders, enabling the generation of diverse grammatical models based on varying actual conditions.

Figure 23. Grammar model of "L"-shaped Tujia dwellings in southeast Chongqing.

Following the traditional construction sequence of Tujia houses, which prioritizes the establishment of the structural framework before detailing, only the elevation and roof details need to be added to the generated framework to produce a complete grammatical model of Tujia houses. This grammar not only provides a systematic overview of the construction techniques employed in Tujia dwellings but also clearly reflects their primary architectural forms.

The current practical applications of dwelling grammar are primarily twofold. First, it involves the conservation and restoration of existing Tujia ethnic dwellings. By invoking corresponding materials and grammar from the material library and rule set, this process assists in the reconstruction of specific architectural structures. Second, it pertains to the generative design of new Tujia rural houses. Inputting the relevant site conditions and layout requirements allows for the stepwise application of grammatical rules to produce floor plans and dwelling styles. Furthermore, secondary grammar can be superimposed based on specific conditions to further customize the design to meet the homeowner's needs. The workflow framework is illustrated in the following figure (Figure 24):

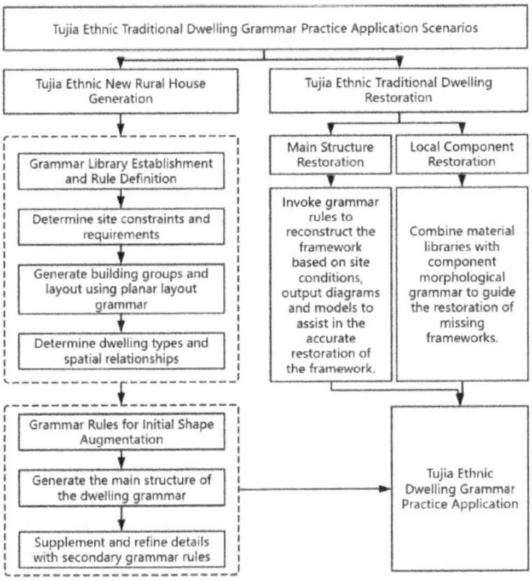

Figure 24. Application scenarios of Tujia ethnic dwelling grammar.

This study primarily uses the "L"-shaped dwelling grammar as an example to explore and analyze the specific rules and logic in the construction of Tujia ethnic dwellings. However, in the actual practice, different types of dwelling constructions are often required based on varying needs. In the actual process of craftsmen building houses, different floor plan configurations can also evolve through the sequential stacking of main rooms and side rooms [46]. This research has parsed the main logical challenges of this evolution during the organization of the "L"-shaped dwelling grammar. Therefore, grammatical models for floor plans such as the "San he shui" (threefold water) and "Si he shui" (fourfold water) can also be obtained through relatively simple replication and transformation operations based on this study. Of course, there are also some corresponding detailed issues in the evolution process of dwelling types, but they do not affect the overall logic; hence, this study focuses on the core research without expansion.

This study selected a relatively complex case of a dwelling in Qiaoqiao Village and compared and deconstructed it with the grammatical rules discussed in this research to verify the practical feasibility of this grammar (Figure 25).

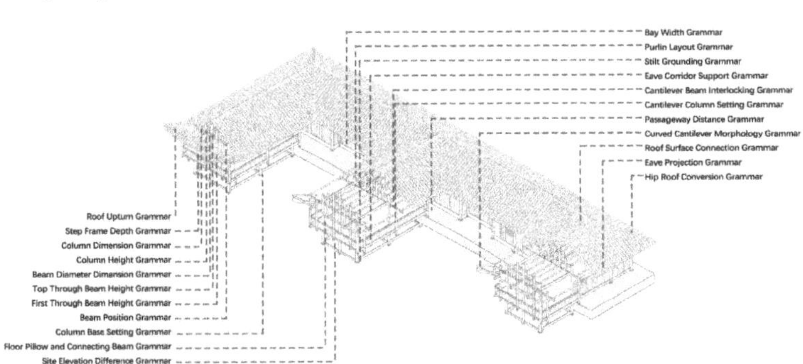

Figure 25. Grammar deconstruction verification of a dwelling in Qiaoqiao Village.

4. Discussion

Traditional dwellings encapsulate the production and lifestyle of a region and an ethnicity, embodying significant cultural and historical value. The perception of regional characteristics and cultural attributes of traditional architecture often stems from morphology. Therefore, it is essential to summarize the stylistic features of traditional dwellings from a morphological perspective [28].

The conservation methods for traditional dwellings are continuously evolving, with digital technology making the process more systematic, efficient, and precise. Previous research on Tujia dwellings primarily focused on the theoretical organization of textual and graphical data through field surveys. While this approach can effectively "record", "summarize", and "classify" traditional dwellings and their cultures, it falls short in deeply deconstructing the underlying construction techniques to address issues of preservation and inheritance. By employing Shape Grammar, the fundamental logic and rules of dwelling construction are transformed into geometric transformation relationships between "shapes", which are then translated into computer language and expressed through digital software. This approach allows for a more in-depth and accurate reconstruction of construction techniques, thereby aiding in the protection and inheritance of Tujia dwellings.

During the construction process of Tujia ethnic dwellings, the methodologies of "prototype-to-typology" and "unit-to-module" are demonstrated. The column-and-tie construction framework serves as a prototype, which is combined and varied in various scenarios, under different owners, and across different periods to generate a rich array of dwelling styles. This study proves the feasibility of using Shape Grammar to restore traditional dwellings and their construction techniques, which can be applied to extract and classify the morphology and stylistic elements of dwellings; meanwhile, the parameterized grammar retains the flexibility and diversity of the generative outcomes.

In the current translation of dwelling grammar, there are still some challenges. On one hand, the construction wisdom and empirical practices of craftsmen have largely been reflected in a set of relatively clear construction rules. On the other hand, due to the traditional self-construction model of traditional dwellings, they possess a great degree of freedom, thus exhibiting a high degree of morphological complexity, with many details handled freely and flexibly. It is difficult to express all the behavioral logic of the construction process in grammatical rules. For instance, regarding the Tujia dwelling wisdom of "adapting to materials", this study analyzed the basis for setting up curved cantilevers and completed the digitalization of this craftsman wisdom through custom morphological grammar. However, on the other hand, the arrangement of certain components, such as the placement of penetrating beams, is almost random, determined not only by variable materials and complex real-world conditions but also by the homeowner's personal preferences, which is hard to frame with unified grammatical rules.

In response to this issue, firstly, this study referred to a multitude of authoritative scholars' preliminary analyses and viewpoints, extracting the widely recognized practices and representative stylistic features of Tujia dwellings, and subsequently compared these with the material library to ensure the scientific validity and effectiveness of the grammar.

Secondly, this grammatical model is based on parameterized regulation, with a series of core variables reserved at the input end of the program serving as control interfaces, allowing for real-time adjustment of the specific style of the generated grammatical model. This approach can provide a certain degree of flexibility in response to non-geometric factors in the construction process.

Furthermore, this study provides a universally adaptable dwelling-construction grammar. To address the aforementioned limitations, this grammatical model can serve as the backbone, with secondary grammars superimposed according to actual changes to meet the individual needs of homeowners. This study has thoroughly analyzed the parameters and rules of each component in the construction process, thus making the superposition of grammar feasible and easily implementable. Of course, the handling of this issue and the research still require further deepening and refinement in the future.

To illustrate the primary operations generated by the data, the charts and program logic diagrams presented in the text have been partially simplified without affecting the outcomes, in order to achieve a clearer presentation. Furthermore, as this grammatical study focuses on the construction logic and building techniques of dwellings, parts that only require selection and addition to the framework, such as lattice windows, ridge decorations, and panel walls, have not been analyzed in detail. Although Shape Grammar holds promising prospects for research, there remain challenges in integrating and compatibility with general computer-aided design tools in the workflow. In the future, the digital grammatical models generated by this study can be further integrated with VR and AR technologies, applying to projects such as digital museums and information model displays in the preservation of traditional villages, promoting the digital conservation of these communities.

5. Conclusions

This study integrates Shape Grammar with digital technology to conserve traditional Tujia dwellings. Through extensive field surveys and sample collections across multiple villages in five districts and counties of Chongqing, the research identifies commonalities and subtle differences in construction techniques from a building-methods perspective. It develops a Shape Grammar set for Tujia dwellings and proposes a construction grammar for the "L"-shaped Tujia dwellings in Southeast Chongqing, generating a digital model. The primary outcomes are threefold:

(1) Comprehensive Case Analysis: This study conducted extensive case surveys and data collections, establishing a rich repository of grammatical materials. It analyzed the unique structural and architectural features of Tujia dwellings, such as the column-riding system, roof frame transitions, cantilevered beam supports, and stilted eaves. By comparing and analyzing these features, the study extracted commonalities and established a Shape Grammar corpus. This corpus clarifies the value elements of Tujia dwellings, providing references for future conservation and design.

(2) Construction Grammar Development: By analyzing and extracting the prototypes of elements in the corpus and consulting relevant literature, this study developed a construction grammar collection for Tujia dwellings in Southeast Chongqing, including sets of labels and rules. The grammar rules, which reflect the geometric transformation relationships between shapes, include the following: (1) connection and transformation rules between components, creating functions such as translation, mirroring, copying, rotation, extrusion, and distance determination; (2) morphological grammar of individual components, such as using two cosine function curves to define the morphological generation grammar of the ox-horn cantilever components. The outcomes of the grammar rules include descriptive grammar composed of text and formulas, as well as computer logic and grammar toolkits translated from programs. These rules rigorously reflect the construction logic of Tujia dwellings, summarizing and preserving their building techniques.

(3) Digital Generation of Grammar Model: Based on the Shape Grammar outcomes, this study proposed a generation grammar for "L"-shaped Tujia dwellings and translated it into digital outcomes using Grasshopper (Based on Rhino 7.35). A three-dimensional grammar model was established, incorporating variable factors such as "Bu Chi", "bay width", and "height difference". These variables reflect the practical choices faced during the construction of Tujia dwellings and integrate the explicit and implicit building rules and techniques through programming languages. Therefore, the grammar model follows the traditional construction logic of Tujia craftsmen, accurately recording and reflecting the construction techniques of Tujia dwellings. The translated grammar model can generate various Tujia dwelling styles that meet requirements through parameter adjustments and rule constraints, providing new support for the conservation, renewal, and design of Tujia dwellings.

Author Contributions: Conceptualization, Q.W., Y.Z.; methodology, Q.W.; software, Y.Z.; validation, Y.Z., X.H. and Q.W.; formal analysis, G.W.; investigation, Y.Z.; resources, Q.W.; data curation, X.H.; writing—original draft preparation, Y.Z.; writing—review and editing, Q.W.; visualization, Y.Z.; supervision, Q.W.; project administration, Q.W. All authors have read and agreed to the published version of the manuscript.

Funding: The National Natural Science Foundation of China: Identification and Evolution Mechanism of the Technical Genealogy of Vernacular Architecture in the Multi-ethnic Southwest Region (52108005). the Postgraduate Scientific Innovation Fund of Chongqing JiaoTong University, project number: CYS240496.

Data Availability Statement: Due to the nature of this research, participants of this study did not agree for their data to be shared publicly, so supporting data are not available.

Conflicts of Interest: The authors declare no conflicts of interest.

References

1. Kang, Y.; Zhou, H. Research Progress of Traditional Chinese Dwellings in the 21st Century. *World Archit.* **2020**, *11*, 18–21.
2. Cai, L. Architecture-Village-Architectural Cultural Area: A Discussion on the Hierarchy and Framework of Traditional Chinese Dwellings Research. *New Archit.* **2005**, *4*, 6–8.
3. Wei, F.; Guo, H.; Tang, X. New Trends in Traditional Dwellings Research: A Review of the 20th Chinese Dwellings Academic Conference. *South. Archit.* **2015**, *1*, 4–7.
4. Tang, X.; Wang, D. Deepening and Expansion of Research on Traditional Chinese Settlements and Dwellings: A Review of the 2015 Chinese Traditional Dwellings Academic Seminar. *New Archit.* **2015**, *6*, 130–131.
5. Zhou, L.; Zhou, T.; Wang, L. Inheritance and Practice of Traditional Chinese Dwellings Research: A Review of the 22nd Chinese Dwellings Architecture Academic Annual Conference. *New Archit.* **2017**, *5*, 114–115.
6. Sun, Y.; Qin, L.; Xia, Z. *Traditional Tujia Dwellings in Southeast Chongqing*, 1st ed.; Chongqing University Press: Chongqing, China, 2004; pp. 001–002.
7. Sun, Y.; Qin, L. Construction Techniques and Artistry of Tujia Ethnic Dwellings in Southeast Chongqing. *J. Chongqing Univ. Technol.* **2006**, *2*, 21–23.
8. Chen, G. Characteristics and Source Analysis of the Traditional Wooden Structure System in the Youshui River Basin. *South. Archit.* **2014**, *1*, 57–61.
9. Guo, S.; Sun, J.; Zhang, Q. Design and Implementation of a Typical Tujia Ethnic Dwelling Exhibition Platform in Guizhou Province. *Inf. Technol. Informatiz.* **2020**, *11*, 46–50.
10. Wang, R. Research on Architectural Design Optimization Based on BIM Quantitative Analysis Techniques. Master's Thesis, Chongqing University, Chongqing, China, 2022.
11. Wang, J.; Fan, W.; Zhao, B.; Zhao, J. Analysis, Transformation, and Generation: A Review and Interpretation of Shape Grammar Research. *Architect* **2022**, *4*, 116–125.
12. Stiny, G.; Gips, J. Shape grammars and the generative specification of painting and sculpture. In Proceedings of the IFIP Congress, Ljubljana, Slovenia, 23–23 August 1971; Volume 2, pp. 125–135.

13. Huang, X.; Zhou, D. A Brief Introduction to Shape Grammar and Its Applications. *Cent. China Archit.* **2004**, *2*, 45–48.
14. Fan, W. Extraordinary Rural Areas: A Study on the Rule Analysis and Design Application of Shape Grammar. Master's Thesis, Shandong Jianzhu University, Jinan, Chian, 2022.
15. Müller, P.; Wonka, P.; Haegler, S.; Ulmer, A.; Van, G. Procedural modeling of buildings. In Proceedings of the ACM SIGGRAPH, 2006, Los Angeles, CA, USA, 10–14 August 2006; pp. 614–623.
16. Veloso, P.; Celani, G.; Scheeren, R. From the generation of layouts to the production of construction documents: An application in the customization of apartment plans. *Autom. Constr.* **2018**, *96*, 224–235. [CrossRef]
17. Yavuz, O.; Sahika, O. Extended prioritizing of store plan alternatives produced with shape grammar using the generalized Choquet integral method. *Environ. Plan. B Urban Anal. City Sci.* **2019**, *46*, 931–947.
18. Wang, J.; Zhou, J.; Zhang, S. Applying shape grammar and BIM for generating the mass modular housing design in ShanLiChen-Jia village, ZhaoYuan, China. *J. Asian Archit. Build. Eng.* **2023**, *24*, 178–198. [CrossRef]
19. Wang, J.; Zhang, S.; Fan, W. Application of shape grammar to vernacular houses: A brief case study of unconventional villages in the contemporary context. *J. Asian Archit. Build. Eng.* **2023**, *23*, 843–859. [CrossRef]
20. Wang, J.; Fan, W.; Zhao, B.; Qi, J.; Yang, Y.; Zhao, J. A Study on the Generation Design Method of Rural Houses Based on Shape Grammar. *Archit. Technol.* **2022**, *28*, 62–65.
21. Stiny, G. Ice-ray: A Note on the Generation of Chinese Lattice Designs. *Environ. Plan. B Plan. Des.* **1977**, *4*, 89–98. [CrossRef]
22. Stiny, G.; Mitchell, W.J. The Palladian Grammar. *Environ. Plan. B Plan. Des.* **1978**, *5*, 5–18. [CrossRef]
23. Knight, T.W. The 41 Steps. *Environ. Plan. B Plan. Des.* **1981**, *8*, 97–114. [CrossRef]
24. Downing, F.; Flemming, U. The Bungalows of Buffalo. *Environ. Plan. B Urban Anal. City Sci.* **1981**, *8*, 269–293. [CrossRef]
25. Flemming, U. The Secret of the Casa Giuliani Frigerio. *Environ. Plan. B Urban Anal. City Sci.* **1981**, *8*, 87–96. [CrossRef]
26. Herbert, T.; Sanders, I.; Mills, G. African Shape Grammar: A Language of Linear Ndebele Homesteads. *Environ. Plan. B Urban Anal. City Sci.* **1994**, *21*, 453–476. [CrossRef]
27. Ali, L.A.; Mustafa, F.A. Mosque Typo-Morphological Classification for Pattern Recognition Using Shape Grammar Theory and Graph-Based Techniques. *Buildings* **2023**, *13*, 741. [CrossRef]
28. Zhang, K.; Zhang, N.; Quan, F.; Li, Y.; Wang, S. Digital Form Generation of Heritages in Historical District Based on Plan Typology and Shape Grammar: Case Study on Kulangsu Islet. *Buildings* **2023**, *13*, 229. [CrossRef]
29. Sun, J. Shape Grammar and Shape Rules. *Chin. J. Comput.* **1986**, *6*, 416–424. [CrossRef]
30. Chiou, S.-C.; Krishnamurti, R. Example Taiwanese traditional houses. *Environ. Plan. B Urban Anal. City Sci.* **1996**, *23*, 191–216. [CrossRef]
31. Li, A.I. A Shape Grammar for Teaching the Architectural Style of the Yingzao Fashi. Ph.D. Thesis, Massachusetts Institute of Technology, Cambridge, MA, USA, 2001.
32. Huang, R.-S. Construction of Song Dynasty Bracket System Using Shape Grammar: A Case Study of Four-Piece Bracket Set with Double Interlocking and Double Angles. Master's Thesis, University of Science and Technology, Taiwan, China, 2002.
33. Yang, G. Representative Plan Design of Rural Dwellings in Zhejiang Based on Shape Grammar Theory. Master's Thesis, Zhejiang A&F University, Hangzhou, China, 2015.
34. Zhang, Y. Research on the Construction of Huizhou Architecture Based on Shape Grammar. Master's Thesis, Hefei University of Technology, Hefei, China, 2017.
35. Gu, Z. Generation of Traditional Courtyard Residential Design Based on Shape Grammar. Master's Thesis, Tianjin University, Tianjin, China, 2021.
36. Zhang, K.; Zhang, N.; Li, Y. *Digital Generation Method of Modern Western-Style Buildings in Gulangyu Based on Shape Grammar. Proceedings of the 2022 National Conference on Digital Technology in Architecture*; Huazhong University of Science and Technology Press: Wuhan, China, 2022; pp. 554–559.
37. Chen, Y.-X.; Shu, B.; Chang, H.-T. Exploring Architectural Shapes Based on Parametric Shape Grammars: A Case Study of the "Three Lanes and Seven Alleys" Historic District in Fuzhou City, China. *Buildings* **2023**, *13*, 2063. [CrossRef]
38. Liu, Z. *Chinese Architectural Types and Structures*; China Architecture & Building Press: Beijing, China, 2000; p. 56.
39. Chao, C. The Art and Culture of Tujia Stilt Houses: An Ethno-Artistic Interpretation of Tujia Architecture. *Archit. J.* **2024**, *7*, 125.
40. Liu, X.; Li, B.; Li, X. Basic Forms and Wisdom of Tujia Dwellings in Southeast Chongqing. *New Archit.* **2007**, *4*, 35–38.
41. Luo, Q. Regional Characteristics of the Ba-Shu Chuan-Dou Structure. *J. West. China's Hum. Settl.* **2017**, *32*, 14–20.
42. Mostafaei, H.; Mostofinejad, D.; Ghamami, M.; Wu, C. Fully automated operational modal identification of regular and irregular buildings with ensemble learning. *Structures* **2023**, *58*, 105439. [CrossRef]
43. Yuan, X.; Zhang, X. Traditional Dwellings in Miao and Tujia Villages under the Influence of the "Zhang-Ba-Ba" System. *Anc. Archit. Landsc. Technol.* **2024**, *3*, 8–13.

44. Zhou, L. Research on Tujia Dwellings and Traditional Construction Techniques in Southeast Chongqing. Master's Thesis, Chongqing University, Chongqing, China, 2005.
45. Pan, W. A Study on the Construction Characteristics and Folk Building Techniques of Tujia Timber Frame Buildings in Southwest Hubei. Master's Thesis, Huazhong University of Science and Technology, Wuhan, China, 2012.
46. Chen, S.; Li, X.; Qiao, J. The Chronicles of Tea Village Houses: A Historical Transformation of a Tujia Ethnic Tradi-tional Dwelling. *Architect* **2024**, *5*, 62–70.

buildings

MDPI

Article

New Insights into Traditional Construction Behind Sibe Dwellings with Swastika Kang for Space Heating in North China

Menglong Zhang, Zhiyuan Shang, Keqian Luo and Kai Xie *

School of Architecture, Southwest Minzu University, Chengdu 610225, China; z18768972650@163.com (M.Z.); 13503702479@163.com (Z.S.); luokeqian@swun.edu.cn (K.L.)
* Correspondence: 80300177@swun.edu.cn; Tel.: +86-136-8409-6927

Abstract: Due to massive urbanization and industrialization, modern constructions tend to be designed as technique-dependent, at the cost of high consumption and emissions for indoor environment control such as heating ventilation and air conditioning. Space heating accounts for about 40% of total building energy usage in northern China in winter. This calls for self-reflection and tracing of local traditional architectural wisdom. In this paper, Sibe Traditional Houses were chosen as a typical illustrative example to reveal the building mechanisms behind such local-adaptive traditional constructions. Based on the field investigation in Shifosi Village, a traditional Sibe settlement in Shenyang City, northern China, thermal modeling and indoor heating effects are studied in Sibe Traditional Houses with unique building spatial patterns. The indoor thermal environment is comparatively analyzed for both passive envelope insulation and active heating considerations. Preliminary results indicate that enhancing roof thermal insulation enhancement is the key passive strategy for improving indoor thermal comfort in winter. It also suggests that a space-heating configuration that combines the traditional "kang" with the architectural layout has a more significant effect on the enhancement of indoor thermal comfort in Sibe dwellings. This paper can provide methodological support and an application reference for the improvement of indoor thermal environment of traditional village dwellings.

Keywords: built environment; historical buildings; Sibe dwelling; heating; thermal comfort; optimization design

1. Introduction

In recent years, with the continuous progress of urbanization, China has shifted its focus towards rural revitalization. The rural living environment stands as a crucial foundation for this revitalization effort, playing a pivotal role in establishing sustainable spaces and creating beautiful countryside landscapes [1]. The issue of rural residents has been a focal point of concern for governments at all levels in the country, given its direct impact on the living standards of farmers [2]. As the living standards of rural inhabitants improve, their demand for thermal comfort indoors is steadily increasing [3]. The indoor thermal environment significantly affects human health and comfort, as adverse indoor thermal conditions are detrimental to human health [4]. Indoor thermal comfort is generally predicted based on environmental parameters such as temperature, humidity, air velocity, and personal parameters including activity levels and clothing resistance [5].

Nowadays occupant thermal comfort has been a research focus for people's increasing indoor time and its potential influence on productivity [6]. To enhance indoor thermal comfort, analysis can be bifurcated into passive strategy and active heating. Buildings have been regarded as one of the major energy consumption sources [7]. If no attention is paid to savin energy in buildings, a 70% increase in building consumption can be expected by 2050 [8]. Passive design measures aim to improve thermal comfort without consuming energy during heating season through reasonable building orientation, improved thermal performance of windows, increased glazing area, enhanced heat capacity, and insulation of building constructions [9]. Passive strategies are more energy-efficient technologies for application in buildings. These have a much longer history associated with their use in buildings than active strategies. Early passive technologies, such as cave dwellings, kang (For a detailed explanation, please refer to Nomenclature), hypocaust, etc., have existed for thousands of years [10]. However, studies from Sweden indicate that achieving passive house standards requires wall insulation of approximately 335 mm (U-value of 0.10 W/m^2K) and roof insulation of about 500 mm (U-value of 0.067 W/m^2K), surpassing the thickness of both exterior walls and roofs beyond 500 mm. This increase in thickness incurs costs far beyond just those of insulating materials [11]. On the other hand, the active heating system relies on special heating equipment, such as the boiler heating system, air source heat pump system, and so on, to meet the requirements of building heating [9]. Different countries, regions, and ethnicities exhibit diverse choices in heating facilities. For instance, central heating systems are prevalent in European and American countries, while stove heating methods are common in Asian and African regions. Underfloor heating systems are popular in North America and some Asian areas. These heating facilities notably elevate indoor thermal comfort.

Both passive insulation and active heating systems can contribute to improving the indoor thermal environment. However, the effectiveness of passive insulation alone in enhancing indoor thermal conditions is often limited, while relying solely on active heating tends to increase building energy consumption.

Therefore, this study focuses on the traditional dwellings of the Sibe ethnic group, employing simulation methods to investigate indoor thermal comfort:

- Validating and analyzing traditional construction wisdom using modern technologies.
- Conducting an in-depth analysis of the "Swastika kang", a traditional heating system (For a detailed explanation, please refer to Nomenclature).
- Exploring optimization strategies by integrating passive insulation with active heating systems.

2. Literature Review

2.1. Indoor Thermal Comfort

The enhancement of indoor thermal comfort has been studied by many scholars. Scholars in China have carried out related research on indoor thermal comfort. Hou JiaWun [4] conducted an optimization study of the indoor thermal environment in winter, and proposed an optimization strategy to improve the thermal performance of the building and the use of solar energy resources by increasing the indoor temperature of the building by up to 18 degrees Celsius in winter. Liu Jiao [12] conducted a study on traditional dwellings in Jia County, implementing optimization measures such as internal insulation for external walls and ceilings with gypsum boards combined with EPS insulation layers. These interventions led to an increase in indoor temperatures during winter and a decrease during summer, without the use of heating or cooling equipment, significantly improving indoor thermal comfort. In China, research on indoor thermal comfort has predominantly focused on the eastern and northern regions, with relatively limited attention given to the southwestern areas. Therefore, Dong et al. [13] evaluated indoor thermal comfort in rural southwestern China, analyzing residents' thermal adaptation behaviors and proposing a thermal neutral temperature of 29.33 °C for local dwellings. Furthermore, they emphasized that passive cooling measures should be more widely adopted in the design and renovation of rural homes. Wang Yan et al. [14] established a new thermo-neutral temperature for residents through methods such as subjective questionnaire surveys and objective data measurements, and proposed different performance enhancement strategies for old and new traditional dwellings: old traditional dwellings should focus on strengthening the thermal performance of the enclosure structure, while the new traditional dwellings should address the problem of low indoor temperatures in winter. Due to the long-standing issue of poor indoor thermal environment quality in Tibetan dwellings on the western Sichuan plateau, He et al. [15] conducted a study on the indoor thermal environment of these residences. Through field measurements, they proposed optimization strategies that include the utilization of both passive and active solar energy, the application of modern insulation materials, and the addition of sunspaces to improve the indoor thermal environment.

Zhang Chen et al. [16] conducted a systematic evaluation of integrated systems with radiant heating/cooling and ventilation, summarized various system configurations, and assessed system performance in terms of thermal comfort and air quality. Due to the issue of frost formation on the outdoor coil of air source heat pumps (ASHP) during heating in low-temperature environments, which interrupts heating supply during defrosting and affects indoor heating performance, Qu et al. [17] developed a novel reverse-cycle defrosting method based on thermal energy storage (TES). Experimental results demonstrated that this approach significantly reduces defrosting time and achieves higher indoor air supply temperatures. Mohammed [11] introduces the basic concept and some illustrative simulated performance results of a new Void Space Dynamic Insulation (VSDI) technology that couples low-cost conventional insulation materials with efficient ventilation to deliver low-loss building envelopes and high indoor air quality in thin wall construction. Joon Are Myhren and Sture Holmberg [18] conducted a study on how different heating systems and their locations affect the indoor climate in exhaust ventilation offices under Swedish winter conditions, and the general conclusion of the study was that low-temperature heating systems can improve the indoor climate and thus reduce the indoor wind speeds and temperature differentials in comparison to conventional high-temperature radiator systems. The disadvantage of low-temperature systems is that they cannot offset the cold of the ventilation supply units. Therefore, the location of heat emitters and the design of the ventilation system proved to be particularly important.

2.2. *Kang*

Research on the kang system has primarily been conducted by Chinese scholars. Zhuang [19] reviewed the basic principles of heat transfer and airflow in traditional Chinese kangs and presented the thermal performance of the kang through field investigations, discussing future research needs. Yu [20] addressed the limited scope of existing kang research by reviewing the detailed structures, flue configurations, and thermal performance of three traditional Chinese kang systems. They also introduced and analyzed recently improved systems, proposing directions for future research. Wei [21] focused on the maze pattern kang, analyzing its cultural significance and extending its interpretation through connections to other cultures, emphasizing the ethnic spirit and cultural value of the kang. Zhang [22] studied traditional fire kangs in Northeast China, testing the thermal performance of full-room and "Swastika kang". Zhang also proposed a method for evaluating thermal performance, simulating heat transfer in the flue system, and providing recommendations for optimal fuel usage and flue design to maximize energy efficiency and maintain reasonable temperatures. Liu [23] compared ground-level and raised kangs in rural Northeast China, finding that raised kangs are a low-cost, high-efficiency heating solution that aligns with sustainability principles.

2.3. Research Status

The current research on indoor thermal comfort in residential homes primarily focuses on passive and active methods aimed at enhancing indoor thermal comfort. Passive methods involve increasing the thickness of building enclosures, using higher-quality insulation materials to improve the thermal properties of the building envelope, or adjusting factors such as building orientation or window-to-wall ratios to enhance indoor thermal comfort. Simultaneously, in the realm of active heating systems, there are explorations into utilizing renewable energy sources like solar power to reduce the energy consumption of building heating and thus improve indoor thermal comfort.

As illustrated in Table 1, current research on indoor thermal comfort predominantly focuses on the impact of either passive or active elements in isolation. Moreover, studies on the influence of active heating systems on the indoor thermal environment often employ modern technologies and advanced materials, thereby overlooking the traditional heating wisdom inherent in many vernacular dwellings. As shown in Table 2, due to the diversity of kang types, a significant number of scholars have conducted research in the form of reviews. Additionally, most studies have focused on the structural aspects of kang and its cultural significance, while relatively few have explored its integration as an active heating device with passive strategies.

Table 1. Literature Review on Indoor Thermal Comfort.

Number	Indoor Thermal Comfort	Passive	Active	
			Modern Technologies	Traditional Wisdom
[4]	✓	✓	✓	
[11]	✓	✓	✓	
[16]	✓	✓		
[17]	✓	✓		
[18]	✓	✓		
[19]	✓	✓		
[20]	✓		✓	
[21]	✓		✓	
[22]	✓		✓	

Table 2. Literature review of kang.

Number	Types			Structure	Indoor Thermal Environment	Cultural
	"Swastika Kang"	Other	Review			
[14]			✓	✓		
[15]			✓	✓		
[23]	✓					✓
[24]	✓			✓		
[25]		✓			✓	

Consequently, this study integrates passive insulation and active heating strategies in traditional northern Chinese residences to validate traditional architectural wisdom and propose optimized strategies for enhancing indoor thermal environments. The study evaluates the potential of passive strategy methods to improve indoor thermal comfort by enhancing the thermal properties of the building envelope. The study also delves deeper into the active heating system embodying traditional wisdom known as the 'kang'. By simulating the heating systems with modern technology, the study explores the impact of this traditional wisdom on indoor thermal comfort, affirming the traditional construction wisdom of the residents in cultural village.

3. Methods

3.1. Geographic and Climatic Information

Positioned along the southern bank of the Liao River, Shifosi Village is situated in Shenyang's Shenbei New District, at the foothills of Qixing Mountain (Figure 1). The village layout exemplifies the classic Feng Shui pattern of ancient northern Chinese villages, surrounded by natural barriers on three sides: water on two sides and mountains on the other.

Figure 1. Case study of Shifo Sibe traditional village location in northern China.

Shifosi Village is located in the temperate continental monsoon climate of the northern zone. Summers are characterized by high temperatures and frequent rainfall, while winters are cold and dry, falling within an intensely cold region. As shown in Figure 2, the hourly temperature variation and daily average temperature of Shifosi Village in 2023 reveal the climatic conditions of the area. The data indicate that summer temperatures in Shifosi Village can reach as high as 30 °C, although the season is relatively short, spanning only from July to August. In contrast, the winter is harsh and prolonged, with approximately five months experiencing temperatures below 0 °C, and the minimum temperature approaching −30 °C.

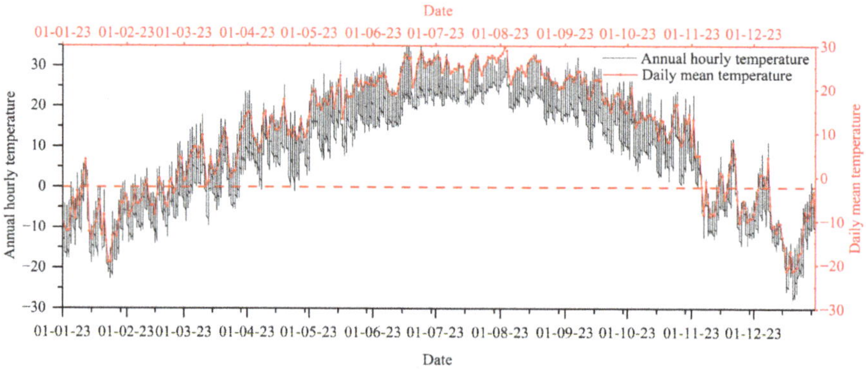

Figure 2. Temperature information of Shifosi Village (Data sources: https://xihe-energy.com/#climate, accessed on 16 November 2024).

3.2. Building Thermal Model

The Village of Shifosi is located in a cold region, and according to the code, building design in this area must fully satisfy the requirements for winter insulation, and can generally be disregarded for summer heat protection [24]. Thermal performance of external walls and other enclosure structures play an important role in building winter heat preservation. The heat transfer coefficient, heat storage coefficient, and thermal inertia of the envelope are usually used to measure the thermal performance of the envelope. For a detailed explanation of the thermal performance parameters, please refer to Nomenclature.

3.3. Folk Houses and Kang

Based on the sixth national census of the People's Republic of China in 2010, 70.2% of the Sibe people reside in Liaoning Province. Within this province, Shenyang's Shenbei New District originally comprised two Sibe townships and one Sibe town, which underwent reorganization in 2017 and were redesignated as Shifosi Street [25]. Shifosi Village was designated as one of the first historical and cultural villages in Liaoning Province in 2007 and was included in the seventh batch of national historical and cultural villages in 2019. In the same year, it was listed in the fifth batch of China's traditional villages. In addition, Shifosi Village has preserved a rich national intangible cultural heritage, especially the traditional culture of the Sibe people. Cultural heritage is "the expression of the way of life produced from social development and passed down across generations" [26]. Therefore, the Sibe residential houses have become the typical representatives of the traditional architecture of the Sibe people.

This study takes Shifosi Village as a case study to analyze traditional Sibe dwellings. As mentioned earlier, winters in Shifosi Village are extremely harsh, with minimum temperatures approaching −30 °C. Therefore, Sibe people's dwellings must thoroughly consider requirements for winter insulation, cold resistance, and frost prevention (Figure 3).

Based on the actual conditions, the building structure is input into the simulation software. Utilizing the thermal parameter calculation formulas provided in Table 3, the thermal parameters of the building materials can be derived. Table 4 illustrates the composition of the outer protective structure of traditional dwellings of the Sibe people, primarily consisting of the roof, walls, and external windows. The roof's structure, from top to bottom, comprises concrete tiles of 20 mm, lime mortar of 20 mm, a layer of clay mixed with grass of 20 mm, and a wooden board of 20 mm. The total thickness of the roof is 80 mm, providing a total thermal resistance of 0.426 ($m^2 \cdot K/W$), a thermal inertia index of 1.301, and a total heat transfer coefficient of 1.74 $W/(m^2 \cdot K)$. The wall primarily employs bricks, with a thickness of 450 mm, offering a thermal resistance of 1.698 ($m^2 \cdot K/W$) and a thermal inertia index of 16.981. The external windows are double-glazed, constructed with wood and plastic, and have a heat transfer coefficient of 2.50 $W/(m^2 \cdot K)$.

Figure 3. Typical Sibe traditional dwelling structure in Shifosi Village.

The following are the formulas for heat transfer coefficient, thermal inertia index, and thermal storage coefficient [27]:

Table 3. Tables of equations.

Number		Equation	Define
(1)	Material Layer Thermal Resistance	$R = \delta/\lambda$	R: Material Layer Thermal Resistance ($m^2 \cdot K/W$). δ: Material layer thickness (m). λ: Material thermal conductivity [$W/(m \cdot k)$].
(2)	Heat transfer resistance of the envelope	$R_0 = R_i + R + R_e$	R_0: Enclosure heat transfer resistance ($m^2 \cdot K/W$). R_i: Internal surface heat transfer resistance ($m^2 \cdot K/W$). R: Enclosure flat wall thermal resistance ($m^2 \cdot K/W$). R_e: External surface heat transfer resistance ($m^2 \cdot K/W$).
(3)	Heat transfer coefficient	$K = \frac{1}{R_0}$	K: Heat transfer coefficient of the enclosure [$W/(m^2 \cdot K)$].
(4)	Heat storage capacity	$S = \sqrt{\frac{2\pi\lambda c\rho}{3.6T}}$	S: Heat storage capacity [$W/(m^2 \cdot K)$]. c: Specific heat capacity [$KJ/(kg \cdot K)$]. ρ: Densities (kg/m^3). T: Temperature fluctuation period (h), generally T = 24 h. π: The circular ratio, π = 3.14.
(5)	Thermal inertness index	$D = R \cdot S$	D: Thermal inertness index.

Table 4. Thermal-physical properties of Sibe building external envelopes.

Material	Thic Thickness (mm)	Thermal Conductivity (λ) W/(m·K)	Heat Storage Coefficient (S) W/(m^2·K)	Correction Factor α	Resistance (R) (m^2·K/W)	Thermal Inertia D = R × S	Heat Transfer Coefficient K = 1/(0.15 + ∑R) W/(m^2·K)
Concrete Tile	20	0.93	10.583	1	0.022	0.228	1.58
lime mortar	40	0.81	10.07	1	0.049	0.497	
Sagebrush clay	40	0.58	7.723	1	0.069	0.533	
Planks	20	0.058	1.627	1	0.345	0.561	
Brickyard	450	0.265	10	1	1.698	16.981	0.54
Wood-plastic windows							2.5

Due to the severe cold climate in Shenyang, relying solely on passive strategy in residential building enclosures is insufficient to meet indoor heating demands during winter. Therefore, residents in the Shenyang area often install active heating facilities indoors to ensure warmth, including a combined stove-kang-chimney structure known as "kang". As an ancient home technology, a typical Chinese kang consists of a stove, a kang body (similar to a bed), and a chimney. It allows at least four different home functions of cooking, bed, domestic heating, and ventilation to be integrated into one system [19]. The operation of the kang can be divided into three phases: heating, stabilization, and cooling. During the heating phase, the surface temperature of the kang rapidly rises through the combustion of wood and other fuels. In the stabilization phase, the surface temperature remains constant and ceases to increase. In the cooling phase, fuel combustion is halted, and the surface temperature gradually decreases. A schematic of these operational phases is shown in Figure 4.

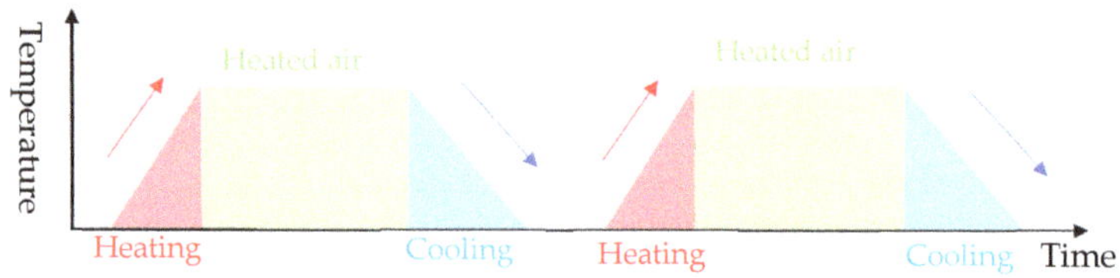

Figure 4. Operation process and construction of Swastika kang.

Additionally, the Sibe ethnic group residing in Shifosi Village developed their unique form of active heating, called the "Swastika kang", which is installed on either side of the bedroom within the traditional three-room layout, evolving over time in the course of history. Chinese kang, which serves as the main space-heating equipment in rural residences of cold regions of China, has received increasing attention due to people's increasing requirements of thermal comfort and concerns about indoor air environment in recent years [20].

3.4. Indoor Thermal Comfort Evaluation

Thermal comfort is defined as a person's subjective satisfaction with the thermal and humid environment. Thermal comfort is mainly affected by six parameters, including two human factors, i.e., the amount of human activity and clothing, and four environmental requirements, i.e., the effects of air temperature, air humidity, air speed, and radiation on human thermal comfort. Different scholars have proposed different evaluation criteria for thermal comfort, and in this paper, we will introduce the PMV-PDD [17] evaluation model as well as the Anticipated Adaptive Thermal Sensory Metrics (APMV). PMV is the predicted mean thermal sensory index and PDD is the percentage of predicted dissatisfiers. The relationship between the two is shown in the following equation [28]:

$$PPD = 100 - 95\exp\left[-\left(0.03353PMV^4 + 0.2179PMV^2\right)\right] \quad (6)$$

The specification [29] states that in a humid-heat environment with natural heating and cooling sources, the expected adaptive mean thermal sensory index (APMV) should be used as the basis for evaluation, and the APMV should be calculated according to the following formula [30]:

$$APMV = \frac{PMV}{(1 + \lambda \cdot PMV)} \quad (7)$$

APMV: Expected adaptive mean thermal sensory index
λ: Adaptive coefficient
PMV: Projected average thermal sensory indicators

This paper will also evaluate the thermal environment of residential houses according to the evaluation standard of the time proportion of indoor thermal environment parameters of the main functional rooms of the building in the adaptive thermal comfort zone in the specification. The indoor thermal comfort time ratio refers to the proportion of hours throughout the year during which the indoor hourly temperature falls within the thermal comfort temperature range.

The range of the indoor thermal comfort temperature zone for the dwelling is determined as illustrated in Figure 5, with the monthly average outdoor temperature derived from outdoor meteorological data.

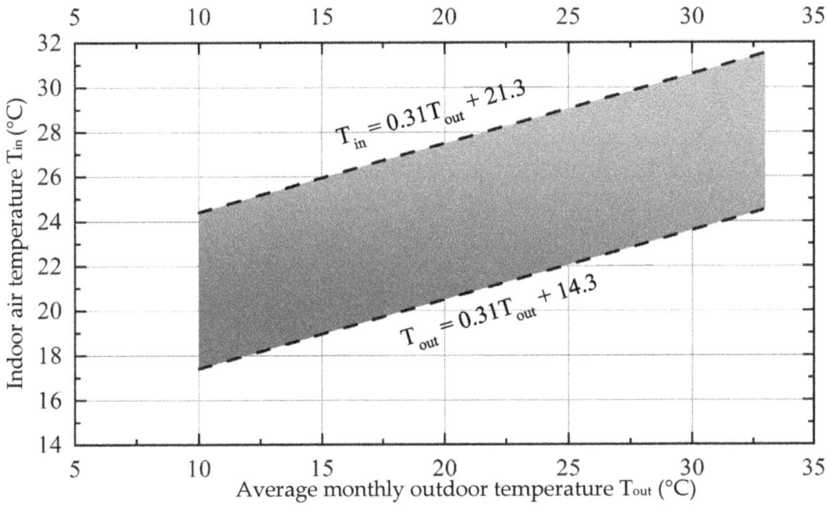

Figure 5. Indoor comfort temperature range for naturally ventilated or composite ventilated buildings.

Table 5 presents the relationship between outdoor temperatures and the corresponding range of indoor thermal comfort temperatures. The outdoor monthly average temperatures are based on typical year averages. Thus, the initial conditions for subsequent simulation modeling are set according to these temperature values.

Table 5. Indoor thermal comfort temperature range.

Months	Average Monthly Outdoor Temperature (°C)	Indoor Thermal Comfort Temperature Range (°C)
1	−11.5	17.4–24.4
2	−6.5	17.4–24.4
3	1.7	17.4–24.4
4	10	17.4–24.4
5	16.7	19.5–26.5
6	21.5	21.0–28.0
7	25.7	22.3–29.3
8	23.2	21.5–28.5
9	17.2	19.6–26.6
10	10.3	17.5–24.5
11	1.1	17.4–24.4
12	−7.5	17.4–24.4

3.5. Building Simulation

The residence adopts a traditional three-room layout with bedrooms on either side to meet the occupants' daily activity needs, while the central area serves as a living room cum kitchen. The building has a width of 16,700 mm and a depth of 8000 mm. In terms of its construction, the non-transparent parts consist of 450-mm-thick walls made of green bricks, and the roof is composed of materials layered from top to bottom: 20 mm of concrete tiles, 20 mm of lime mortar, 20 mm of clay mixed with straw (ρ = 1400), and a 20-mm template. The transparent sections are constructed using wooden and plastic double-glazed windows, with a spacing of 100 to 140 mm between the two panes of glass.

The study utilized Sware ITES2023 (20220401) to investigate the impact of the building envelope of Shifosi traditional dwellings on the indoor thermal environment. Simulation experiments were conducted on the external walls, roof, and windows of the dwellings as follows:

1. External Walls: Two sets of experiments were performed:

Maintaining a constant wall thickness, exploring the influence of different wall constructions on the indoor thermal environment.

Keeping the wall construction unchanged, investigating the effect of varying wall thicknesses on the indoor thermal environment.

2. Roof: Three sets of experiments were conducted:

Increasing the thickness of each roof layer.

Maintaining a constant roof thickness, examining the impact of different construction materials on the indoor thermal environment.

Adding an insulation layer and incrementally increasing its thickness.

3. Windows: three sets of experiments were carried out using different modern window constructions to study their effects on the indoor thermal environment.

Figure 6 presents a simplified floor plan of the residence and a schematic diagram of the preliminary simulation model.

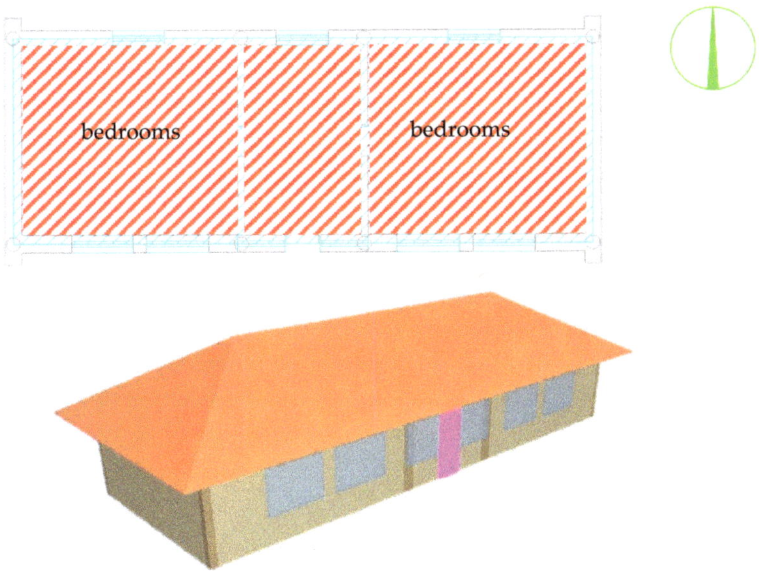

Figure 6. Floor layout of typical Sibe house for case study.

The month with the highest average temperature is July, with an average temperature of 25.7 °C. During this period, the corresponding indoor thermal comfort temperature range is between 22.3 °C and 29.3 °C. Conversely, the month with the lowest average temperature is January, recorded at −11.5 °C, where the corresponding indoor thermal comfort temperature range is between 17.4 °C and 24.4 °C. The passive thermal insulation design of the building envelope is evaluated based on the proportion of time the indoor temperature complies with the thermal comfort standards.

Figure 7 illustrates the model setup of the active heating system within the traditional residential interior at Shifosi Village. The "Swastika kang", typically constructed using materials like adobe and bricks, incorporates thermal adobe bricks made by mixing wheat straw and thatch with mud. These bricks are exceptionally hardy and proficient in retaining heat [31]. The surface of the "kang" is usually coated with plaster, utilizing the technique of "inserted plaster": a mixture of one part white lime, three parts loess, and one part hemp blade, followed by a 1 mm-thick layer of yellow mud containing dry straw, then further smoothed by inserted plaster using a hemp blade [15]. The surface of the "kang" is usually coated with plaster, utilizing the technique of "inserted plaster": a mixture of one part white lime, three parts loess, and one part hemp blade, followed by a 1 mm-thick layer of yellow mud containing dry straw, then further smoothed by inserting plaster using a hemp blade.

For the kang, an active heating device, the study selected Airpak3.0.16 software as the simulation tool. In the simulation model, the parameters for the building envelope of the dwelling were set based on the configurations described earlier. The outdoor temperature was set to −11.5 °C. Considering that the surface of the kang is usually covered with blankets, bedding, and other materials, it is recommended that the comfortable temperature of the surface of the fire bed fluctuates between 24 °C and 28 °C, with the maximum temperature not exceeding 40 °C for human comfort. Therefore, the temperature variation on the "Swastika kang" surface is set within the range of 20 °C to 45 °C. In the simulation process, considerations were made regarding the temperature and area of the "Swastika kang" and their impact on indoor thermal comfort. Therefore, the simulation model for the kang does not consider its internal structure but focuses on its surface temperature and the area of the kang surface.

As the human body is typically seated or standing above the "Swastika kang," the main torso region is usually positioned at approximately 1.2 m above the indoor floor level. Therefore, the simulation experiments primarily investigate temperature and PMV variations at a height of 1.2 m above the floor, with the indoor ground level set as the zero-elevation reference. This plane effectively represents the thermal environment at the torso level when a person is either sitting on the kang or standing on the ground.

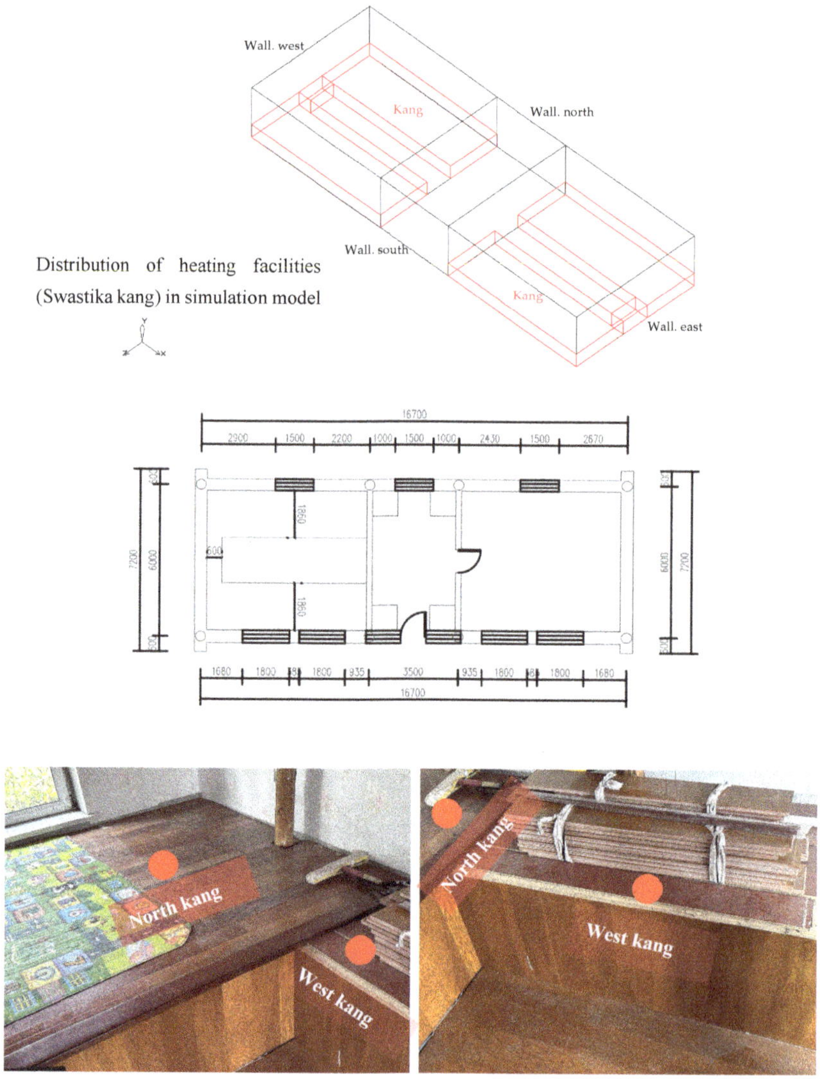

Figure 7. Building simulation model of Sibe house with Swastika kang for indoor space heating.

4. Results

4.1. External Wall Impact

Table 6 presents the simulation results for optimizing residential wall structures. Using 450 mm green bricks in the construction of traditional Sibe ethnic dwellings as the benchmark, the wall's heat transfer coefficient is 0.54, with a thermal inertia index of 16.981. Under these conditions, the proportion of time within the thermal comfort zone reaches 24.03%.

Table 6. Influence of external wall construction on indoor thermal comfort.

	Analogue Serial Number	Materials (From Outside to Inside)	Thicknesses (mm)	Heat Transfer Coefficient (k) $K = 1/(0.15 + \sum R)$ ($W/m^2 \cdot k$)	Thermal Inertia $D = R \times S$	Thermal Comfort Time Percentage
External wall	Standard	Brickyard	450	0.54	16.981	24.03%
	1	Brickyard	400	0.6	15.094	24.03%
		Brickyard	500	0.49	18.868	24.13%
	Total thickness of the wall remains the same, changing the wall construction					
	1	Brickyard (430 mm) Extruded polystyrene foam (with skin) (20 mm)	450	0.41	16.453	24.03%
		Extruded polystyrene foam (with skin) (20 mm), Brickyard (430 mm)	450	0.41	16.453	24.03%
	2	Brickyard (410 mm), Extruded polystyrene foam (with skin) (20 mm), lime mortar (20 mm)	450	0.42	15.944	24.03%
	Changing the total thickness of the wall					
	1	Brickyard (450 mm), Extruded polystyrene foam (with skin) (20 mm)	470	0.40	17.208	24.03%
	2	Brickyard (450 mm), Extruded polystyrene foam (with skin) (50 mm)	500	0.29	17.548	24.03%

Firstly, adjusting wall thickness:

- The wall thickness was increased and decreased by 50 mm to 500 mm and 400 mm, respectively. The overall heat transfer coefficients were 0.49 and 0.60, while the thermal inertia index was 18.868 and 15.094, respectively. Simulation results indicated that whether increasing or decreasing wall thickness, the proportion of time meeting the criteria for thermal comfort remained at 24.03%, similar to the baseline case.

Secondly, maintaining the total wall thickness but adjusting wall construction:

- Maintaining a constant wall thickness, adding 20 mm extruded polystyrene foam insulation for both interior and exterior walls resulted in overall heat transfer coefficients of 0.41 and 0.42, with thermal inertia indices of 16.453 and 15.944, respectively. At this point, the proportion of time meeting the criteria for thermal comfort remained at 24.03%.
- Maintaining a constant wall thickness, adding 20 mm extruded polystyrene foam and 20 mm lime mortar for external wall insulation led to a heat transfer coefficient of 0.42 and a thermal inertia index of 15.944. The results showed that the proportion of time meeting the criteria for thermal comfort remained at 24.03%.

Finally, altering the total wall thickness:

- By maintaining consistent insulation materials and thickness while increasing the total wall thickness, constructing interior and exterior walls with 450 mm bricks and 20 mm extruded polystyrene foam for the simulation, the overall heat transfer coefficients and thermal inertia indices reached 17.208. Under these conditions, the proportion of time meeting the criteria for thermal comfort remained at 24.03%.

- Adjusting wall thickness and increasing the insulation material thickness, with walls composed of 450 mm bricks and 50 mm extruded polystyrene foam, yielded a heat transfer coefficient of 0.29 and a thermal inertia index of 17.548. The results showed that the proportion of time meeting the criteria for thermal comfort remained at 24.03%.

According to the simulation results for the walls, neither changing the wall thickness, altering the insulation material, nor combining both approaches led to significant improvements in the indoor thermal environment. Therefore, in further optimization efforts, the impact of walls on the indoor thermal environment can be disregarded. This also highlights the wisdom embedded in the construction of traditional dwellings.

4.2. Roof Impact

Table 7 presents the optimized simulation results of residential roof structures. Taking the traditional roof structure of the Sibe ethnic group as the baseline, the structure comprises layers of 20 mm concrete tile, 20 mm lime mortar, 20 mm mixed clay (ρ = 1400), and 20 mm template, totalling 80 mm in thickness, with a heat transfer coefficient of 1.74 and a thermal inertia index of 1.301. At this stage, the proportion of time within the thermal comfort zone reaches 24.03%.

Table 7. Influence of roof construction on indoor thermal comfort.

	Analogue Serial Number	Materials (From Outside to Inside)	Thicknesses (mm)	Heat Transfer Coefficient (k) $K = 1/(0.15 + \sum R)$ ($W/m^2 \cdot k$)	Thermal Inertia $D = R \times S$	Thermal Comfort Time Percentage
Roof	Standard	Concrete tiles (20 mm), Lime mortar (1) (20 mm), Grass-filled clay (ρ = 1400) (20 mm), Wooden boards (20 mm)	80	1.74	1.301	24.03%
	1	Concrete tiles (30 mm), Lime mortar (1) (30 mm), Grass-filled clay (ρ = 1400) (30 mm), Wooden boards (30 mm)	120	1.27	1.951	24.03%
	Changing the roof structure while keeping the total thickness of the roof structure unchanged					
	1	Concrete tiles (20 mm), Aerated concrete, foam concrete (ρ = 700) (20 mm), Grass-filled clay (ρ = 1400) (20 mm), Wooden boards(20 mm)	80	1.51	1.399	24.14%
	2	Concrete tiles (20 mm), Lime mortar (1) (20 mm), Rigid-foam polyurethane sheet PUR (ρ ≥ 35) (20 mm), Wooden boards (20 mm)	80	0.73	4.612	24.03%
	Changing the total thickness of the roof and the thickness of the insulation					
	1	Concrete tiles (20 mm), Lime mortar (20 mm), Rigid-foam polyurethane sheet PUR (ρ ≥ 35) (1) (60 mm), Wooden boards (20 mm)	120	0.33	11.770	24.35%

Firstly, variations are made in the total roof thickness: All roof materials were increased by 10 mm, resulting in a total roof thickness of 120 mm, yielding a total heat transfer coefficient of 1.27 and a thermal inertia index of 1.951. Under these conditions, the proportion of time within the thermal comfort zone remained at 24.03%.

Secondly, while maintaining the roof thickness unchanged, adjustments were made to the roof structure to investigate its impact on the indoor thermal environment of the dwelling.

- With the total roof thickness unchanged and other roof materials held constant, the 20 mm lime mortar was replaced with the same thickness of aerated concrete and foam concrete (ρ = 700), resulting in a heat transfer coefficient of 1.51 and a thermal inertia index of 1.399. In this scenario, the proportion of time within the thermal comfort zone increased to 24.14%, an increment of 0.11% compared to the baseline roof structure.

- Maintaining the total roof structure thickness and other roof materials unchanged, the 20 mm mixed clay was replaced with 20 mm rigid polyurethane foam board PUR ($\rho \geq 35$), resulting in a heat transfer coefficient of 0.93 and a thermal inertia index of 4.612. Under these conditions, the proportion of time within the thermal comfort zone remained at 24.03%, consistent with the baseline roof structure.

Finally, adjustments were made to both the total roof thickness and the thickness of insulating materials. Keeping other roof materials constant, the 20 mm lime mortar was replaced with 60 mm rigid polyurethane foam board PUR ($\rho \geq 35$), resulting in a heat transfer coefficient of 24.35% and a thermal inertia index of 11.770. Under these conditions, the proportion of time within the thermal comfort zone reached 24.35%, representing an increase of 0.32% compared to the baseline roof structure.

In conclusion, for traditional residential roofs, altering thickness has a minimal impact on enhancing thermal comfort, whereas adjusting the insulating materials in residential roofs significantly improves thermal comfort. Therefore, in further studies, it is recommended to improve the indoor thermal environment by adding insulation to the roofs.

4.3. External Window Impact

Table 8 presents the optimization of residential windows. Using the traditional window structure of the Sibe ethnic group as the baseline, consisting of wooden and plastic double-layered windows (double glass spacing of 100 to 140), with a heat transfer coefficient of 2.50, the proportion of indoor thermal comfort time is 24.03%.

- Replacing the window structure with 6 mm + LE35AMARL film glass, with a heat transfer coefficient of 4.60, resulted in a thermal comfort time ratio of 22.83%, representing a decrease of 1.2% compared to the baseline structure.
- Substituting the window structure with plastic + 6Low-E + 12A + 6 mm transparent hollow glass, with a heat transfer coefficient of 1.90, yielded a thermal comfort time ratio of 23.31%, exhibiting a decrease of 0.72% compared to the baseline structure.
- Changing the window structure to [5 mm + 9A (air) + 5 mm] nanometer-coated glass (HJ-N-series) + 9A (air) + 5 mm white glass (warm edge seal), with a heat transfer coefficient of 1.56, resulted in a thermal comfort time ratio of 24.09%, indicating an increase of 0.06% compared to the baseline structure.

Table 8. Influence of external window construction on indoor thermal comfort.

	Analogue Serial Number	Materials (From Outside to Inside)	Thicknesses (mm)	Heat Transfer Coefficient (k) $K = 1/(0.15 + \sum R)$ ($W/m^2 \cdot k$)	Thermal Inertia $D = R \times S$	Thermal Comfort Time Percentage
External wall	standard	Wooden, plastic-double-glazed windows (double-glazed spacing (100–140)		2.50		24.03%
	1	6 mm + LE35AMARL film glass		4.60		22.83%
	2	Plastic +6Low-E + 12A + 6 mm white transparent insulating glass		1.90		23.31%
	3	[5 mm + 9A (air) + 5 mm] nano-coated (HJ-N-series) + 9A (air) + 5 mm white glass (warm edge sealing)		1.56		24.09%

It is evident that altering the window structure has a minimal effect on enhancing indoor thermal comfort and might even diminish it. Furthermore, for rural residences, investing in superior window structures is economically unfavourable due to the relatively insignificant improvement in indoor thermal comfort. This reaffirms the wisdom behind traditional residential construction methods

Based on the previous analysis of the walls, roof, and windows of the dwelling, and taking into account the economic constraints in rural areas, optimizing the roof structure proves to be the most effective strategy for passively improving indoor thermal comfort in traditional Sibe ethnic dwellings through structural modifications. In this scenario, the roof structure comprises, from top to bottom, 20 mm concrete tiles, 20 mm lime mortar, 60 mm hard polyurethane foam board PUR ($\rho \geq 35$), and 20 mm templates.

Through the study of the building envelope, it is evident that even though traditional dwellings employ simple envelope structures and rudimentary materials, they provide excellent indoor insulation, effectively ensuring thermal comfort within the residence. This demonstrates the wisdom inherent in traditional construction practices. Consequently, the subsequent discussion will delve into enhancing indoor thermal comfort through the utilization of traditional construction-based active heating methods.

5. Discussion: Swastika Kang Heating Effect

5.1. Surface Temperature Impact

Figure 8 illustrates the location of the kang surface references in the experimental program. The blue area is the upper surface of the kang. Figure 9 illustrates the variation in temperature distribution at a height of 1.2 m above the floor as the upper surface temperature of the "Swastika kang" increases. Considering the placement of the Swastika kang in the eastern and western sections of the residence, which are the primary activity areas, our discussion primarily focuses on the impact of the Swastika kang on these two rooms, specifically emphasizing the temperature and PMV changes at a height of 1.2 m above the ground.

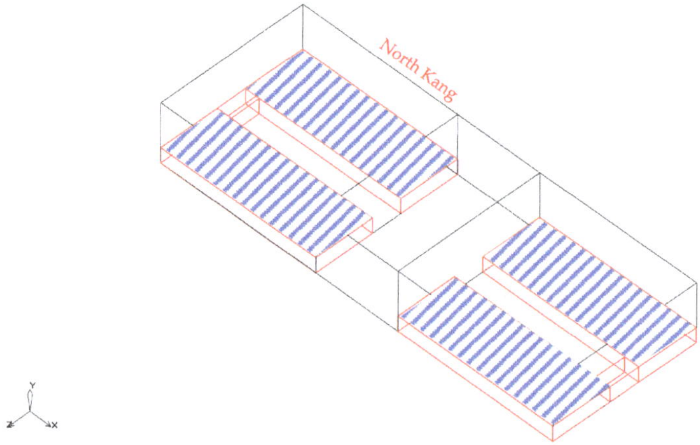

Figure 8. Schematic diagram of the upper surface of the kang.

Figure 9 shows the simulation of the temperature distribution in the plane at 1.2 m in the room under different kang surface temperature conditions, which clearly indicates the distribution of the temperature in the plane and provides support for exploring the temperature trend in Figure 10.

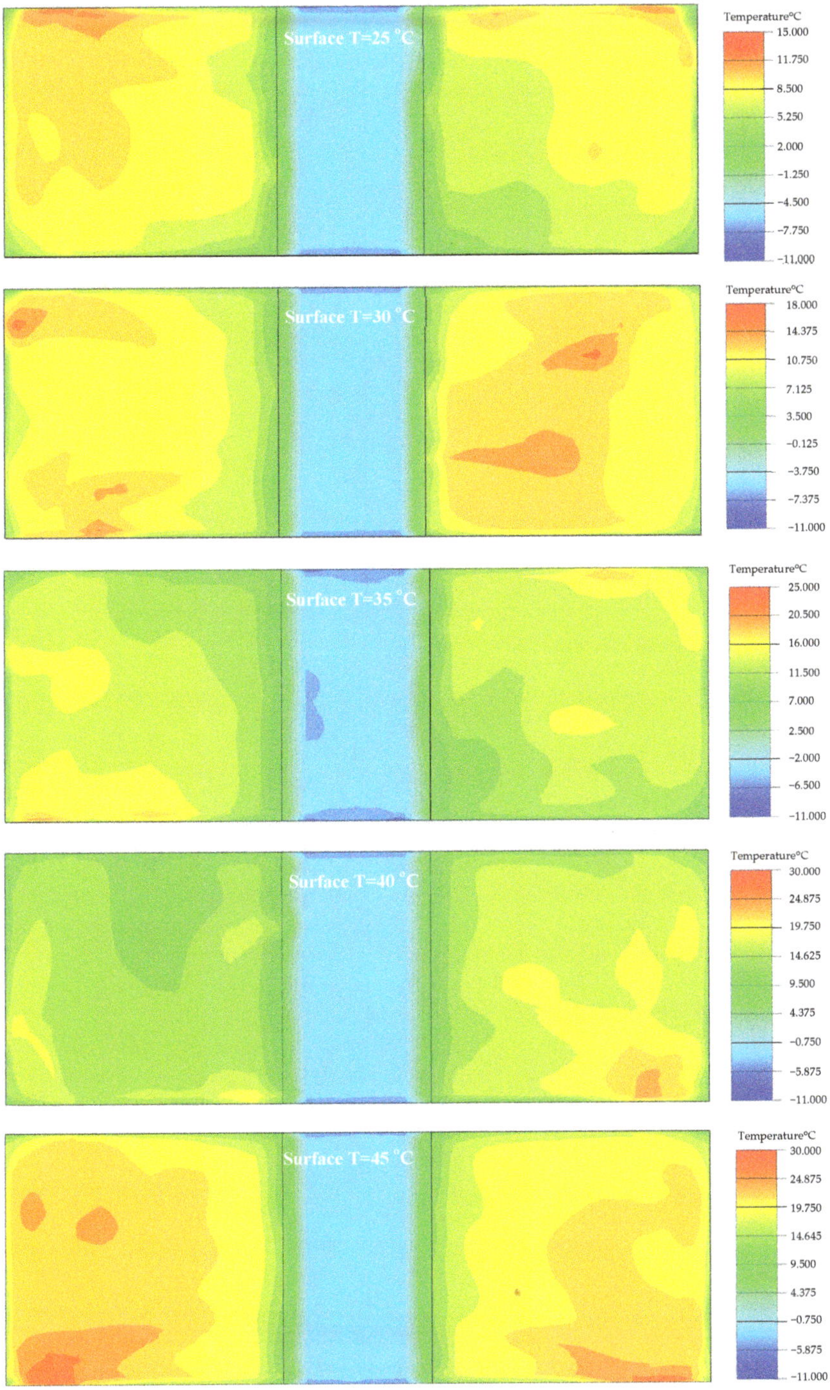

Figure 9. The influence of the surface temperature of the "Swastika kang" on the temperature distribution at a height of 1.2 m.

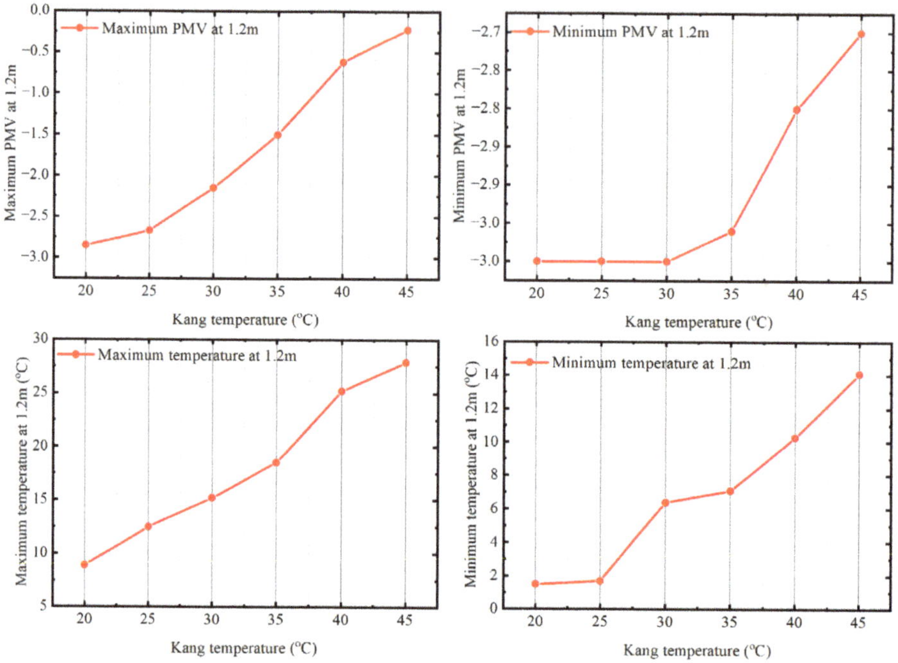

Figure 10. The influence of the upper surface temperature of the kang on the temperature distribution and PMV variations at a height of 1.2 m.

Initially, as the temperature increases, both the maximum and minimum indoor temperatures and PMV values demonstrate an upward trend. At a height of 1.2 m, the maximum temperature increases by 19 °C, rising from 8.9 °C to 27.9 °C, while the minimum temperature increases from 1.5 °C to 14.1 °C, marking a rise of 12.6 °C. The temperature variation between the maximum and minimum values amounts to 6.4 °C. Additionally, within the temperature range of 35 °C to 40 °C, there is a noticeable increase in the slope of both the maximum and minimum temperature values, whereas between 40 °C to 45 °C, the slope decelerates or remains stable, further confirming the conclusion that 40 °C serves as the optimal temperature for the Swastika kang surface.

Furthermore, concerning the indoor PMV, the trends in the maximum and minimum values at 1.2 m align closely with the temperature variations, displaying an upward trend. The maximum PMV at 1.2 m rises from −2.85 to −0.22, an increase of 2.63. Conversely, the minimum PMV within this plane increases marginally from −3 to −2.7, representing a rise of 0.3. In summary, with the elevation of the Swastika kang surface temperature, both the maximum temperature and PMV at a height of 1.2 m show a notable increase, whereas the rise in the minimum temperature and PMV remains relatively gradual.

5.2. Heating Area Impact

Considering that the main components of the "Swastika kang" include the north-south main kang and the west kang, this section separately discusses the impact of the areas of the north-south kang and the west kang on the indoor thermal environment, as shown in Figure 11.

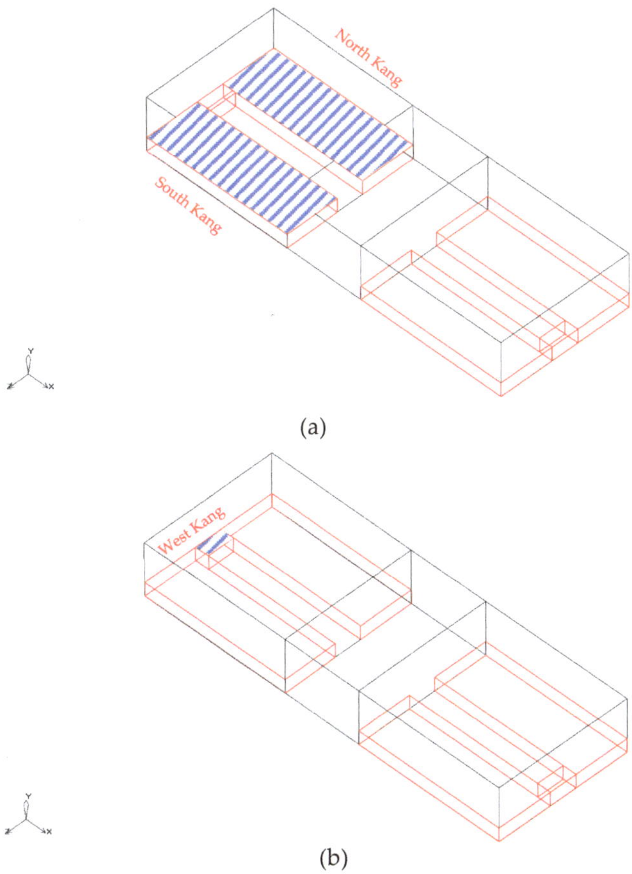

Figure 11. Schematic of the simulation strategy. (**a**) North-south kang. (**b**) West kang.

Figure 12a illustrates the impact of the surface area of the main north-south bed (the primary part of the Swastika kang) on the thermal comfort within the indoor space, focusing specifically on the temperature and PMV (Predicted Mean Vote) changes at a height of 1.2 m in the east-west rooms. As depicted in the graph, an increase in the bed's surface area results in an ascending trend in both the maximum and minimum temperatures at the 1.2-m height plane. The maximum temperature rises from 25.2 °C to 27.5 °C, marking a 2.3 °C increase, while the minimum temperature increases from 10.3 °C to 14 °C, representing a 3.7 °C rise. The range of change in the minimum temperature is 1.4 °C higher than that of the maximum temperature.

Regarding the PMV values at a height of 1.2-m, both the maximum and minimum PMV values also display an upward trend. The maximum PMV value elevates from −0.62 to −0.26, a rise of 0.36, while the minimum PMV value climbs from −2.8 to −2.2, showing a 0.6 increase. It is apparent from Figure 12a that the variations in temperature and PMV values at the 1.2-m height plane exhibit a relatively gradual trend. Therefore, the conclusion drawn is that enlarging the surface area of the main north-south bed contributes to improving indoor thermal comfort. However, the enhancement in thermal comfort within the indoor space due to the expanded bed area does not appear significant in comparison to the reduction in the available space for indoor activities.

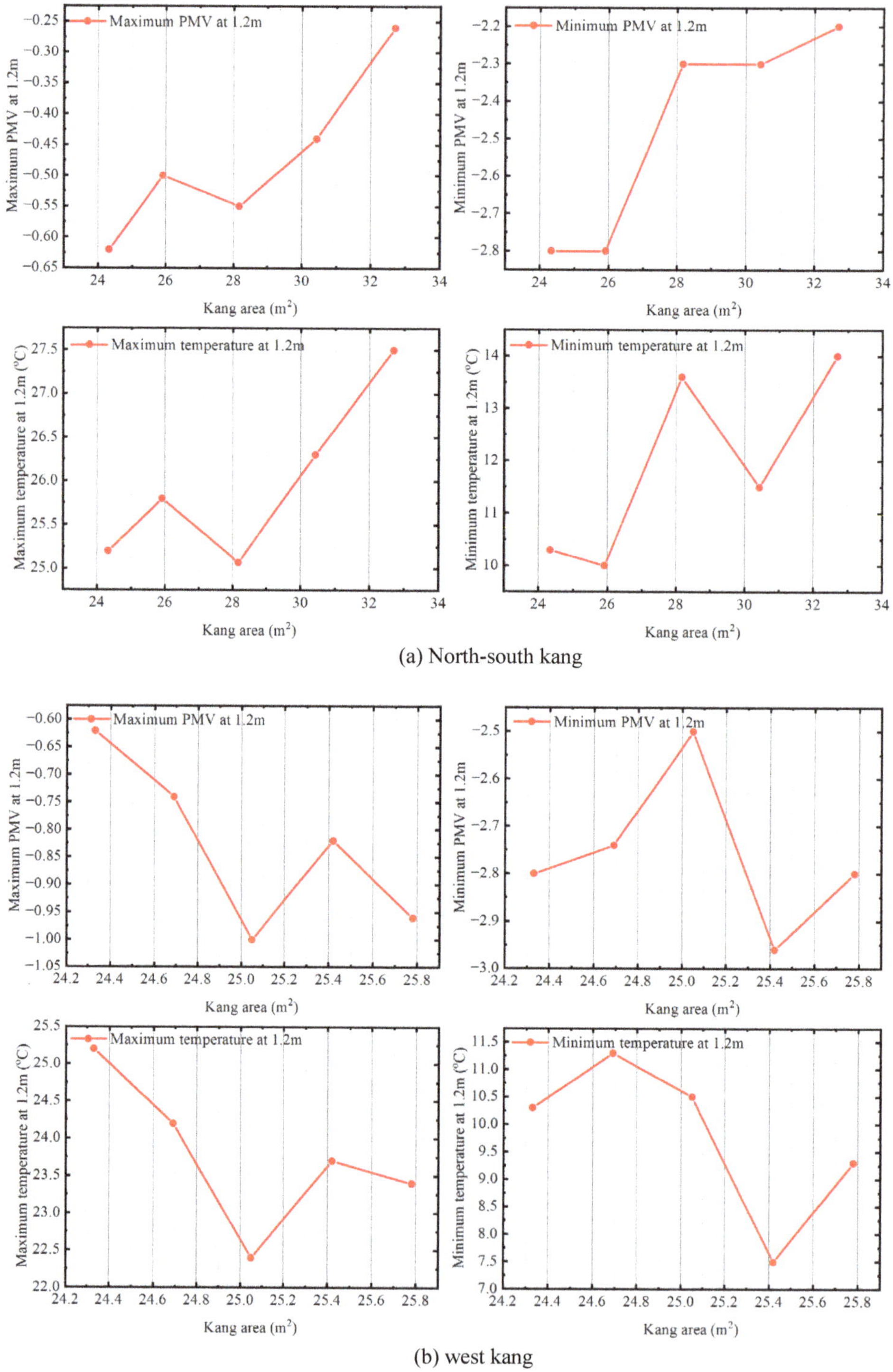

Figure 12. The influence of changes in the upper surface area at different locations of the kang on the changes in temperature and PMV at a height of 1.2 m.

When increasing the surface area of the west kang, Figure 12b illustrates the variation in indoor thermal comfort, considering the temperature and PMV changes at a height of 1.2 m of the east-west two-room configuration. As depicted in the graph, an increase in the surface area of the west kang leads to an unstable and relatively minor fluctuation in both the maximum and minimum values of temperature and PMV at a height of 1.2 m. Given the initially small size of the west kang, its impact on indoor thermal comfort appears relatively modest. Considering the traditional cultural context of the Sibe ethnic group, it can be inferred that the necessity of increasing the surface area of the west kang is not significantly substantial. It is clear that compared to the effect of changing the kang temperature on indoor thermal comfort, the effect of changing the kang area on indoor thermal comfort is smaller, so for the active heating device with traditional construction wisdom of the Sibe dwellings, the indoor thermal comfort can be further improved by raising the kang surface temperature.

6. Conclusions

Indoor thermal comfort and living environments have become increasingly prominent topics of concern. Scholars are dedicating more effort to enhancing indoor thermal comfort and proposing various methods and suggestions across different fields to improve living conditions. However, for traditional rural dwellings, advanced improvement methods are not always applicable due to factors like economic conditions and cultural traditions.

In the case of Sibe ethnic dwellings, economic factors significantly limit the enhancement of indoor thermal comfort. Through investigations, it was found that the Sibe employ passive strategy methods by reinforcing the thermal performance of structural enclosures. Additionally, they utilize traditional construction techniques to install active heating devices—known as "kang"—to combat harsh cold weather conditions. The introduction of these structural enclosures and "kang" has notably improved the indoor thermal comfort of their dwellings. Over time, "kang" has evolved into an indispensable heating facility in Sibe homes. This study focuses on two methods within Sibe dwellings—passive strategy and active heating—to enhance indoor thermal comfort, aiming to affirm traditional construction techniques while laying the groundwork for future research. The main conclusions of this paper are as follows:

1. Simulations of residential structural enclosures revealed that enhancing roof insulation had the most pronounced effect on indoor thermal comfort compared to improving wall or window insulation. However, simulations also indicated that despite continuous enhancements, the effectiveness of passive strategy methods on indoor thermal comfort within structural enclosures is limited and less applicable to Sibe dwellings.
2. The configuration of the traditional "kang" combined with the architectural layout of the Sibe residence is better suited for space heating
3. Regarding the simulation research on the active heating device—"kang"—it was found that raising "kang" temperature significantly improves indoor thermal comfort, while increasing the "kang" area has a comparatively small effect. Therefore, future improvement measures could emphasize temperature-related factors.

By analyzing the structural enclosures and active heating devices of traditional Sibe dwellings, this study explicitly acknowledges the traditional construction techniques of the Sibe, guiding the direction for enhancing indoor thermal comfort in Sibe dwellings. Moreover, the methods and approaches used in this research offer insights into addressing thermal comfort issues in traditional rural dwellings in other regions. However, this study specifically focuses on a particular type of dwelling influenced by geographical and cultural factors, gradually forming its present style. Other regions and distinctive rural dwellings are influenced by varying geographical and cultural factors, resulting in different construction techniques. Therefore, when studying traditional dwellings in other regions, these diverse influences need consideration. Additionally, this study only conducts preliminary research on structural enclosures and active heating devices without further extensive improvements. Hence, future research could start by enhancing active heating devices to better improve the indoor thermal environment of Sibe dwellings.

Author Contributions: Conceptualization, M.Z.; methodology, Z.S.; software, K.L.; writing—original draft preparation, M.Z.; writing—review and editing, K.X.; supervision, K.X.; funding acquisition, K.L. All authors have read and agreed to the published version of the manuscript.

Funding: This research was Supported by Sichuan Science and Technology Program, grant number 2025ZNSFSC1302 and Scientific research startup fund subsidy project for talent introduction, Southwest Minzu University, grant number RQD2022030.

Data Availability Statement: The raw climate data provided in the study are available on the XIHE-ENERGY.COM at https://www.xihe-energy.com, accessed on 25 February 2025. The raw data supporting the conclusions of this article will be made available by the authors on request.

Conflicts of Interest: The authors declare no conflicts of interest.

Nomenclature

Nomenclature	**Definition**
Heat storage coefficient (S) $W/(m^2 \cdot K)$	The ability of a material to store heat is defined as its thermal capacity. The greater this value, the better the thermal stability of the material.
Heat transfer coefficient $W/(m^2 \cdot K)$	Under steady-state heat transfer conditions, it refers to the amount of heat transferred per unit time through a unit area of a building envelope when there is a temperature difference of 1 degree (K or °C) between the air on either side.
Kang	In northern China, a kang is a sleeping platform constructed from bricks or adobe, featuring hollow spaces underneath that are connected to a chimney. It can be heated by burning fuel, providing warmth for sleeping and living areas.
Resistance (R) $(m^2 \cdot K/W)$	When heat is transferred through an object, the ratio between the temperature difference across the object and the power of the heat source is defined as the thermal resistance.
"Swastika kang"	In the bedroom, a continuous kang is built along the north and south walls, with a narrower kang constructed on the west side. In some cases, the west kang is of the same width as the south and north kangs, connecting with them to form a "π"-shaped structure. The chimney extends through the wall to the outside.
Thermal conductivity (λ) $W/(m \cdot K)$	The measure of a material's ability to conduct heat.
Thermal inertia	The thermal inertia index is a measure of how quickly temperature fluctuations on one side of an object's surface attenuate within the object when subjected to periodic thermal effects.

References

1. Zhao, H.; Jiang, Y. Research on Strategies for Improving Rural Human Settlements in the Context of Rural Revitalization—A Case Study of Zhejiang Province. *Small Town Constr.* **2019**, *37*, 9–14+93. (In Chinese)
2. Ma, B. Research on Energy-Saving Design Solutions for Rural Houses in Northern China from the Perspective of Rural Revitalization. Master's Thesis, North China University of Science and Technology, Tangshan, China, 2021. (In Chinese)
3. Bao, C.; Zhang, C.; Bi, L. Research on Optimization Strategies for Indoor Thermal Comfort in Rural Houses in Severe Cold Regions. *Archit. Cult.* **2023**, 26–27. (In Chinese) [CrossRef]
4. Hou, J. Research on Optimization of Indoor Thermal Environment in Traditional Tibetan Dwellings on the Northwest Sichuan Plateau during Winter. Master's Thesis, Sichuan Agricultural University, Ya'an, China, 2019. (In Chinese)
5. Kaynakli, O.; Kilic, M. Investigation of Indoor Thermal Comfort under Transient Conditions. *Build. Environ.* **2005**, *40*, 165–174. [CrossRef]
6. Cheng, C.; Deng, X.; Zhao, X.; Xiong, Y.; Zhang, Y. Multi-Occupant Dynamic Thermal Comfort Monitoring Robot System. *Build. Environ.* **2023**, *234*, 110137. [CrossRef]
7. Ma, K.; Yu, Y.; Yang, B.; Yang, J. Demand-Side Energy Management Considering Price Oscillations for Residential Building Heating and Ventilation Systems. *IEEE Trans. Ind. Inform.* **2019**, *15*, 4742–4752. [CrossRef]
8. Deng, X.; Zhang, Y.; Zhang, Y.; Qi, H. Toward Smart Multizone HVAC Control by Combining Context-Aware System and Deep Reinforcement Learning. *IEEE Internet Things J.* **2022**, *9*, 21010–21024. [CrossRef]
9. Wang, X.; Mai, X.; Lei, B.; Bi, H.; Zhao, B.; Mao, G. Collaborative Optimization between Passive Design Measures and Active Heating Systems for Building Heating in Qinghai-Tibet Plateau of China. *Renew. Energy* **2020**, *147*, 683–694. [CrossRef]
10. Li, X.; Shen, C.; Yu, C.W.F. Building Energy Efficiency: Passive Technology or Active Technology? *Indoor Built Environ.* **2017**, *26*, 729–732. [CrossRef]
11. Imbabi, M.S.-E. A Passive–Active Dynamic Insulation System for All Climates. *Int. J. Sustain. Built Environ.* **2012**, *1*, 247–258. [CrossRef]
12. Liu, J. Research on Optimization Strategies for Traditional Dwellings in Jia County Based on Thermal Comfort Improvement. Master's Thesis, Zhengzhou University, Zhengzhou, China, 2022. (In Chinese)
13. Wei, D.; Zhao, G.; Liu, S.; Yang, L. Indoor Thermal Comfort in a Rural Dwelling in Southwest China. *Front. Public Health* **2022**, *10*, 1029390. [CrossRef]
14. Wang, Y.; Dong, Q.; Guo, H.; Yin, L.; Gao, W.; Yao, W.; Sun, L. Indoor Thermal Comfort Evaluation of Traditional Dwellings in Cold Region of China: A Case Study in Guangfu Ancient City. *Energy Build.* **2023**, *288*, 113028. [CrossRef]
15. He, Q.; Liu, D.; Zhu, X.; Yang, L.; Liu, J. Field Study on Indoor Thermal Environment of Tibetan Dwellings in the Western Sichuan Plateau. *J. Xi'an Univ. Archit. Technol. Nat. Sci. Ed.* **2015**, *47*, 402–406. (In Chinese) [CrossRef]
16. Zhang, C.; Pomianowski, M.; Heiselberg, P.K.; Yu, T. A Review of Integrated Radiant Heating/Cooling with Ventilation Systems-Thermal Comfort and Indoor Air Quality. *Energy Build.* **2020**, *223*, 110094. [CrossRef]
17. Minglu, Q.; Liang, X.; Deng, S.; Yiqiang, J. Improved Indoor Thermal Comfort during Defrost with a Novel Reverse-Cycle Defrosting Method for Air Source Heat Pumps. *Build. Environ.* **2010**, *45*, 2354–2361. [CrossRef]
18. Myhren, J.A.; Holmberg, S. Flow Patterns and Thermal Comfort in a Room with Panel, Floor and Wall Heating. *Energy Build.* **2008**, *40*, 524–536. [CrossRef]
19. Zhuang, Z.; Li, Y.; Chen, B.; Guo, J. Chinese Kang as a Domestic Heating System in Rural Northern China—A Review. *Energy Build.* **2009**, *41*, 111–119. [CrossRef]
20. Yu, K.; Tan, Y.; Zhang, T.; Zhang, J.; Wang, X. The Traditional Chinese Kang and Its Improvement: A Review. *Energy Build.* **2020**, *218*, 110051. [CrossRef]
21. Wei, L. Research on the Cultural Value of the Manchu "Swastika Kang". Master's Thesis, Jilin University of Arts, Changchun, China, 2022. (In Chinese)
22. Zhang, Q. Simulation Analysis of Flow and Heat Transfer of Flue Gas in Traditional Kang of Northeast China. Master's Thesis, Harbin Institute of Technology, Harbin, China, 2017. (In Chinese)
23. Liu, Y.; Gu, Y.; Li, Y. Comparative Analysis of Energy-Saving Design between Ground Kang and Elevated Kang in Rural Houses in Northeast China. *J. Anhui Agric. Sci.* **2014**, *42*, 4326–4328. (In Chinese) [CrossRef]
24. *GB 50352-2019*; Uniform Standard for Design of Civil Buildings. Ministry of Housing and Urban-Rural Development of the People's Republic of China: Beijing, China, 2019. (In Chinese)
25. Xiao, H.; Jiang, N. Investigation and Conservation Strategies for the Living Environment of Traditional Xibo Ethnic Villages—A Case Study of Shifosi Village, Shenbei New District, Shenyang, Liaoning Province. *Urban Hous.* **2020**, *27*, 98–102. (In Chinese)

26. Lin, H.; Wang, J.-B.; Zhang, X.; Hu, F.; Liu, J.; Hong, X.-C. Historical Sensing: The Spatial Pattern of Soundscape Occurrences Recorded in Poems between the Tang and the Qing Dynasties amid Urbanization. *Humanit. Soc. Sci. Commun.* **2024**, *11*, 730. [CrossRef]
27. *GB 50176-2016*; Code for Thermal Design of Civil Buildings. Ministry of Housing and Urban-Rural Development of the People's Republic of China: Beijing, China, 2016. (In Chinese)
28. Wang, Y.; Zhang, X.; Zhang, H. Thermal Comfort Analysis of Loader Cab Based on PMV-PPD. *J. Taiyuan Univ. Sci. Technol.* **2024**, *45*, 38–43. (In Chinese)
29. *GB/T 50785-2012*; Evaluation Standard for Indoor Thermal Environment in Civil Buildings. Ministry of Housing and Urban-Rural Development of the People's Republic of China: Beijing, China, 2012. (In Chinese)
30. Yao, R.; Li, B.; Liu, J. A Theoretical Adaptive Model of Thermal Comfort—Adaptive Predicted Mean Vote (aPMV). *Build. Environ.* **2009**, *44*, 2089–2096. [CrossRef]
31. Shi, J. Research on the Interpretation of the Character "Kang" and the Development Process and Origin of Kang. *Pop. Art Lit.* **2016**, *266*. (In Chinese)

buildings

MDPI

Article

Natural Ventilation in Building Buffer Spaces of Traditional Qiang Dwellings: Field Study in Western China

Ying Zhao [1], Kun Li [1], Meng Han [1], Jianwu Xiong [2,*] and Yifan Zhang [1]

[1] School of Architecture, Southwest Minzu University, Chengdu 610225, China; zhaoying0704@163.com (Y.Z.); a19822962185@163.com (K.L.); menghan0406@163.com (M.H.); zhangyfans@163.com (Y.Z.)

[2] AI + Arch Lab, Southwest Minzu University, Chengdu 610225, China

* Correspondence: 80300151@swun.edu.cn

Abstract: As China's rural revitalization progresses, the green and sustainable development of traditional dwellings has gained significant attention. Traditional Qiang dwellings in western Sichuan, located in high-altitude cold regions, often feature thick walls and small windows, limiting air circulation. Buffer spaces play a key role in improving indoor air quality and reducing energy consumption through natural ventilation. This study explores traditional Qiang dwellings in western Sichuan through field research, architectural analysis, and numerical simulations. The study analyzes three buffer space types and compares natural ventilation in dwellings with and without buffer spaces. The results show better air circulation in buildings with buffer spaces. The simulation further shows that when the courtyard's aspect ratio is 1.3 and the width-to-height ratio is 0.9, ventilation and air renewal rates are optimized. Based on this, the study proposes natural ventilation optimization strategies to reduce reliance on mechanical systems and improve energy efficiency. The study provides a scientific basis for low-carbon Qiang dwelling design and offers practical strategies for improving living environments, supporting sustainable rural development.

Keywords: Qiang ethnic minority; traditional dwellings; buffer space; natural ventilation; energy saving potentials; sustainable built environment

1. Introduction

Global warming and energy shortages have become a major challenges facing the world today, and the construction industry is the main source of greenhouse gas emissions, accounting for 30% of the total global emissions [1]. In developed countries, the proportion of emissions from the building sector is even higher, at approximately 40%. In China, carbon emissions from residential buildings throughout their life cycle account for 51.2% of the country's total carbon emissions [2]. With the ongoing urbanization and rural revitalization processes, carbon emissions from rural and mountainous areas in China account for 15% of the national total [3]. In response to this issue, China has committed to achieving peak carbon emissions by 2030 and carbon neutrality by 2060, with a goal of reducing residential energy consumption to below 1% [4–6]. To achieve this, the country is accelerating the low-carbon transformation of rural housing, optimizing traditional buildings through passive energy-saving strategies, improving indoor thermal environments, enhancing air quality, and reducing energy consumption [7]. In recent years, natural ventilation has been widely applied as a passive design strategy to improve indoor air quality and thermal comfort, becoming one of the key measures for energy conservation and emissions reduction [8,9].

The energy efficiency and thermal comfort of traditional buildings have garnered widespread attention [10]. In the field of architecture, the ecological design principles of traditional buildings are considered fundamental to sustainable architectural design, providing effective solutions to contemporary architectural challenges [11]. This "integration of old and new" design philosophy holds significant advantages in achieving building sustainability, as traditional buildings have successfully facilitated the harmonious coexistence of humans and the natural environment, which is one of the core elements of sustainable development [12]. In China, many traditional dwellings demonstrate the characteristic of "warm in winter and cool in summer", a design advantage stemming from the accumulated experience of local craftsmen over centuries and their deep understanding of the natural environment. Through this long-term process of summarization and experimentation, traditional dwellings have continually evolved in their adaptation to the local climate and environment [13].

Specifically, the Qiang ethnic group, a minority in southwestern China, is known for its traditional dwellings that are characterized by their construction on mountain slopes and the use of stone as a primary material. Based on materials and structural forms, Qiang architecture can be categorized into three types as follows: stone masonry, rammed earth, and timber houses [14]. In cold climatic conditions, Qiang dwellings have developed a design system closely aligned with the natural environment, particularly demonstrating unique ingenuity in a spatial layout [15,16]. These dwellings incorporate multiple buffer spaces, which not only reflect rich cultural features but also effectively promote natural ventilation and climate regulation, thereby significantly enhancing indoor thermal comfort and energy efficiency. Compared to traditional dwellings in other cold climates, the architectural design strategies of Qiang dwellings exhibit both commonalities and distinctions. For example, Tibetan dwellings typically feature thick stone walls combined with small windows, which minimize heat loss. Moreover, Tibetan dwellings place particular emphasis on the spatial layout and structural design of the interior, often incorporating buffer spaces and internal–external isolation walls to enhance heat exchange and effectively control indoor temperature fluctuations [17]. A similar design philosophy is also found in the wooden houses of Scandinavia. In response to the harsh climate, these wooden houses prioritize airtightness and efficient thermal insulation while strategically positioning ventilation openings to promote natural ventilation [18].

In recent years, research on natural ventilation has been increasing. Deng et al. [19] analyzed the impact of building cluster designs on wind environments and found that semi-enclosed courtyards perform better in natural ventilation than fully enclosed courtyards, particularly when the openings are aligned with prevailing wind directions. Kajjoba et al. [9] studied the effects of natural ventilation (NV) and outdoor environments (OE) on indoor air quality (IAQ) and the health of residents in low-income housing in Kampala, Uganda, highlighting the role of natural ventilation in improving air quality and health. Mateus et al. [20] addressed the growing demand for energy-efficient and well-ventilated buildings by studying natural ventilation in large spaces, confirming the effectiveness of this method in analyzing ventilation in expansive areas. Toe et al. [21] investigated the potential of passive cooling technologies in improving natural ventilation and thermal comfort in modern brick row houses in Malaysia. For traditional dwellings, Zhong et al. [22] used CFD models to study the natural ventilation of courtyards in Southern Jiangnan residential buildings, providing quantitative evidence for the effectiveness of courtyard natural ventilation strategies under diurnal and seasonal conditions. Xu et al. [23] studied attic spaces in traditional dwellings in Southern Shaanxi, finding that attics extend comfort periods and significantly reduce energy consumption in both winter and summer, thereby improving the indoor thermal environment. Transition spaces, serving as buffers between the interior and exterior of buildings, also play a significant role in improving the thermal environment and saving energy. Xu et al. [24] also studied the effect of corridor spaces in traditional dwellings in regions with hot summers and cold winters on thermal environment regulation. They conducted a quantitative analysis using Grasshopper and Energy Plus platforms, confirming the effectiveness of corridor spaces in enhancing indoor comfort, thus providing support for adaptive design in traditional residences in Southern Shaanxi. Chen et al. [25] studied the gray spaces in traditional Jiangnan gardens, classifying five common types of gray spaces and their climatic responses. Through wind environment simulations, they determined the optimal climate adaptation strategies for each type.

Existing research has predominantly focused on natural ventilation systems in modern architecture, with limited attention given to traditional dwellings, especially in cold regions. Furthermore, most studies concentrate on hot and humid climates, overlooking the advantages of traditional dwellings in cold regions concerning climate adaptation. As modernization and urbanization continue to advance, these traditional buildings have increasingly been marginalized. This study specifically examines the natural ventilation mechanisms of buffer spaces in Qiang traditional dwellings in the cold regions of western Sichuan. It proposes optimization strategies to improve the comfort and health of the living environment, providing theoretical support for the sustainable development of traditional dwellings in this region.

2. Field Investigation and Geomatic Study

2.1. Geographical Information

The study area, Dongmen Village in Wenchuan County, Sichuan Province, lies along the Zagunao and Longxi Rivers in a mountainous valley (Figure 1). Covering 1.34 km^2 with elevations from 1300 to 1800 m, it is a traditional Qiang settlement with houses built at varying elevations along the mountainside, preserving its ethnic character.

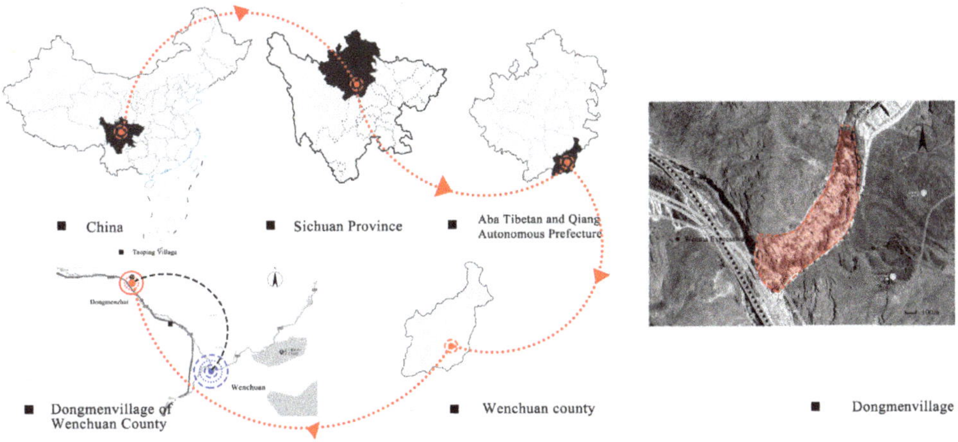

Figure 1. Geographic location information map.

The climatic condition of Dongmen village belongs to the north subtropical arid valley climate, with a high altitude, strong sunlight, dry and cold climate, less precipitation, more sunny days, and a large temperature difference between day and night. The average annual sunshine is 1042–1694 h, the average annual outdoor temperature is 14.1 °C, the annual precipitation is 491.7 mm, the average annual outdoor relative humidity is 67.3%, and the average annual wind speed is 1.95 m/s [26]. Due to the terrain, defensive needs, and insulation requirements, Qiang dwellings feature very small windows, limiting airflow and resulting in poor indoor air quality [14]. Buffer spaces are essential in these traditional dwellings. This study focuses on how the buffer spaces in Qiang traditional homes influence ventilation performance.

2.2. Types and Functions of Buffer Spaces

This study, through field research, identified that Qiang traditional dwellings primarily consist of four types, namely watchtowers, fortified houses, board houses, and mud houses, each with its distinct characteristics. These buildings commonly include functional spaces such as living rooms, kitchens, bathrooms, bedrooms, and sun terraces. The combinations of these spaces vary according to the dwelling type, creating diverse buffer spaces. The primary forms of buffer spaces include courtyards, eaves spaces, and overhanging floors. The selection of these types is based on their ability to regulate the local climate and improve the comfort of the inhabitants. The exclusion of other potential buffer spaces (such as semi-enclosed corridors) is based on the architectural characteristics of Qiang dwellings, which are primarily designed with fully enclosed corridors for warmth and protection against harsh weather. In the region's high altitude and dry climate, temperature fluctuations and strong winds prevail, and semi-enclosed corridor spaces cannot effectively mitigate extreme conditions. Previous studies [27] have shown that such spaces cannot provide sufficient climate regulation under harsh conditions, which is why they were excluded from this study. The reasons for selecting these specific buffer spaces (courtyards, eaves spaces, and overhanging floors) are detailed as follows:

(1) Courtyards: Qiang dwellings feature two types of courtyards (Figure 2), which are enclosed and open. The enclosed courtyard is surrounded by walls, typically 2 to 2.8 m high, forming a closed space that serves both defensive and climate-modulating functions. It effectively mitigates external climate fluctuations, maintaining stability in the indoor microclimate, which is especially advantageous in high-altitude areas with significant diurnal temperature variation. Studies have shown [28,29] that enclosed courtyards offer significant advantages in temperature regulation and climate adaptation, making them a common buffer space in Qiang dwellings.

Figure 2. Actual photograph of the courtyard.

(2) Eaves Spaces: The eaves spaces (Figure 3) of Qiang traditional dwellings are diverse, mainly including overhanging eaves and eaves corridors. Overhanging eaves are typically located at the main entrance, canopy rooms, or the edges of the roof. Research [24] indicates that eaves spaces are particularly effective in reducing heat exposure and mitigating the impact of intense sunlight at high altitudes and in dry climates. Acting as a transitional buffer zone, eaves spaces reduce the effects of wind and sun, enhance the regulation of solar radiation, and improve thermal comfort. Therefore, eaves spaces play a crucial role

not only in physically separating indoor and outdoor spaces but also in buffering climate factors, making them an important form of buffer space.

Figure 3. Actual photograph of the eaves spaces.

(3) Overhanging Floors: The overhanging floor (Figure 4) is one of the most distinctive spaces in Qiang traditional dwellings, consisting of a protruding wooden platform, which can be enclosed on its outer side by wooden sticks or boards, forming a balcony structure. Overhanging floors are classified into enclosed and open types, with the enclosed version enclosed by wooden boards, while the open version is protected by railings. The typical projection of an overhanging floor is about 0.9–1.2 m, and some larger overhanging floors are supported by additional columns underneath, evolving into stilted houses. These overhanging floors interact directly with the external environment, initially influenced by climatic factors such as temperature, wind, and precipitation. The buffering effect of overhanging floors moderates these external conditions, allowing them to be mitigated before reaching the indoor space. Research [30] indicates that in high-altitude regions, overhanging floors provide significant shelter from intense sunlight and strong winds, thereby enhancing indoor comfort. Given their key role in climate regulation, overhanging floors are considered an important form of buffer space.

Figure 4. Actual photograph of the overhanging floors.

2.3. Study Subject

This study conducted a field survey to analyze 28 traditional Qiang dwellings in Dongmen Village, selecting two representative types of traditional Qiang dwellings as the research subjects (Figure 5). The choice of these two dwellings was based primarily on their similarities in function and traditional characteristics. Specifically, dwelling 1 was chosen because its owner had renovated the original structure by enlarging the windows and courtyard to enhance ventilation while still retaining the core architectural features of a Qiang dwelling. It represents an adaptation of traditional architecture to modern needs. Dwelling 2, on the other hand, is an old traditional Qiang dwelling that has not undergone any renovations and retains its original architectural style, serving as a comparison building. These two dwellings share significant similarities in both function and traditional characteristics, particularly in terms of spatial organization and external structure. Therefore, they can be effectively compared in terms of their natural ventilation and indoor comfort.

Figure 5. Location of dwelling 1 and dwelling 2.

Dwelling 1 (Figure 6) is a relatively modern representative of Qiang dwellings, featuring an L-shaped layout with a southwest–northeast orientation. The site is flat and open. This dwelling serves not only as a residence but also as a small shop for the village, providing a space for community interaction and leisure. The building encloses a courtyard where villagers typically gather for tea. The structure consists of two floors; the first floor contains daily living spaces, including a living room, kitchen, bathroom, and bedroom, while the second floor serves as guest rooms, open to visitors during harvest season. The second floor is connected to a drying platform. The two floors are linked by an exterior staircase. The walls are primarily constructed with concrete, and the exterior is decorated with wooden sticks. The wall thickness ranges from 270 to 600 mm, with no tapering. The roof is made of concrete, with a total thickness of approximately 160 mm. The door frame is approximately 1.9 m in height, and the windows are large, with the maximum size being approximately 1500 mm by 1100 mm. These windows have been modified by the owner to improve natural light and ventilation. The larger window openings were introduced to address the limited airflow and light intake that often characterizes traditional Qiang dwellings.

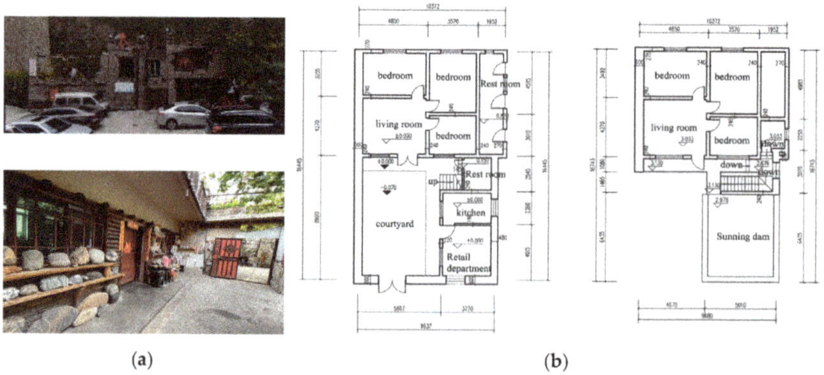

(a) (b)

Figure 6. Status and floor plan of the dwelling 1. (**a**) Current status of residential buildings. (**b**) Plane diagram.

Dwelling 2 (Figure 7) is a more typical representative of Qiang dwellings, with a rectangular shape and a southwest–northeast orientation. The site is flat and adjacent to a pond. The building consists of two floors; the first floor houses daily living spaces, including the main hall, kitchen, bathroom, and bedroom, while the second floor is used for storage and is connected to two drying platforms. The two floors are linked by an interior staircase. The walls are mainly constructed using abundant local materials, including crushed stone and yellow clay sourced from the mountain, with the exterior decorated with wooden sticks. The wall thickness is approximately 500 mm, with no tapering. The roof is composed of a composite material of sand, gravel, and earth, with an estimated thickness of 300 mm. The doorframe is relatively low, about 1.7 m high, and the windows have two sizes of 400 mm by 500 mm and 900 mm by 600 mm. Both sizes are relatively small and are usually kept closed. According to previous studies on traditional Qiang architecture [31,32], windows of this size are considered small, reflecting a traditional design approach aimed at minimizing heat loss in cold climates. The residents have built a firepit, which is currently used as a storage room.

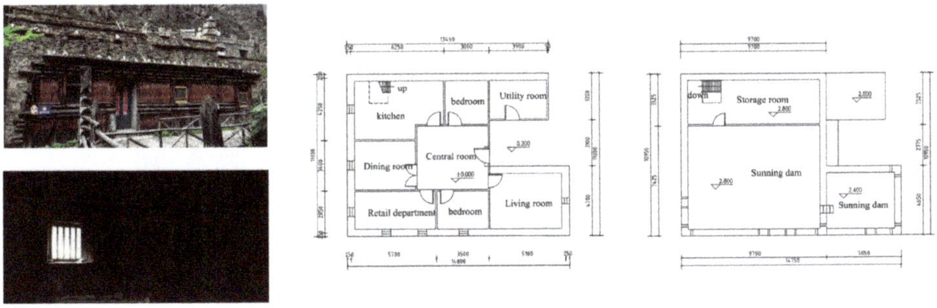

(a) (b)

Figure 7. Status and floor plan of the dwelling 2. (**a**) Current status of residential buildings. (**b**) Plane diagram.

3. Methodology

3.1. Natural Ventilation Mechanism

Thermal pressure ventilation (Figure 8) occurs as a result of natural convection caused by temperature differences in the air. Warmer air, having a lower density, rises and exits the building, while cooler and denser air descends and enters the interior [33]. Open buffer spaces, such as courtyards, facilitate this process by providing a larger volume and open pathways for air circulation. The open nature of courtyards allows hot air to rise and escape more easily, while cooler outdoor air can flow into the building through lower openings. Additionally, the height difference between the interior and the courtyard increases the pressure differential, enhancing the natural ventilation effect and improving airflow efficiency [34–36].

Figure 8. Principles of thermal pressure ventilation.

3.2. CFD Model

Before solving the problem, it is necessary to first establish the governing equations. Typically, in the absence of heat exchange, the continuity equation and momentum equation can be directly used as governing equations. For steady-state problems, initial conditions are not required but boundary conditions must be specified. These boundary conditions describe the variations in variables or their derivatives in space and time [27]. After establishing the governing equations, grid discretization is needed to discretize the equations in the spatial domain. This study uses Sware VENT2024 software and employs a second-order upwind scheme for the discretization of the equations. This scheme provides sufficient accuracy for conventional fluid simulations and has been widely used in the simulation studies of natural ventilation and airflow [37–40], ensuring high reliability and accuracy. The project adopts the k-ε turbulence model recommended by the "Green Building Evaluation Technical Guidelines" for indoor airflow calculations. This model is suitable for simple industrial flow fields and heat exchange simulations, with no large pressure gradients, separation, or strong curvature flows, making it suitable for initial parameter studies.

This study primarily focuses on the natural ventilation effect in the summer. In the summer, the outdoor temperature is higher than the indoor temperature, but the temperature difference is relatively small. As a result, the heat transfer effect has a weak influence on the ventilation process. Huang et al. [41] proposed that buoyancy-driven natural convection dominates when the indoor temperature is significantly higher than the outdoor temperature ($\Delta T > 4\ °C$), and the building has strong thermal insulation with poor thermal conductivity. Therefore, it is considered that the heat transfer effect in this study can be neglected, and natural ventilation driven mainly by wind speed plays a dominant role in air movement.

The wind field boundary is divided into the inlet boundary, outlet boundary, side boundary, ground boundary, and top boundary. Different boundaries have different boundary conditions (Figure 9).

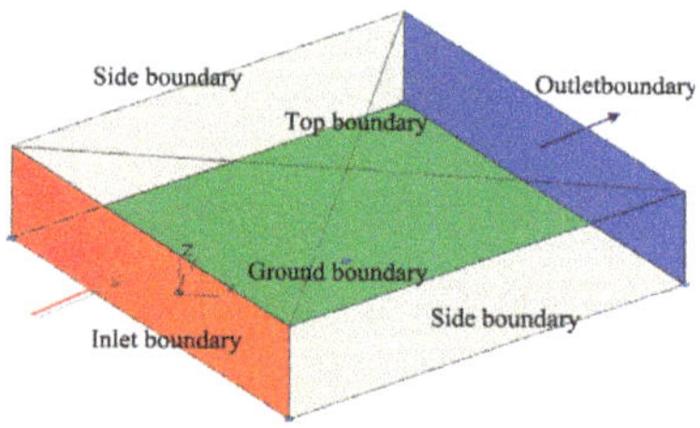

Figure 9. Wind field boundary diagram.

Inlet Wind Speed Gradient: The inlet boundary conditions are typically associated with wind speed, wind direction, and other factors under different conditions. In this study, a gradient wind model is applied to the wind speed at the inlet:

$$\nu = \nu_R \left(\frac{u}{u_M} \right)^{\alpha} \quad (1)$$

where ν, z are the average wind speed and height at any given point; V_m, Z_R are the average wind speed at the standard height and the standard height value. According to the "Load Code for the Design of Building Structures" [42], he standard height for natural wind fields is 10 m, and the average wind speed corresponds to the inlet wind speed setting. α is the ground roughness index, which is 0.22 for this project.

This project adopts a free outflow as the outlet boundary condition. The two side boundaries and the top boundary of the wind field are set as slip wall surfaces, meaning that airflow is assumed to be unaffected by wall friction, simulating the real outdoor wind flow. The ground boundary of the wind field is set as a no-slip wall surface, meaning that the airflow is influenced by ground friction.

This project uses Sware VENT2024 software to solve the wind field. Specifically, mass conservation, momentum conservation, and energy conservation equations are established within the analyzed computational domain to form the governing mathematical equations. The general form of these equations is shown as follows:

$$\frac{\partial(\rho\phi)}{\partial t} + \nabla\cdot(\rho U\phi) = \nabla\cdot(\Gamma_\phi \nabla\phi) + S_\phi \quad (2)$$

In this equation, ϕ can represent physical quantities such as velocity, turbulent kinetic energy, the turbulent dissipation rate, and temperature, as referenced in Table 1 below.

Table 1. Governing equations of fluid dynamics.

Term	Variable	Γ_ϕ	S_ϕ
Continuity Equation	1	0	0
x Velocity	μ	$\mu_{eff} = \mu + \mu_t$	$-\frac{\partial P}{\partial x} + \frac{\partial}{\partial x}\left(\mu_{eff}\frac{\partial u}{\partial x}\right) + \frac{\partial}{\partial y}\left(\mu_{eff}\frac{\partial v}{\partial x}\right) + \frac{\partial}{\partial z}\left(\mu_{eff}\frac{\partial w}{\partial x}\right)$
y Velocity	v	$\mu_{eff} = \mu + \mu_t$	$-\frac{\partial P}{\partial y} + \frac{\partial}{\partial x}\left(\mu_{eff}\frac{\partial u}{\partial y}\right) + \frac{\partial}{\partial y}\left(\mu_{eff}\frac{\partial v}{\partial y}\right) + \frac{\partial}{\partial z}\left(\mu_{eff}\frac{\partial w}{\partial y}\right)$
z Velocity	ω	$\mu_{eff} = \mu + \mu_t$	$-\frac{\partial P}{\partial z} + \frac{\partial}{\partial x}\left(\mu_{eff}\frac{\partial u}{\partial z}\right) + \frac{\partial}{\partial y}\left(\mu_{eff}\frac{\partial v}{\partial z}\right) + \frac{\partial}{\partial z}\left(\mu_{eff}\frac{\partial w}{\partial z}\right) - \rho g$
Turbulent Kinetic Energy	κ	$\alpha_k \mu_{eff}$	$G_k + G_B - \rho\varepsilon$
Turbulent Dissipation Rate	ε	$\alpha_\varepsilon \mu_{eff}$	$C_{1\varepsilon}\frac{\varepsilon}{k}(G_k + C_{3\varepsilon}G_B) - C_{2\varepsilon}\rho\frac{\varepsilon^2}{k} - R_\varepsilon$
Temperature	T	$\frac{\mu}{Pr} + \frac{\mu_t}{\sigma_T}$	S_T

The constants in the above table are as follows:

$G_k = \mu_t S^2$, $S = \sqrt{2S_{ij}S_{ij}}$, $S_{ij} = \frac{1}{2}\left(\frac{\partial u_j}{\partial x_i} + \frac{\partial u_i}{\partial x_j}\right)$, $G_B = \beta_T g \frac{\mu_t}{\sigma_T}\frac{\partial T}{\partial y}$, $\mu_t = \rho C_\mu \frac{k^2}{\varepsilon}$, $C_\mu = 0.0845$, $C_{1\varepsilon} = 1.42$, $C_{2\varepsilon} = 1.68$, $C_{3\varepsilon} = \tanh\left|\frac{v}{\sqrt{u^2+w^2}}\right|$, and $\sigma_T = 0.85$, $\sigma_C = 0.7$, calculated by $\alpha_k = \alpha_\varepsilon$, $\left|\frac{\alpha-1.3929}{\alpha_0-1.3929}\right|^{0.6321}\left|\frac{\alpha+2.3929}{\alpha_0+2.3929}\right|^{0.3679} = \frac{\mu}{\mu_{eff}}$, and $\alpha_0 = 1.0$. If $\mu \ll \mu_{eff}$, then $\alpha_k = \alpha_\varepsilon \approx 1.393$.

$R_\varepsilon = \frac{C_\mu\rho\eta^3(1-\eta/\eta_0)}{(1+\beta\eta^3)} \times \frac{\varepsilon^2}{k}$, where $\eta = \frac{Sk}{\varepsilon}$, $\eta_0 = 4.38$ and $\beta = 0.012$. In this project, the SIMPLE algorithm is employed to solve the aforementioned system of equations.

To simulate the impact of natural ventilation on indoor air quality in traditional Qiang dwellings of Dongmen Village using Sware VENT2024, the first step is to establish the outdoor wind field computational domain, as shown in Figure 10. The domain size was chosen based on the scale of the study area to ensure sufficient space for accurately capturing the wind flow dynamics. The windward direction has a dimension of 405 m, the width is 536 m, and the height is 122 m. Automatic grid generation was performed using Sware VENT2024 software. In regions with higher wind speeds, such as around door and window openings, local grid refinement was manually applied to improve grid quality and resolution. This approach allows for a more accurate representation of airflow in critical areas. Additionally, for the ground grid, higher levels of refinement were applied to the near-field region, where surface details are crucial for capturing local airflow dynamics, while the far-field ground grid distribution was sparser, reflecting the less critical nature of these areas for the simulation. A total of 344,038 grids were generated. The choice of grid sizes aims to balance computational efficiency with the required accuracy to simulate airflow patterns around the building, and grid quality was carefully assessed to ensure it meets the simulation requirements.

The meteorological data for Wenchuan comes from the "Code for Design of Heating, Ventilation, and Air Conditioning in Civil Buildings", which provides wind speed and direction for both winter and summer. The data are based on the observation statistics of the coldest three months and the hottest three months over multiple years. In winter, the wind direction is set as northwest with a wind speed of 3.3 m/s, and in summer, the wind direction is also set as northwest with a wind speed of 3.1 m/s (Table 2).

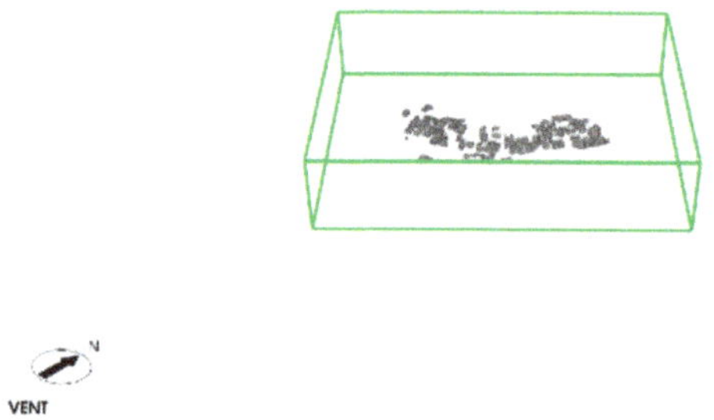

Figure 10. Calculation domain of outdoor wind field.

Table 2. Winter outdoor evaluation criteria of the Green Building Evaluation Standard.

Reason	Wind Speed (m/s)	Wind Direction	Wind Direction (°)
Winter	3.30	NW	135.0
Summer	3.10	NW	135.0

This study mainly investigates the effect of natural ventilation in buffer spaces in traditional Qiang dwellings. The simulations are divided into the following three scenarios:

(1) Site Wind Environment Analysis: A detailed analysis of wind speed and wind pressure in the Dongmen Village area to assess the ventilation performance of the overall site design.

(2) Comparison of Buffer Space Presence: A comparison between winter and summer seasons to analyze the differences in wind speed and air age between dwellings with and without buffer spaces, evaluating the role of buffer spaces in natural ventilation.

(3) Impact of Courtyard Dimensions on Ventilation Performance: By setting different aspect ratios (length-to-width and height-to-width), the study investigates the impact of courtyard size variations on indoor wind speed and air age, exploring the optimization effect of courtyard morphology on natural ventilation efficiency.

3.3. *CFD Validation Through Grid Sensitivity Analysis*

To assess the impact of grid resolution on the simulation results, this study conducted a grid sensitivity analysis to verify the stability of the model [43]. Three grid resolutions were set as follows: coarse grid (172,019 grid cells), medium grid (344,038 grid cells), and fine grid (516,057 grid cells). The horizontal wind speed distribution along the windward direction on the leeward side of the building was compared for each resolution. Through this analysis, we confirmed that the wind speed variation trend was consistent across different grids (Figure 11), and with the increase in grid resolution, the accuracy of the simulation results improved. The grid sensitivity analysis provided an effective validation of the stability and rationality of the model results. Furthermore, future research can further validate the model's accuracy and applicability by incorporating field measurement data.

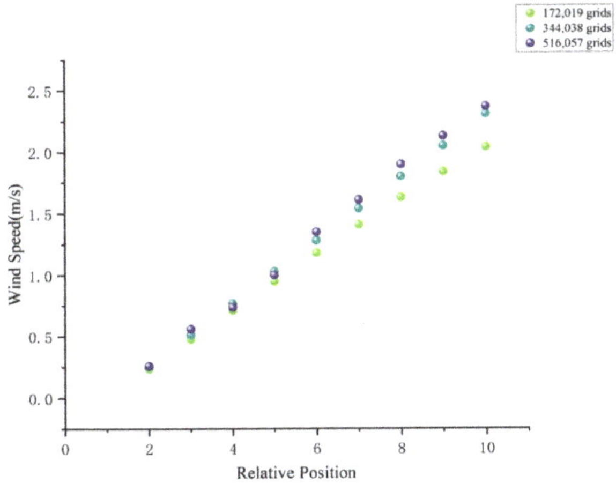

Figure 11. Horizontal velocity distributions of three different grids.

3.4. Evaluation Criteria

The evaluation criteria for the outdoor wind environment in this study are based on the "Green Building Evaluation Standard" [44], while the indoor wind environment assessment criteria consider both the comfort of wind speed and air age. The specific evaluation standards are presented in Tables 3–6:

Table 3. Winter outdoor evaluation criteria of the Green Building Evaluation Standard [44].

Project Evaluation	Standard Requirements
Wind Speed (m/s)	The wind speed at 1.5 m above the ground in pedestrian areas around the building should be <5 m/s, and in outdoor resting and children's play areas, the wind speed should be <2 m/s
Wind Speed Amplification Factor	The outdoor wind speed amplification factor should be <2
Windward/Leeward Building Facades	Except for the first row of buildings on the windward side, the surface wind pressure difference between the windward and leeward facades should not exceed 5 Pa

Table 4. Summer outdoor evaluation criteria of the Green Building Evaluation Standard [44].

Project Evaluation	Standard Requirements
Windless Zone	No windless zones, defined as areas where the wind speed is below 0.2 m/s, should occur in the human activity areas within the site
Vortex Zone	No vortexes zones should occur in the human activity areas within the site
Wind Pressure Difference Between the Interior and Exterior Surfaces of Openable Windows	For more than 50% of openable windows, the wind pressure difference between the interior and exterior surfaces should be >0.5 Pa

Table 5. Assessment of indoor wind speed effects on human activity [45].

Wind Speed (m/s)	Impact on Human Activity
0–0.25	Difficult to Perceive
0.25–0.5	Comfortable and Pleasant
0.5–1.0	Relatively Comfortable, Requires Measures to Prevent Papers from Blowing Away
1.0–1.5	Some Discomfort, Papers on the Desk May Be Dispersed
More than 1.5	Causes Discomfort

Table 6. Standards for evaluating indoor air age [46].

Air Age T (s)	Assessment of Air Freshness
T < 225	Fresh Air
225 ≤ T < 400	Fairly Fresh Air
T ≥ 400	Stale Air

4. Results

4.1. Outdoor Wind Environment Analysis

Dongmen Village is located in Bazhou Town, and the slope differences in the surrounding terrain may have a significant impact on the natural ventilation effect. The windward slope of Bazhou Town is 33.65°, while the leeward slopes are relatively gentle, measuring 22.7° and 22.81° [47]. According to related studies, steeper slopes accelerate wind flow, thereby increasing wind speed, which helps enhance natural ventilation [48]. Therefore, the windward side of Dongmen Village may experience stronger ventilation, improving air movement. In contrast, the gentler slope of the leeward side may cause wind speed to slow down, obstructing air movement and resulting in localized air stagnation, which affects air exchange and ventilation efficiency. Furthermore, the surrounding mountain slopes of Bazhou Town are 25.89°, suggesting that the terrain in this area may lead to asymmetrical local wind flows, further influencing the wind speed and ventilation effects in Dongmen Village [48]. Nevertheless, considering that the main objective of this study is to evaluate the influence of buffer spaces on natural ventilation, the selection of the computational domain focuses on the airflow path around the building and does not account for terrain factors in detail. The size of the simulation domain and grid settings are sufficiently precise to capture key areas of airflow dynamics around the building.

To evaluate the outdoor winter wind environment of the site, this study conducted a series of simulations. Figure 12a illustrates the winter wind speed contour plot at a height of 1.5 m above the ground level within the computational domain. No areas exceeding the prescribed limits are identified in the figure, with the maximum wind speed recorded in pedestrian activity zones being 2.21 m/s. This value is below the 5 m/s threshold specified in the green building standards, indicating compliance with the relevant criteria. Figure 12b presents the winter wind speed amplification coefficient contour plot at the same height of 1.5 m. The analysis of the data reveals that the maximum wind speed amplification coefficient in pedestrian areas is 1.02, well below the 2.0 limit set by the green building standards, thus meeting the required standard. Figure 12c,d displays the wind pressure difference between the windward and leeward sides of the building. The computed data show that the average wind pressure difference between the windows on the windward and leeward sides does not exceed 5 Pa, complying with the prescribed requirements.

Therefore, it can be concluded that the building fully meets the relevant green building standards for outdoor winter wind conditions.

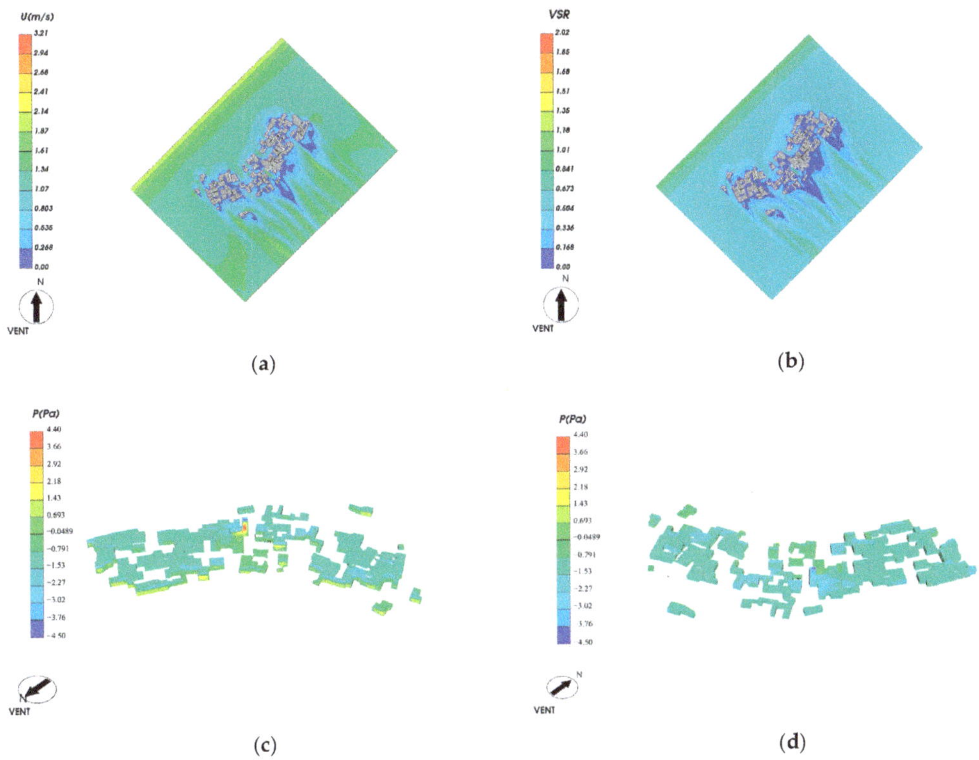

Figure 12. Winter outdoor wind environment simulation of the site. (**a**) Winter wind speed contour map. (**b**) Winter wind speed amplification factor contour map. (**c**) Wind pressure contour map on the building's windward side. (**d**) Wind pressure contour map on the building's leeward.

This study conducted a simulation analysis of the summer outdoor wind environment of the site. Figure 13a presents the summer wind speed contour at a 1.5 m height level within the computational domain. It can be observed from the figure that no areas with wind speeds lower than 0.2 m/s are identified within the human activity zones, and no significant vortices appear across the entire computational domain, in compliance with green building standards. This indicates that the current building layout is rational and exhibits favorable wind environment characteristics in terms of natural ventilation. Figure 13b,c shows the window wind pressure difference analysis for two residential buildings. For dwelling 1, there is a total of 16 operable windows, 11 of which exhibit an indoor-to-outdoor pressure difference exceeding 0.5 Pa, resulting in a compliance rate of 68.75%, in accordance with the "Green Building Evaluation Standard". Dwelling 2 has four operable windows, all of which have an indoor-to-outdoor pressure difference exceeding 0.5 Pa, achieving a compliance rate of 100%, fully meeting the relevant requirements of the standard.

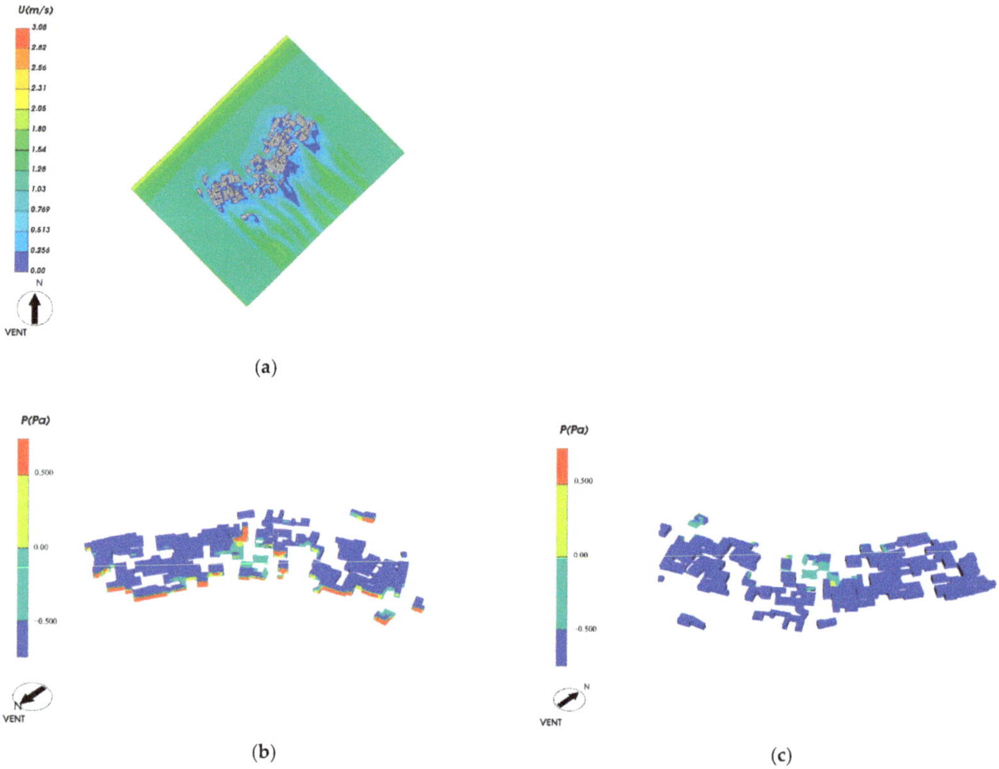

Figure 13. Summer outdoor wind environment simulation of the site. (**a**) Wind speed contour map. (**b**) Wind pressure contour map on the building's windward side. (**c**) Wind pressure contour map on the building's leeward side.

4.2. Ventilation Comparison for Building Buffer Spaces

To systematically explore the impact of buffer spaces on natural ventilation performance, this study conducted numerical simulations of indoor wind speed and air age for dwellings 1 and 2 during both winter and summer. Given that the second floor of dwelling 2 is used as a storage room and does not involve actual living spaces, it was excluded from the analysis. For comparison purposes, the analysis of dwelling 1 is limited to the indoor ventilation conditions on the first floor.

Under the influence of northwesterly winds during winter, the wind speed and air age distributions for both the first and second floors of dwelling 1 were analyzed at a reference height of 1.2 m. The choice of 1.2 m as the reference plane corresponds to the breathing zone of a seated adult [49]. This height was selected to more accurately assess the ventilation performance within human activity areas, particularly to quantify ventilation quality and air freshness within the living space.

As shown in Figure 14a, the indoor wind speed distribution indicates that the wind speed in dwelling 1 ranges from 0.00793 to 1.21 m/s, with an average wind speed of 0.21 m/s. The maximum wind speeds occur in the small shop, where wind speeds exceed 1 m/s, potentially causing slight discomfort. According to the indoor air age distribution, the maximum air age in dwelling 1 is 84.8 s, with an average air age of 39.7 s, reflecting good ventilation and relatively fresh air quality. Although in some areas, particularly in the enclosed spaces on the first floor, air movement is somewhat restricted, leading to a slight increase in air age, overall, the ventilation in dwelling 1 is efficient and the air quality remains high.

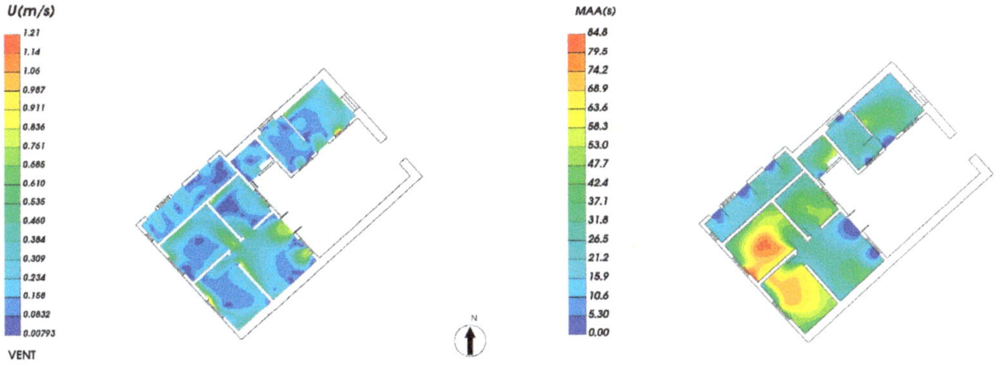

(a)

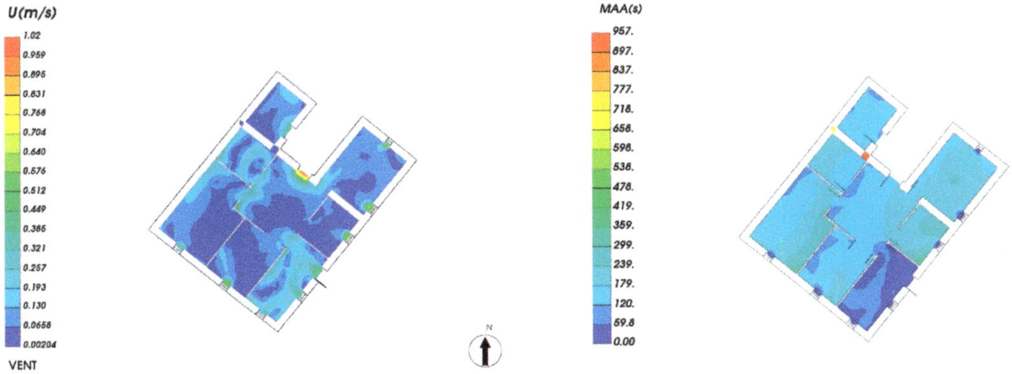

(b)

Figure 14. Winter simulation results. (**a**) Wind speed cloud map and air age map of dwelling 1. (**b**) Wind speed cloud map and air age map of dwelling 2.

As shown in Figure 14b, the indoor wind speed contour map reveals that the wind speed on the first floor of dwelling 2 ranges from 0.00204 to1.02 m/s, with the highest wind speed occurring at the entrance of the dwelling, an average wind speed of 0.11 m/s. According to the analysis of the air age map, the maximum air age in dwelling 2 exceeds 400 s, primarily concentrated in the kitchen, bedroom, and living room areas, indicating poor air circulation in these zones, resulting in stale air. The overall average air age is 195.5 s. Overall, the winter natural ventilation performance in the bedroom, living room, and kitchen of dwelling 2 is poor, while the ventilation conditions in other rooms are relatively better.

Under the influence of the northwestern wind during summer, the wind speed and air age maps for dwelling 1 are shown in Figure 15a. According to the wind speed contour map, the indoor wind speed in dwelling 1 ranges from 0.0052 m/s to 1.25 m/s, with an average wind speed of 0.17 m/s. According to the air age distribution map, the maximum air age in dwelling 1 is 122 s, with an average air age of 67.98 s. This result indicates that despite some ventilation, the indoor air movement during summer remains insufficient, with low air renewal, which may affect the comfort of the occupants.

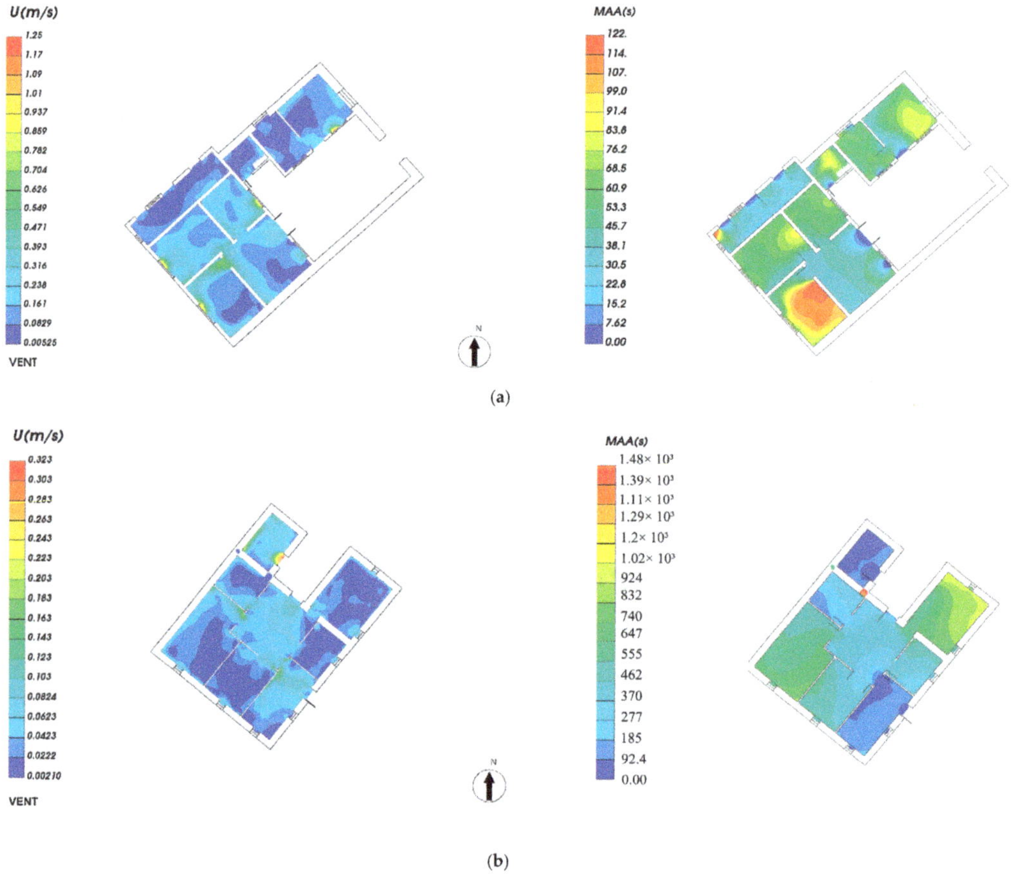

Figure 15. Summer simulation results. (**a**) Wind speed cloud map and air age map of dwelling 1. (**b**) Wind speed cloud map and air age map of dwelling 2.

Similarly, under the influence of the northwestern wind during summer, the wind speed and air age maps for dwelling 2 are shown in Figure 15b. The indoor wind speed in dwelling 2 ranges from 0.0021 m/s to 0.323 m/s, with an average wind speed of approximately 0.043 m/s. The overall wind speed is relatively low, which may result in a stuffy indoor environment. According to the air age distribution map, the maximum air age in dwelling 2 is 1110 s, with an average air age of approximately 621 s, indicating poor ventilation during summer and insufficient air movement. Consequently, the occupants may experience an uncomfortable and stuffy environment.

In summary, the addition of buffer spaces in traditional Qiang dwellings significantly enhances their natural ventilation performance. Specifically, both dwellings 1 and 2 maintain wind speeds within reasonable ranges. However, the air age in dwelling 1 remains consistently below 225 s during both winter and summer, and although it slightly increases in the summer, the indoor air quality remains relatively fresh. In contrast, dwelling 2, which lacks buffer spaces, demonstrates poorer ventilation performance, especially in the summer, where the maximum air age reaches 1110 s and the average air age is 621 s, indicating a clear stagnation of airflow. The winter simulation results further highlight that the air circulation in dwelling 2 is suboptimal, with a maximum air age of 419 s and an average air age of 195.5 s, both of which are significantly higher than the ventilation performance in dwelling 1 (Figure 16). The simulation results further indicate that buffer spaces play a crucial role in improving indoor air quality and reducing dependence on mechanical ventilation systems. By promoting natural ventilation, buffer spaces can effectively support a more energy-efficient and livable environment.

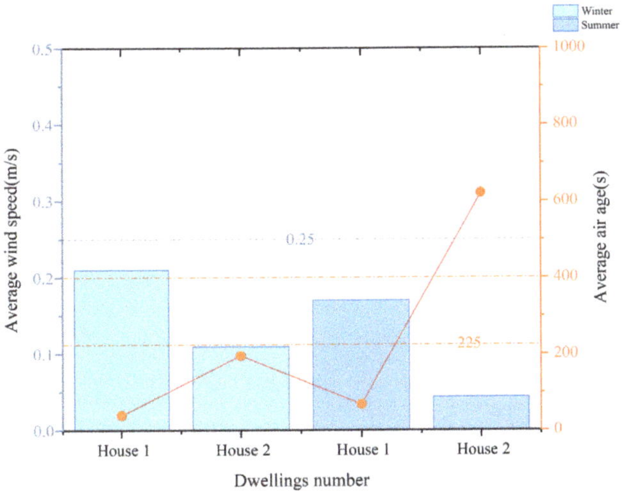

Figure 16. Comparison of seasonal wind speed and air age in two traditional dwellings.

4.3. Impact of Courtyard Dimensions on Ventilation Performance

Through simulation analysis, it was found that the ventilation performance of dwelling 1, which includes a buffer space, is significantly better than that of dwelling 2, which lacks a buffer space, with the winter ventilation performance being superior to that of summer. To further investigate the impact of the buffer space on indoor ventilation, this study sets the central hall of dwelling 2 as a buffer space courtyard and conducts indoor ventilation simulations by controlling the variations in its dimensions. According to the field survey data, the original dimensions of the main hall are 4.8 m × 4.41 m × 2.6 m, with a length-to-width ratio of approximately 1.1 and a height-to-width ratio of approximately 0.6. To investigate the effects of different scales on ventilation performance, simulations were conducted with five length-to-width ratios (0.7, 0.9, 1.1, 1.3, 1.5) and five height-to-width ratios (0.6, 0.9, 1.2, 1.5). These ratios were selected based on the measurements of typical courtyard dimensions in Qiang traditional dwellings from the field survey, as well as the courtyard dimension ranges found in other studies [50–52]. In this study, the changes in the length-to-width ratio and height-to-width ratio of the courtyard space were considered independently. To clearly assess the impact of each ratio on indoor ventilation, one ratio was kept constant during each test while the other was adjusted. This approach ensured that the observed differences in ventilation performance were due to changes in the ratios.

Figure 17 presents the impact of varying the length-to-width ratio of the courtyard, with values of 0.7, 0.9, 1.1, 1.3, and 1.5 under different length-to-width ratio control conditions. The simulation examines the effect of changes in the courtyard space dimensions on indoor wind speed and air age during summer operating conditions for dwelling 2.

Length-to-Width Ratios	Wind Speed Cloud Map	Air Age Map
0.7	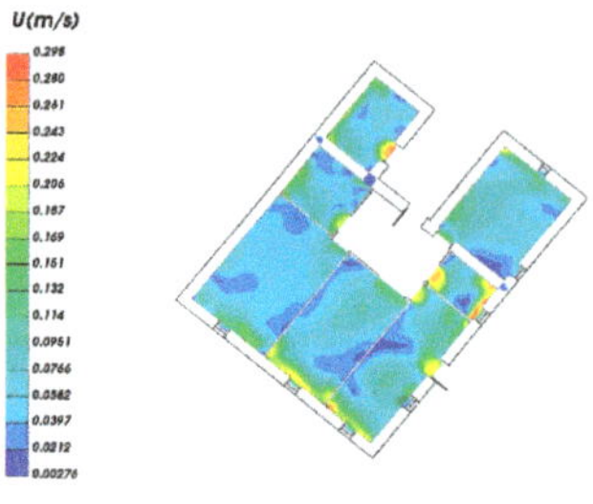	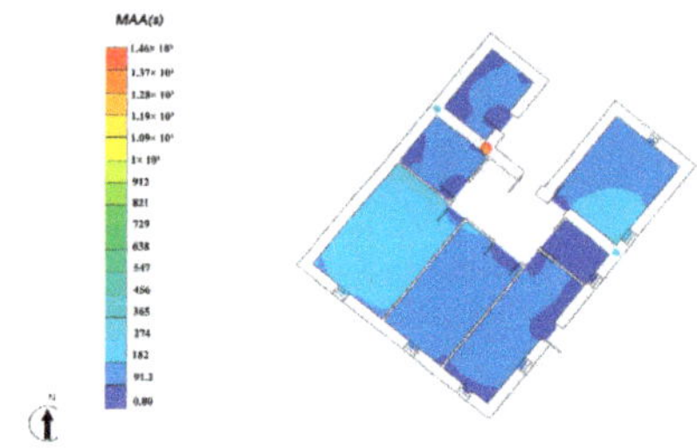
0.9	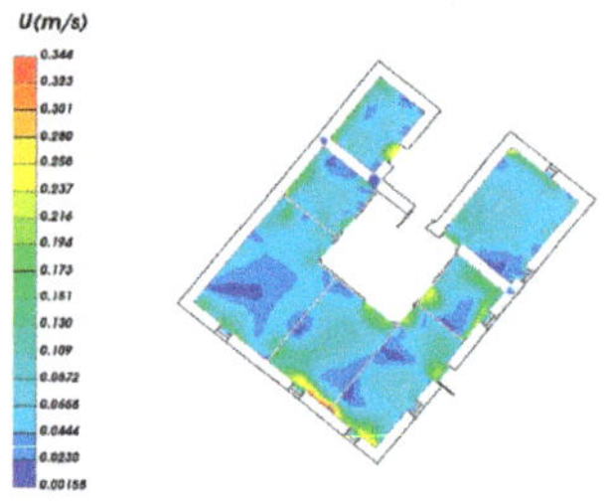	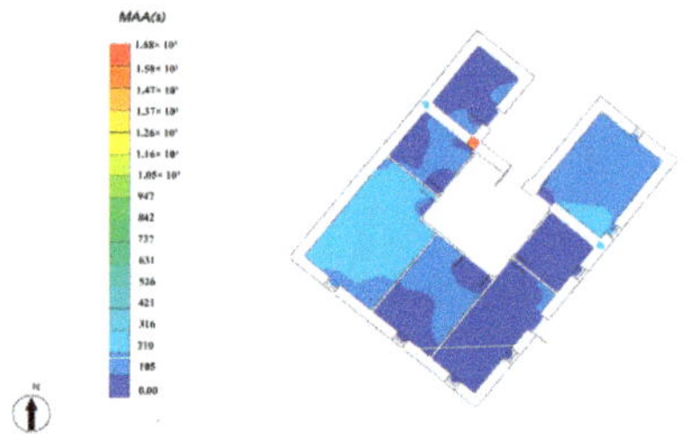
1.1	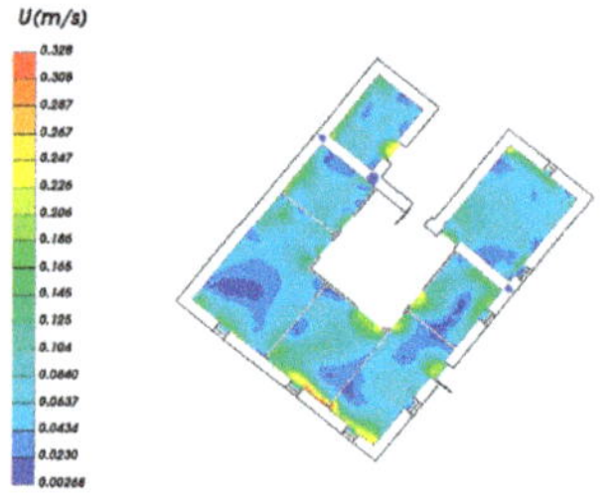	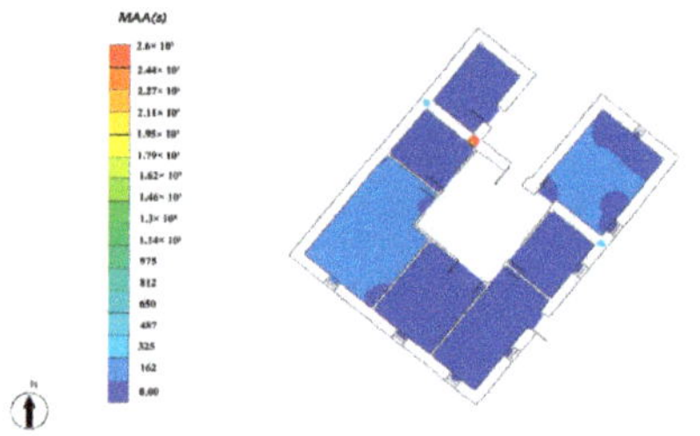
1.3	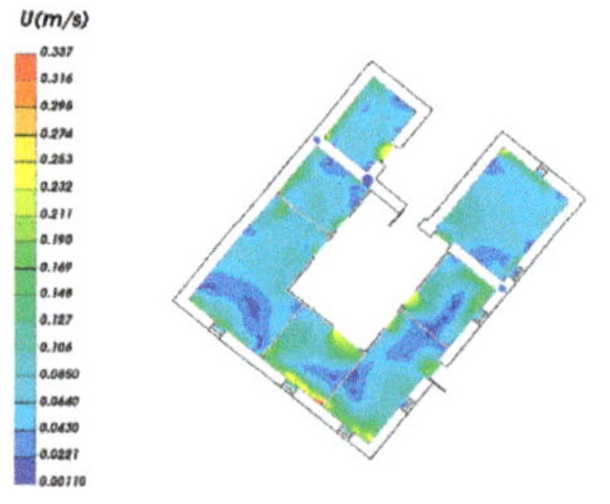	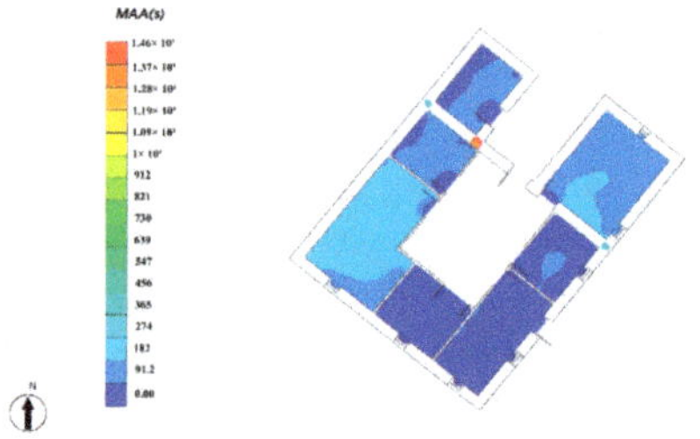
1.5	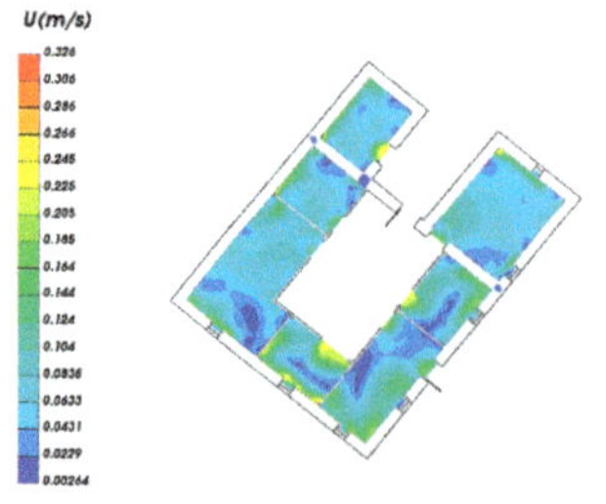	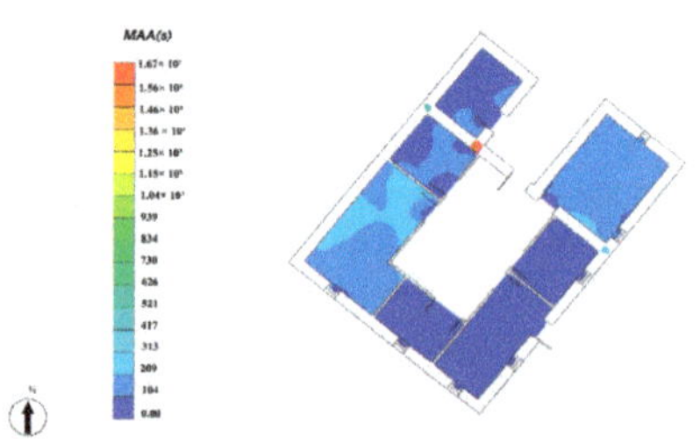

Figure 17. The impact of different courtyard length-to-width ratios on indoor ventilation.

As shown in Figure 17, when the courtyard's length-to-width ratio is 0.7, the maximum air age inside the house occurs in the living room and kitchen, with values ranging from 182 to 365 s, while the minimum air age is found in the bedroom, ranging from 0 to 91.2 s. The air age in other rooms typically falls between 91.2 and 182 s, and the overall average air age is 196 s. The maximum wind speed in the house occurs at the entrance of the storage room, the window of the small shop, and the window of the bedroom, with a speed of approximately 0.3 m/s. The minimum indoor wind speed is 0.002 m/s, with the majority of rooms having wind speeds ranging from 0.02 to 0.114 m/s; the overall average wind speed is 0.08 m/s. When the courtyard's length-to-width ratio increases to 0.9, the maximum air age in the house is still observed in the kitchen and living room, with values ranging from 105 to 336 s. The minimum air age is found in the bedroom, small shop, dining room, and storage room, with values ranging from 0 to 210 s. The air age in other rooms is mostly concentrated between 105 and 210 s; the overall average air age is 188 s. The maximum wind speed in the house is observed at the window openings of the dining room and small shop, with a speed of approximately 0.34 m/s. The minimum wind speed inside the rooms occurs at the corners, where the wind speed is 0.0015 m/s, while the majority of rooms have wind speeds ranging from 0.02 to 0.109 m/s, where the overall average wind speed is 0.083 m/s. When the courtyard's length-to-width ratio is 1.1, which corresponds to the original dimensions of the central hall, the maximum air age in the house remains in the kitchen and living room, with values ranging from 162 to 325 s, while the air age in other rooms is mostly 0 to 162 s and the overall average air age is 218 s. The maximum wind speed in the house is observed at the window openings of the dining room and small shop, with a speed of approximately 0.33 m/s. The minimum wind speed inside the rooms is 0.0027 m/s, with the majority of rooms having wind speeds ranging from 0.02 to 0.125 m/s, and the overall average wind speed is 0.079 m/s. When the courtyard's length-to-width ratio is 1.3, the maximum air age in the house is found in the kitchen and living room, with values ranging from 91.2 to 274 s, while the minimum air age is observed in the bedroom, small shop, and dining room, with values ranging from 0 to 91.2 s. The air age in other rooms is mostly concentrated between 0 and 162 s, and the overall average air age is 162 s. The maximum wind speed in the house occurs at the window openings of the dining room, with a speed of approximately 0.34 m/s. The minimum wind speed inside the rooms is 0.0011 m/s, while the majority of rooms have wind speeds ranging from 0.02 to 0.169 m/s; the overall average wind speed is 0.07 m/s. When the courtyard's length-to-width ratio is 1.5, the maximum air age in the house is observed in the kitchen, with values ranging from 104 to 313 s, while the minimum air age is found in the bedroom, small shop, and dining room, with values ranging from 0 to 104 s. The air age in other rooms is mostly concentrated between 0 and 209 s, and the overall average air age is 165 s. The maximum wind speed in the house occurs at the window openings of the dining room and the door of the storage room, with a speed of approximately 0.245 m/s., with the majority of rooms having wind speeds ranging from 0.02 to 0.185 m/s; the overall average wind speed is 0.085 m/s.

In summary, as the length-to-width ratio of the courtyard increases, the indoor air age of dwelling 2 initially decreases, then increases. When the length-to-width ratio is 0.7, the overall air age is relatively high, with the maximum value being 365 s, and the average value is 196 s. When the length-to-width ratio is 1.3, the overall air age of the dwelling is at its minimum, with the maximum value being 274 s, and the average value is 188 s. When the length-to-width ratio exceeds 1.3, the air age increases, with the maximum value being 313 s, and average value is 218 s. This indicates that the indoor air age of dwelling 2 decreases initially and then increases as the courtyard's length-to-width ratio increases, with the indoor air being relatively freshest when the dimensions are 5.7 m by 4.41 m. According to the air age diagram, the rooms with the lowest air freshness are consistently the living room and kitchen. This suggests that changes in the courtyard's length-to-width ratio do not significantly affect the balancing of indoor air age. Meanwhile, the average wind speed shows a relatively stable trend; this indicates that changes in the courtyard's length-to-width ratio have a negligible effect on indoor wind speed (Figure 18).

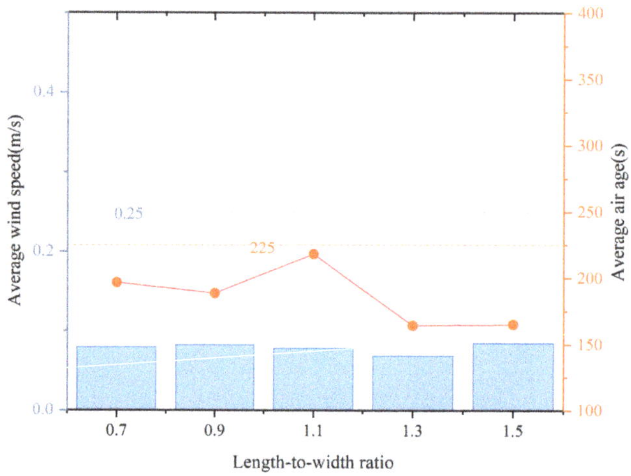

Figure 18. Impact of length-to-width ratio variation on indoor wind speed and air age.

Figure 19 presents the impact of varying the length-to-width ratio of the courtyard, with values of 0.6, 0.9, 1.2, and 1.5 under different height-to-width ratio control conditions. The simulation examines the effect of changes in the courtyard space dimensions on indoor wind speed and air age during summer operating conditions for dwelling 2.

As illustrated in Figure 19, when the height-to-width ratio of the courtyard is 0.6, corresponding to the original height of the hall, the maximum indoor air age in dwelling 2 occurs in the living room and kitchen, with a range of 162 to 325 s. The air age in other rooms generally ranges from 0 to 162 s, and the overall average air age is 218 s. The highest wind speed in the dwelling is observed at the windows of the dining room and convenience store, at approximately 0.33 m/s. The minimum indoor wind speed is 0.002 m/s, with most rooms exhibiting wind speeds between 0.02 and 0.125 m/s, and the overall average wind speed is 0.079 m/s. When the height-to-width ratio of the courtyard increases to 0.9, the maximum air age in the dwelling remains in the kitchen, ranging from 102 to 170 s. The minimum air age is found in the convenience store, ranging from 0 to 68.1 s. The air age in other rooms generally varies between 34 and 136 s, and the overall average air age is 104 s. The highest wind speed is recorded at the windows of various rooms, at approximately 1.3 m/s. The minimum wind speed inside the rooms is 0.005 m/s, primarily in the kitchen and bedroom, with the wind speed in other rooms ranging from 0.113 to 0.976 m/s, and the overall average wind speed is 0.258 m/s. At a height-to-width ratio of 1.2, the maximum air age in the dwelling is observed in the bedroom and storage room, with a range of 228 to 308 s. The minimum air age is recorded in the dining room, bedroom, and convenience store, ranging from 0 to 152 s. The air age in other rooms primarily falls between 76.2 and 305 s, and the overall average air age is 132 s. The highest wind speed is again recorded at the windows of various rooms, at around 1.2 m/s. The minimum wind speed inside is 0.001 m/s, observed mainly in the bedroom and storage room, while the wind speed in other rooms generally ranges from 0.102 to 0.710 m/s, and the overall average wind speed is 0.26 m/s. When the height-to-width ratio of the courtyard increases to 1.5, the maximum air age in dwelling 2 is found in the kitchen, ranging from 214 to 356 s. The minimum air age is observed in the bedroom and convenience store, ranging from 0 to 71.3 s. The air age in other rooms primarily falls between 71.3 and 214 s, and the overall average air age is 183 s. The highest wind speed is measured at the windows and door openings of the rooms, at around 1.6 m/s. The minimum wind speed is 0.002 m/s, occurring mainly in the kitchen and bedroom, with the wind speed in other rooms ranging from 0.104 to 1.12 m/s, and the overall average wind speed is 0.275 m/s.

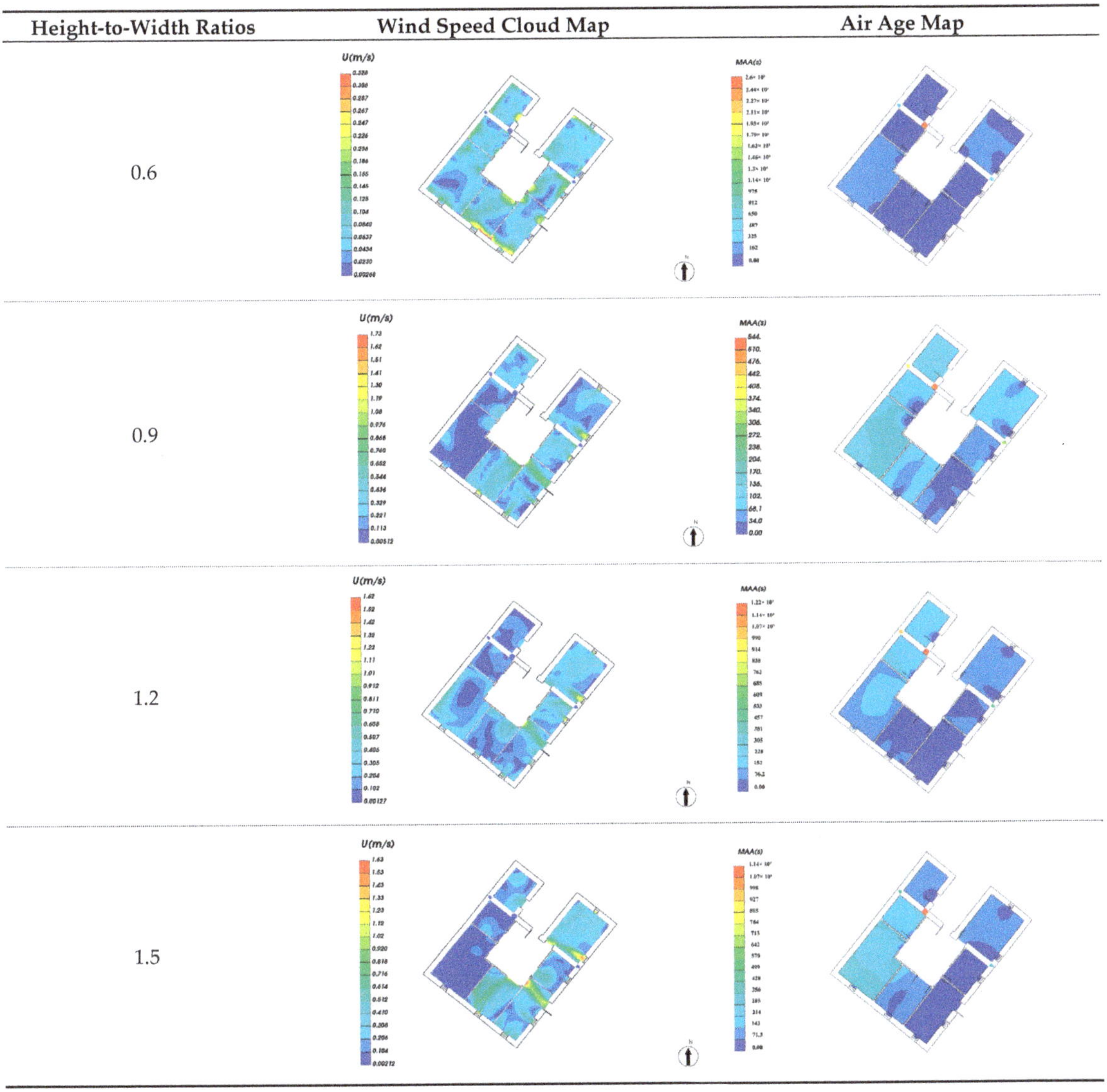

Figure 19. The impact of different courtyard height-to-width ratios on indoor ventilation.

In summary, as the height-to-width ratio of the courtyard increases, the indoor average air age of the dwelling exhibits a trend of initially decreasing and then increasing. When the height-to-width ratio is 0.9, the dwelling achieves the minimum air age, with a maximum value of 170 s and an average value of 104 s. As the height-to-width ratio increases to 1.5, the air age reaches a relatively maximum value, with a maximum of 356 s and an average of 183 s. This suggests that a skylight height of approximately 4 m helps improve indoor air circulation. According to the air age distribution, in four of the scenarios, except for the one with a 1.2 height-to-width ratio, the room with the maximum air age is either the bedroom or the storage room, while in other scenarios, the kitchen exhibits the highest air age. Notably, the room with the minimum air age in all scenarios is consistently the small shop. This indicates that the variation in the courtyard height-to-width ratio does not significantly affect the balance of indoor air age. Furthermore, the indoor wind speed increases with the height-to-width ratio of the courtyard. As the height-to-width ratio increases, the maximum wind speed across all scenarios rises from 0.125 m/s to 1.12 m/s. This indicates that an increase in courtyard height has a positive effect on the variation in indoor wind speed, contributing to enhanced ventilation efficiency within the residence (Figure 20).

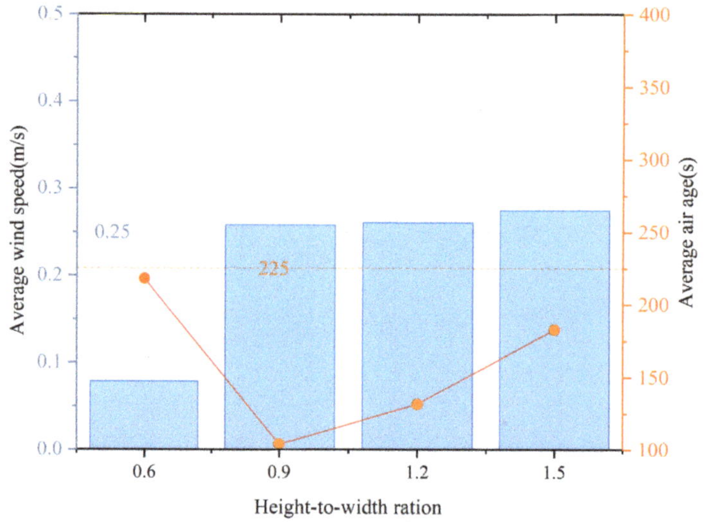

Figure 20. Impact of height-to-width ratio variation on indoor wind speed and air age.

5. Discussion

This study investigates the impact of courtyard dimensions on natural ventilation performance across various architectural layouts and climate conditions. The results reveal that the length-to-width and height-to-width ratios of courtyards significantly influence indoor air age, with optimal ratios being affected by climate, building layout, and external environmental factors. In hot summer–cold–winter zones, increasing the courtyard height helps lower indoor and outdoor air temperature in lower spaces, with a height-to-depth (H/D) ratio not less than one, enhancing air circulation and thermal comfort [53]. In contrast, in cold climates, such as Beijing's siheyuans, a courtyard length-to-width ratio close to one and a height-to-width ratio between 1.2 and 1.4 effectively balances summer ventilation needs with winter wind protection [29]. However, excessively large ratios in extremely cold climates may lead to heat loss, compromising thermal comfort [54]. Therefore, the study recommends using a length-to-width ratio of 1.3 and a height-to-width ratio of 0.9 for Qiang dwellings in cold climates, ensuring good ventilation and meeting green building standards for wind speed while effectively expelling stale air to maintain indoor air freshness.

Traditional Tibetan and Scandinavian dwellings, located in similar cold climate zones, share many similarities with Qiang dwellings. In recent years, with advancements in building technology, the thermal and mechanical properties of window glazing have significantly improved. Tibetan dwellings have begun adopting larger window-to-wall ratios, a change that not only enhances indoor lighting but also effectively mitigates poor ventilation and indoor air pollution issues [55]. Scandinavian homes promote natural ventilation by strategically placing vents and designing multiple buffer spaces. These spaces act as "spatial filters", gradually mitigating the cold external airflow through elements such as windbreaks, vestibules, and sunrooms, thereby maximizing solar gain and minimizing heat loss [18,56]. Overall, these buildings demonstrate a profound understanding of natural principles in adapting to cold climates. Specifically, they commonly use heavy materials like stone, rammed earth, or wood to increase thermal mass and reduce heat loss. Additionally, their spatial layouts typically include buffer or transitional spaces, such as inner courtyards, vestibules, or sunrooms, to regulate the indoor–outdoor temperature difference and facilitate natural ventilation. Therefore, the simulation results of courtyard dimensions in Qiang dwellings not only provide a valuable reference for architectural optimization in the region but also offer useful insights for building design in cold climates. In terms of natural ventilation, Qiang dwellings can draw on passive ventilation strategies from other cold climate regions to enhance indoor ventilation efficiency and optimize indoor environmental quality.

The results of this study provide empirical support for ventilation optimization strategies. Based on the simulation results, it is recommended that in practical architectural design, the length-to-width and height-to-width ratios of courtyards be flexibly adjusted according to regional climate and microclimate conditions. Additionally, to achieve optimal indoor air circulation and thermal comfort, the building design should be optimized in consideration of external environmental factors. For example, the building orientation should take into account the prevailing wind direction, and excessive window openings should be avoided in unfavorable wind directions to minimize the negative impact of wind pressure differences on indoor ventilation.

6. Conclusions

This study employs field research and CFD simulation methods to analyze the typical buffer space types in traditional Qiang dwellings in western China. It explores the impact of natural ventilation in these buffer spaces on indoor air age and wind speed comfort, proposing corresponding optimization strategies for natural ventilation. The aim of this research is to reduce the reliance on mechanical ventilation systems and enhance overall energy efficiency. The findings not only provide a theoretical basis for the construction of modern low-carbon Qiang dwellings but also offer practical guidance for improving existing living environments and advancing the construction of beautiful rural areas. The specific conclusions are as follows:

(1) This study centers on buffer spaces, selecting Dongmen Village as the research area. It mapped typical Qiang traditional dwellings and identified three main types of buffer spaces, namely courtyard space, eaves space, and overhanging eaves space. The study also explored the proportions and functions of these spaces. The findings

provide theoretical support for the preservation and sustainable development of Qiang traditional dwellings and promote the integration of traditional culture with modern lifestyles.

(2) This study demonstrates that the addition of buffer spaces significantly enhances the natural ventilation performance of traditional Qiang dwellings. Through simulation analysis, it was found that the average wind speed of dwelling 1, which includes a buffer space, increased by 90.9% in winter and 295.3% in summer compared to dwelling 2, which does not have a buffer space. Moreover, the wind speed in dwelling 1 meets green building standards in both seasons. Meanwhile, the maximum air age in dwelling 1 was approximately 79.7% lower in winter and 89.0% lower in summer compared to dwelling 2. These results indicate that buffer spaces play a critical role in improving ventilation efficiency and optimizing indoor air quality, effectively reducing reliance on mechanical ventilation systems.

(3) Optimizing the design parameters of buffer spaces, such as adjusting the aspect ratio of the courtyard space, can effectively enhance natural ventilation. The study found that changes in the aspect ratio of the skylight have limited effects on balancing indoor air age, and the impact on indoor wind speed is also not significant. However, when the aspect ratio of the courtyard space is approximately 1.3, the indoor air age reaches its optimal state, with the maximum air age decreasing to 274 s, a reduction of 75.32% from the previous maximum, significantly improving indoor air freshness. This finding provides theoretical support for further optimizing the natural ventilation design of Qiang traditional dwellings.

(4) Adjusting the height-to-width ratio of the courtyard space can also effectively improve natural ventilation performance. The study shows that changes in the skylight height-to-width ratio have a limited effect on balancing indoor air age, but increasing the skylight height positively influences indoor wind speed. When the aspect ratio of the courtyard space is approximately 0.9, the indoor air age reaches its optimal state, with the maximum air age being 170 s, a decrease of 84.69% from the previous maximum, significantly enhancing indoor air freshness. This finding provides strong theoretical evidence for further optimizing the natural ventilation design of Qiang traditional dwellings.

(5) Enhancing the functionality of buffer spaces is especially important in the cold regions of western Sichuan, as buffer spaces not only improve ventilation but also regulate the temperature difference between the indoors and outdoors. By optimizing the layout of buffer spaces, such as planting low-maintenance plants in the courtyard and eaves spaces, wind speed can be slowed down, providing shade in the summer while preventing excessive cold air infiltration in the winter. Additionally, the ground of buffer spaces can be made of permeable materials (such as stone), but considerations regarding thermal insulation and cold resistance are essential to reduce heat loss in winter.

The buffer space optimization strategies proposed in this study are not only of significant importance for the sustainable design of traditional Qiang dwellings but also provide valuable insights for the architectural optimization of buildings in other cold regions. These design solutions are highly applicable and offer considerable reference value, presenting new ideas for the future development of traditional green buildings. However, this study also has certain limitations. Firstly, due to the lack of field measurements for validation, the CFD simulation results are based on theoretical assumptions and simplifications, which may not fully reflect the complex wind patterns and climatic variations in real-world environments. Furthermore, the study primarily focuses on the design of typical dwellings in a specific region. While optimization strategies have strong potential for broader application, building types in different geographical locations and climate conditions may require further validation and adjustments. Future research could integrate field data from a broader range of regions to conduct a more extensive verification and optimization of buffer space designs under varying climatic contexts.

Author Contributions: Conceptualization, Y.Z. (Ying Zhao); methodology, Y.Z. (Ying Zhao); software, M.H. and K.L.; validation, Y.Z. (Ying Zhao) and K.L.; formal analysis, M.H.; investigation, K.L.; resources, Y.Z. (Ying Zhao); data curation, Y.Z. (Ying Zhao); writing—original draft preparation, Y.Z. (Ying Zhao) and M.H.; writing—review and editing, J.X. and Y.Z. (Ying Zhao); visualization, Y.Z. (Yifan Zhang); supervision, J.X. All authors have read and agreed to the published version of the manuscript.

Funding: This study was funded by the Chengdu Social Science Research Program (no. 2024CS081) and the Project of China-Portugal Joint Laboratory of Cultural Heritage Conservation Science (no. SDYY2403).

Data Availability Statement: The original contributions presented in this study are included in the article. Further inquiries can be directed to the corresponding author.

Acknowledgments: The Fundamental Research Funds for the Central Universities, Southwest Minzu University.

Conflicts of Interest: The authors declare no conflicts of interest. The funders had no role in the design of the study; in the collection, analyses, or interpretation of data; in the writing of the manuscript; or in the decision to publish the results.

References

1. Mostafavi, F.; Tahsildoost, M.; Zomorodian, Z. Energy efficiency and carbon emission in high-rise buildings: A review (2005–2020). *Build. Environ.* **2021**, *206*, 108329. [CrossRef]
2. Zhou, Z.; Xu, Y.; Ouyang, C.; Gui, M.; Jiang, W.; Zhou, C.; Ma, K.; Zhang, J.; Huang, J. Climate Adaptability Research of Vernacular Dwellings in Jiangxi Based on Numerical Simulation—An Example from Nanfeng County. *Buildings* **2024**, *14*, 2211. [CrossRef]
3. Han, J.; Zhang, L.; Li, Y. Spatiotemporal analysis of rural energy transition and upgrading in developing countries: The case of China. *Appl. Energy* **2022**, *307*, 118225. [CrossRef]
4. Qi, Y.; Stern, N.; He, J.-K.; Lu, J.-Q.; Liu, T.-L.; King, D.; Wu, T. The policy-driven peak and reduction of China's carbon emissions. *Adv. Clim. Change Res.* **2020**, *11*, 65–71. [CrossRef]
5. Zhang, F.; Shi, L.; Liu, S.; Shi, J.; Ma, Q.; Zhang, J. Climate Adaptability Based on Indoor Physical Environment of Traditional Dwelling in North Dong Areas, China. *Sustainability* **2022**, *14*, 850. [CrossRef]
6. National Development and Reform Commission. The Fourth Anniversary of the Major Declaration on Peak Carbon and Carbon Neutrality: The "Top Ten Actions for Carbon Peaking" Have Achieved Positive Results. Available online: https://www.ndrc.gov.cn/fggz/202409/t20240923_1393135.html (accessed on 23 September 2024).
7. Wang, Z.; Gadi, M. Optimal HVAC setpoints for energy efficiency and thermal comfort in Chinese residential buildings. *J. Build. Eng.* **2024**, *96*, 110286. [CrossRef]
8. Jarrahi, A.; Aflaki, A.; Khakpour, M.; Esfandiari, M. Enhancing indoor air quality: Harnessing architectural elements, natural ventilation and passive design strategies for effective pollution reduction—A comprehensive review. *Sci. Total Environ.* **2024**, *954*, 176631. [CrossRef]
9. Kajjoba, D.; Kasedde, H.; Kirabira, J.B.; Wesonga, R.; Mugwanya, R.; Lwanyaga, J.D.; Olupot, P.W. Impact of natural ventilation and outdoor environment on indoor air quality and occupant health in low-income tropical housing. *Energy Rep.* **2024**, *12*, 4184–4194. [CrossRef]
10. Zhao, M.; Mehra, S.-R.; Künzel, H.M. Energy-saving potential of deeply retrofitting building enclosures of traditional courtyard houses—A case study in the Chinese Hot-Summer-Cold-Winter zone. *Build. Environ.* **2022**, *217*, 109106. [CrossRef]
11. Asadi, S.; Fakhari, M.; Sendi, M. A study on the thermal behavior of traditional residential buildings: Rasoulian house case study. *J. Build. Eng.* **2016**, *7*, 334–342. [CrossRef]
12. Alabsi, A.A.N.; Song, D.X.; Garfield, W.H. Sustainable Adaptation Climate of Traditional Buildings Technologies in the Hot Dry Regions. *Procedia Eng.* **2016**, *169*, 150–157. [CrossRef]

13. Xie, L.; Li, Z.; Li, J.; Yang, G.; Jiang, J.; Liu, Z.; Tong, S. The Impact of Traditional Raw Earth Dwellings' Envelope Retrofitting on Energy Saving: A Case Study from Zhushan Village, in West of Hunan, China. *Atmosphere* **2022**, *13*, 1537. [CrossRef]
14. Xiao, Y. Climate Adaptability Study of Traditional Qiang Residences in Sichuan. Master's Thesis, Southwest University of Science and Technology, Mianyang, China, 2017.
15. Mo, Y.; Guo, P.; Lu, J.; Wu, Z.; Yang, B.; Jiang, Z.; Chen, F.; Jiang, W. Study on the Influence of Window Openings on Seismic Performance of Stone Walls of Tibetan and Qiang Dwellings. *Buildings* **2024**, *14*, 3829. [CrossRef]
16. Lin, L.; Gui, Y. Spatial characteristics and ecological wisdom of Amdo Tibetan traditional dwellings in western Sichuan. *J. Asian Archit. Build. Eng.* **2024**, 1–16. [CrossRef]
17. Ding, D. Solar passive features in vernacular Gyalrong Tibetan houses: A quantitative investigation. *Ain Shams Eng. J.* **2022**, *13*, 101525. [CrossRef]
18. Bhattacharjee, S.; Lidelöw, S.; Schade, J. Performance Evaluation of a Passive House in Sub-arctic Climate. In *Sustainable Buildings in Cold Climates*; Springer International Publishing: Berlin/Heidelberg, Germany, 2019; pp. 145–157.
19. Deng, J.-Y.; Xia, Y.; Lao, H.; Ye, Y.; Wang, Z.; Jiang, H. Natural ventilation potential of teaching building complexes with different block shapes and layout patterns. *J. Build. Eng.* **2024**, *96*, 110420. [CrossRef]
20. Mateus, R.; Pinto, A.; Pereira, J.M.C. Natural ventilation in large spaces: CFD simplified model validated with full-scale experimental data of Roman Baths. *Build. Environ.* **2024**, *266*, 112077. [CrossRef]
21. Toe, D.H.C.; Kubota, T. Comparative assessment of vernacular passive cooling techniques for improving indoor thermal comfort of modern terraced houses in hot–humid climate of Malaysia. *Sol. Energy* **2015**, *114*, 229–258. [CrossRef]
22. Zhong, W.; Xiao, W.; Zhang, T. Numerical Investigations on Natural Ventilation in Atria of China's Southern Yangtze Vernacular Dwellings. *Sustain. Cities Soc.* **2023**, *89*, 104341. [CrossRef]
23. Xu, J.; Xu, B.; Yang, W.; Zhu, C.; Li, Q. Comparative research on improving the thermal environment of traditional dwellings with attic space in Southern Shaanxi region, China. *Case Stud. Therm. Eng.* **2024**, *55*, 104092. [CrossRef]
24. Xu, J.; Meng, W.; Yang, W.; Lu, Z.; Xie, G.; Wang, X. Impact of transition spaces on the indoor thermal environment: A case study of traditional dwellings in Southern Shaanxi, China. *J. Build. Eng.* **2024**, *95*, 110283. [CrossRef]
25. Chen, H.; Tan, Z.; Sun, P. Research on Wind Environment Simulation in Five Types of "Gray Spaces" in Traditional Jiangnan Gardens, China. *Sustainability* **2024**, *16*, 7765. [CrossRef]
26. China Meteorological Data Service Centre. Available online: https://data.cma.cn/dataService/cdcindex/datacode/A.0012.0001_510/show_value/normal.html (accessed on 24 February 2025).
27. Liu, R.; Wang, Y.; Zhang, Y.; Peng, Z.; Chen, H.; Li, X.; Li, H.; Li, W. Analysis of the city-scale wind environment and detection of ventilation corridors in high-density metropolitan areas based on CFD method. *Urban Clim.* **2025**, *59*, 102274. [CrossRef]
28. Li, M.; Jin, Y.; Guo, J. Dynamic characteristics and adaptive design methods of enclosed courtyard: A case study of a single-story courtyard dwelling in China. *Build. Environ.* **2022**, *223*, 109445. [CrossRef]
29. Li, Y.; Wang, C.; Zhao, Z.; An, J.; Shi, Y.; Hao, S.; Ren, Z.; Zhang, J. Study on Thermal Environment and Energy Consumption of Typical Residential Courtyards in Beijing's Old City. *Energies* **2025**, *18*, 626. [CrossRef]
30. Ge, H.; Wang, R.; Baril, D. Field measurements of hygrothermal performance of attics in extreme cold climates. *Build. Environ.* **2018**, *134*, 114–130. [CrossRef]
31. Southwest Nationality Research Institute of Southwest University for Nationalities. *A Social Survey on Tibetan and Qiang Nationality in Northwest Sichuan Province*; Nationality Publishing House: Beijing, China, 2008; pp. 21–23.
32. Zhang, M.; He, Y.; Huang, L.; Xiong, R.; Zhang, Y. Built-In Environmental Construction Mechanism and Sustainable Renewal Strategies of Traditional Qiang Dwellings in Western China. *Sustainability* **2024**, *16*, 7122. [CrossRef]
33. Kravchenko, I.; Kosonen, R.; Jokisalo, J.; Kilpeläinen, S. Performance of Modern Passive Stack Ventilation in a Retrofitted Nordic Apartment Building. *Buildings* **2022**, *12*, 96. [CrossRef]
34. Jiang, Z.; Gao, W.; Yao, W. Research on the wind environment in an enclosed courtyard: Effect of the opening position, size and wind angle. *Urban Clim.* **2023**, *52*, 101737. [CrossRef]
35. Wang, Z.; Fan, Z.; Liu, J. Thermal environment optimisations of cave dwellings for the elderly in heating season among cold regions of China. *Build. Environ.* **2023**, *244*, 110656. [CrossRef]
36. Shaeri, J.; Mahdavinejad, M.; Vakilinejad, R.; Bazazzadeh, H.; Monfared, M. Effects of sea-breeze natural ventilation on thermal comfort in low-rise buildings with diverse atrium roof shapes in BWh regions. *Case Stud. Therm. Eng.* **2023**, *41*, 102638. [CrossRef]
37. Xie, K.; Xiong, R.; Bai, Y.; Zhang, M.; Zhang, Y.; Han, W. Traditional Architectural Heritage Conservation and Green Renovation with Eco Materials: Design Strategy and Field Practice in Cultural Tibetan Town. *Sustainability* **2024**, *16*, 6834. [CrossRef]
38. Li, B.; Zhao, X.; Luo, Y. Design Strategies for Green Retrofitting of a University Student Dormitory in the Hot-humid and Less-windy Climate. In Proceedings of the 2022 International Conference on Green Building, Civil Engineering and Smart City, Guilin, China, 25–27 July 2022; Springer: Singapore, 2023; pp. 22–31.

39. Lee, Z. Courtyard of Liujia Village Traditional Residence Based on Swire Software Natural Ventilation. Master's Thesis, Hunan University of Science and Technology, Xiangtan, China, 2022.
40. Ji, P. Facade Renovation Design of the Main Building in Nanjing University of Science and Technology. Master's Thesis, Nanjing University, Nanjing, China, 2016.
41. Huang, Z.; Li, Q.; He, Y.; Ding, X.; Dong, Y.; Gao, W. Coupled heat and mass transfer analysis for indoor air quality and thermal comfort in naturally ventilated offices. *Appl. Therm. Eng.* **2025**, *269*, 126019. [CrossRef]
42. *GB 50009-2012*; Load Code for the Design of Building Structures; Ministry of Housing and Urban-Rural Development of the People's Republic of China. China Architecture & Building Press: Beijing, China, 2012.
43. Celik, I.B.; Ghia, U.; Roache, P.J.; Freitas, C.J. Procedure for Estimation and Reporting of Uncertainty Due to Discretization in CFD Applications. *J. Fluids Eng.* **2008**, *130*, 078001. [CrossRef]
44. *GB/T 50378-2019*; Assessment Standard·for Green Building; Ministry of Housing and Urban-Rural Development People's Republic of China. China Architecture & Building Press: Beijing, China, 2019.
45. Soligo, M.J.; Irwin, P.A.; Williams, C.J.; Schuyler, G.D. A comprehensive assessment of pedestrian comfort including thermal effects. *J. Wind Eng. Ind. Aerodyn.* **1998**, *77–78*, 753–766. [CrossRef]
46. Gao, C.F.; Lee, W.L. Evaluating the influence of openings configuration on natural ventilation performance of residential units in Hong Kong. *Build. Environ.* **2011**, *46*, 961–969. [CrossRef]
47. Xu, L. Research on the Shading Effects from Mountainous Terrain on Solar Radiation of Towns in Western Sichuan. Ph.D. Thesis, Sichuan University, Chengdu, China, 2021.
48. Kai, W.; Hong, L.; Yan, Y. Wind field law of different terrain patterns based on computational fluid dynamics simulation and application in wind sensing planning. *Acta Ecol. Sin.* **2021**, *41*, 3499–3511.
49. *ANSI/ASHRAE Standard 55-2020*; Thermal Environmental Conditions for Human Occupancy. American Society of Heating, Refrigerating and Air-Conditioning Engineers: Atlanta, GA, USA, 2020.
50. Azimi, Z.; Shafaat, A. Proposing design strategies for contemporary courtyards based on thermal comfort in cold and semi-arid climate zones. *Build. Environ.* **2024**, *266*, 112150. [CrossRef]
51. Rodríguez-Algeciras, J.; Tablada, A.; Chaos-Yeras, M.; De la Paz, G.; Matzarakis, A. Influence of aspect ratio and orientation on large courtyard thermal conditions in the historical centre of Camagüey-Cuba. *Renew. Energy* **2018**, *125*, 840–856. [CrossRef]
52. Zhu, J.; Feng, J.; Lu, J.; Chen, Y.; Li, W.; Lian, P.; Zhao, X. A review of the influence of courtyard geometry and orientation on microclimate. *Build. Environ.* **2023**, *236*, 110269. [CrossRef]
53. Sun, Q.; Fan, Z.; Bai, L. Influence of space properties of enclosed patio on thermal performance in hot-humid areas of China. *Ain Shams Eng. J.* **2024**, *15*, 102370. [CrossRef]
54. Lin, L.; Gui, Y. The key factors of microclimate in courtyard-style architectural ensembles spaces—Taking the Yi minority Tusi manor as an example. *Build. Environ.* **2025**, *270*, 112553. [CrossRef]
55. Zhang, G. Regional Adaptability of Tibetan Dwelling in East Khams, Ganz. Master's Thesis, Xi'an University of Architecture and Technology, Xi'an, China, 2019.
56. Schauerte, T. Wooden house construction in Scandinavia—A model for Europe. In Proceedings of the Internationales Holzbau-Forum (IHF 2010): Aus der Praxis–Für die Praxis, Doha, Qatar, 17–21 May 2010; Eberl Print GmbH: CH-Biel, Swizerland, 2010; pp. 1–10.

buildings

MDPI

Article

Aging Adaptation Transition of Health Care Buildings for Accessibility Optimization for the Elderly

Chang Yi, Wenyang Han, Yiheng Liu, Yijie Lin and Yicong Qi *

School of Architecture, Southwest Minzu University, Chengdu 610225, China;
240813002005@stu.swun.edu.cn (C.Y.); 230813002016@stu.swun.edu.cn (W.H.);
240813002025@stu.swun.edu.cn (Y.L.); 202031707080@stu.swun.edu.cn (Y.L.)
* Correspondence: 80300254@swun.edu.cn

Abstract: As society develops, the aging population issue is becoming more serious and gaining global attention. Meanwhile, the building industry worldwide is focusing on making buildings more convenient for the elderly. This study focuses on a health care building, analyzing its aging-friendly design. It examines issues related to walking situations and activity spaces and proposes optimization strategies based on relevant codes and actual needs. Through optimization and transformation, the walking distance to the nearest exit for the elderly in the building has been reduced by 36.8%, the walking distance to activity space for the elderly has been reduced by 8.4%, and the average public activity space of each elderly person has been increased by about 23.5%. In addition, the handrails of the accessible stairway have been changed to double handrails, which is more suitable for the different needs of the elderly, and the space of the wheelchair-accessible elevator has been expanded, which is more convenient for the elderly's activities in elevators. This paper explores the feasibility and design direction of the aging-friendly architecture, and it aims to provide a valuable reference for the renovation of aging buildings.

Keywords: health care buildings; building renovation; elderly occupants; aging adaptation transition; inclusive design; walking situation; activity space

1. Introduction

1.1. Backgrounds

Nowadays, China's society and economy are developing rapidly, but at the same time, the problem of an aging society is becoming more and more serious. According to the World Health Organization, China is a country with one of the fastest rates of population aging in the world, and it is forecasted that the population of people over 60 years old will reach 28% by 2040 [1]. One of the problems caused by the aging problem is the increasing burden of chronic diseases. Currently, in China, people over 40 years old suffer from at least one non-communicable disease, and the number of people suffering from non-communicable diseases will continue to increase in the future; among them, about 50% are over 65 years old [2]. According to the China Research Center on Aging, in 2013, 100 million of China's 202 million elderly people suffered from at least one chronic non-communicable disease [3], and many of them even suffered from multiple chronic diseases at the same time. The proportion of deaths due to non-communicable diseases gradually increases with age, and about 80% of elderly people aged 60 years and above will die from chronic non-communicable diseases. The increase in the number of elderly people and the number

 https://doi.org/10.3390/buildings15030379

of people suffering from chronic diseases will also lead to a significant increase in the number of care-dependent elderly people in China, with the dependency rate of China's total population projected to rise from 5.6% to 6% from 2010 to 2050 [4], and about 60% of elderly people will seek institutionalized care when they no longer consider themselves suitable for aging in their own homes [5]. All these points have led to an increasing demand for elderly care institutions in China.

In the face of these problems, the construction of social infrastructure is faced with great challenges, and the government has begun to pay more attention to building facilities that meet the needs of the elderly and enable them to enjoy equal public services. In the building industry, many kinds of architectural design have also begun to pay more attention to aging-friendly design. As the process of urbanization is accelerating, many elderly people are moving to cities, and the pressure of family support is high; in response, many health care buildings for the elderly have been developed. However, as living standards rise, the requirements of elderly people for health care buildings have also gradually increased. According to Maslow's hierarchy of needs, human needs are divided into five categories, including physiological needs, security needs, belonging and love needs, esteem needs, and self-actualization needs; similarly, for the elderly, there is also the need to think about spiritual and emotional needs [6]. At present, the design of many health care buildings is only considered to comply with the national codes, neglecting the higher needs of the elderly. In addition, according to the study, in 2010, 19% of elderly people over 60 years old had difficulties in daily activities [7], while 26% of elderly women and 11.3% of elderly men were unable to walk 1 km or use stairs [8]. This leads to a very high susceptibility to falls, and falls are officially the main cause of injuries among the elderly, so paying attention to the accessibility of the elderly is also very important. All in all, there is a pressing need to optimize and transform these buildings to better serve the elderly population.

1.2. Literature Review

Population aging has become an important global issue, and if measures are not taken promptly, the aging problem will further lead to other social problems, such as labor shortages, GDP decline, and social care burdens. According to the guidelines of the World Health Organization, there are eight aging-friendly domains: outdoor space and architecture, transportation, housing, social participation, respect and social inclusion, civic engagement and employment, communication and information, and community support and health services [9]. As the elderly age, their physical functions decline and their movement is restricted; buildings become important for their activities, so it is very urgent to optimize and retrofit the interior of buildings for the elderly. Nowadays, because of the various practical needs of many families, the elderly will choose to live in health care buildings, and the services provided by health care buildings can be divided into 24 h care and day care, and health care buildings not only take care of the physical health of the elderly but also take care of the mental health of the elderly. If the elderly experience loneliness, they will be more likely to experience psychological problems [10], so the public activity space renovation and walking distance optimization of elderly health care buildings is very necessary, and the optimization of the renovation can provide the elderly with a more convenient opportunity to find companions.

Fallah et al. [11] highlight the increased demand for age- and disability-friendly residential design as the world's population ages. An inclusive design approach allows designers to design products and services that meet the needs of a wider group of users, regardless of age and ability. Liu et al. [12] discuss the problem of currently designed buildings being unsuitable for older people under the concept of age-friendliness, and they present the influencing factors, principles, and problems and solutions for the spatial design of modern buildings for the elderly. Han et al. [9], using a scient metric approach, proposed a theoretical framework that considers the active role of older people in community spaces, as well as the physical and social aspects of community spaces, to provide guidance for the development of age-friendly community policies. Furthermore, in another article, Han et al. [13] explored how the concept of age-friendly cities and communities (AFCC) can be integrated into sustainable urban development (SUD), emphasizing the importance of policy integration and proposing a strategy that uses the silver-haired market as an entry point. A study by Han et al. [14] guides "active aging" community regeneration in modular residential areas based on community spatial characteristics and neighborhood interactions and suggests using existing resources to improve spatial organization and community communication. This article suggests using existing resources, improving spatial organization and community communication, and targeting the increase and optimization of the layout of elderly care facilities. Tao et al. [15] studied five communities in Singapore known for their "aging-in-place" strategies and found that adequate physical facilities and connectivity to the city were critical to older people's satisfaction with the community. According to Ayala et al. [16], the principles of age-friendly home design include a holistic definition of home and its impact on older adults, emphasizing their mobility and physical needs in the home environment. Liu [17] uses computer data analysis software to data mine user behavior in interior spaces to guide the design of more rational spaces. A replicable methodology is proposed for all remodeling processes. Fang et al. [18] further proposed the concept of the co-creation of inclusive spaces and places, emphasizing the importance of intergenerational and age-friendly living ecosystems to enhance public health planning. The study by Luciano et al. [19] introduced a framework for assessing the aging of housing by detecting and identifying physical and non-physical features of home environments through qualitative and quantitative metrics to support older adults' independent living at home. Tao et al. [20] assessed the thermal, light, and acoustic environments of Care and Attention (C&A) homes in Hong Kong through field measurements and occupant surveys. The study found that older adults preferred warmer environments, while actual measurements showed lower indoor temperatures and higher humidity. A study by Jiao et al. [21] explored the thermal environment characteristics of transition spaces in elderly care buildings in Shanghai and their effects on the thermal adaptation of older adults through field surveys and physical measurements. It was found that the temperature difference between the transition space and the indoor environment had a significant effect on the indoor thermal satisfaction of the elderly, with the highest thermal satisfaction at 6 °C in winter and 2 °C in summer. Ideal transition space designs should utilize semi-open exterior porches to help older adults adapt to temperature changes through activities such as sunbathing and exercise.

These studies indicate that age-friendly buildings are not only just technical issues but also involve multiple social, economic, and cultural dimensions, requiring interdisciplinary collaboration and policy support to ensure that the elderly can maintain independent living in a safe and comfortable environment. One of the most important aspects of enabling older people to live independently is to ensure that they are safe and have enough recreational activities during their daily activities, so it is very important to study activity spaces and walking of the elderly.

1.3. Objects and Focus

In order to better study the area of public activity space and walking for the elderly and optimize elderly health care buildings, this paper takes a designed elderly health care building as the object of study and reveals how to enhance the comfort and functionality of senior living environments through careful study of these key elements and then improve their quality of life and social interaction. The research is structured into the following sections: (1) Analyze the existing health care building and clarify the nature and type of the health care building as well as its existing condition. (2) Analyze the relevant building codes to determine the requirements for the access and public activity spaces for the elderly, ensuring that the walking distance, safety facilities, and public activities space in the building meet the requirements of codes, and then conduct further optimization on this basis. (3) Propose feasible optimization strategies to improve the accessibility and safety of the health care building and aim to provide a valuable reference for the aging-friendly renovation of other aging buildings.

2. Materials and Methods

This section first introduces the subject of this paper, a health care building for the elderly located in Sanya, Hainan, China, so that readers can understand the internal situation and geographical location of the subject. Then, it analyzes the relevant codes so that readers can understand the minimum requirements for public activity spaces, walking distances, and accessibility of stairs and elevators in health care buildings for the elderly. These minimum standards are compared with the existing design to find out the problems of the existing buildings and propose optimization strategies according to these problems (Figure 1).

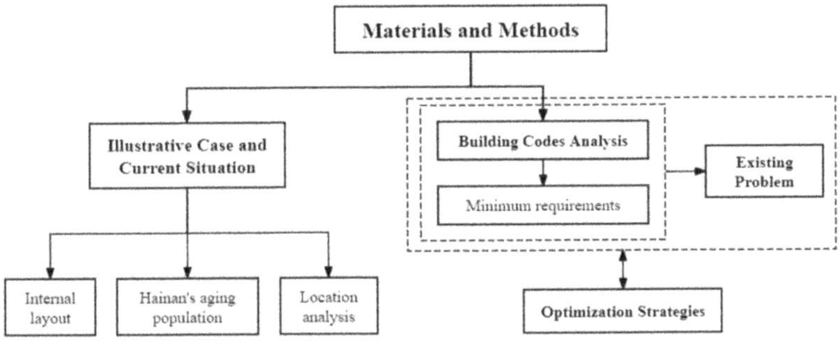

Figure 1. Materials and methods.

2.1. Illustrative Case and Current Situation

A health care building for the elderly is the subject (Figure 2). This health care building is one of the newly designed community centers. The design starts with solving the practical needs of residents and aims to design a neighborhood community center that considers all age groups and covers multiple functions.

Due to the limited research resources, it is not possible to cover multiple cases, so this study chooses to conduct an in-depth excavation study on a single case. This study chose this recreational building for analysis mainly based on its data availability, with complete design information, and the fact that the building faces typical problems such as insufficient public space and a long walking distance, which makes it a good sample for studying optimization strategies.

According to the research, it is known that the health care building for the elderly has a very important and positive effect on the health of the elderly [22]; in today's society, the aging problem is very serious, and it is very important to set up a health care building for the elderly in the community center. This building can provide simple consultation for the elderly's daily health issues as well as a place for routine health check-ups. In addition, this building offers 24 h care for the elderly, with a total of 28 beds, of which there are 16 nursing care beds and 12 non-nursing care beds (Figure 3).

Figure 2. Research target representation.

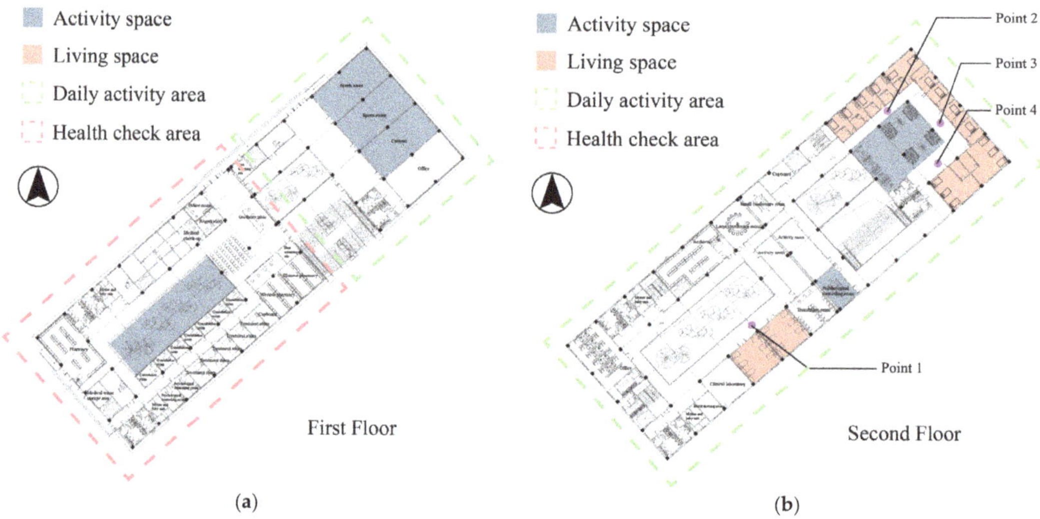

Figure 3. (**a**) First-floor plan of the original design. (**b**) Second-floor plan of the original design.

According to the data of the Seventh National Population Census Bulletin of Sanya City, among the population of Sanya City, aged 0–14 years old are 17.67%; aged 15–59 years old are 71.92%; and aged 60 years old and above are 10.42%, of which aged 65 years old and above are 6.95%, and the proportion of the population aged 60 years old and above increased by 3.41% compared with that in the Sixth National Population Census in 2010, with the proportion of the population aged 65 years old and above increasing by 2.25%

(Figure 4). Compared with the Sixth National Population Census in 2010, the proportion of people aged 60 and above increased by 3.41%, of which the proportion of people aged 65 and above increased by 2.25% [23]. Sanya, as a well-known tourist city, is especially popular in winter, when many elderly people from the cold northern regions come to Sanya for short stays. Therefore, the design of community centers also needs to give full consideration to the needs of the elderly in order to adapt to the needs of all-age living. Therefore, it is necessary to build new health care buildings for the elderly in community centers.

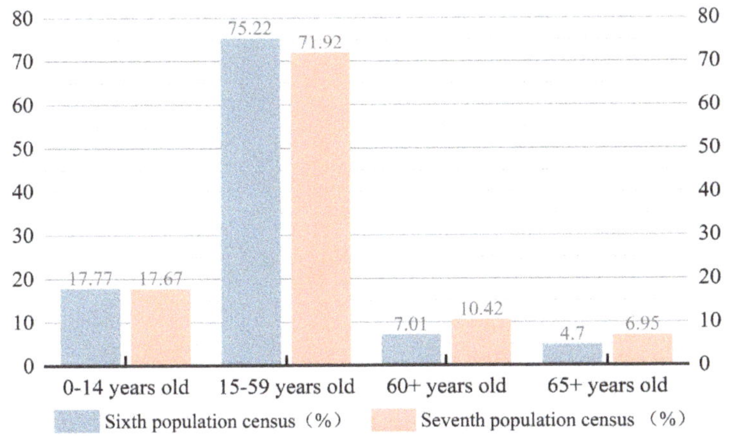

Figure 4. Proportion of population by age in Sanya City. (Data source: The seventh national census).

Elderly people are very likely to have their conditions aggravated because they are unable to seek medical treatment in time; the newly built health care building can avoid this situation, which is usually equipped with professional nursing staff and advanced medical equipment to provide high-quality geriatric care services, so that elderly people can receive timely and comprehensive medical support under the care of professional staff. Furthermore, health care buildings are not only places for the elderly to live and receive treatment but also serve as community activity centers to enhance the interactions between the elderly and the outside world and their sense of social participation. This socialized lifestyle can effectively improve the overall health of the elderly population.

The project is located in the Hailuo area of Jiyang District, Sanya City, Hainan Province. Jiyang District is located in the central and eastern part of Sanya City, it is the seat of the party and governmental organs of the Sanya Municipal Party Committee and City Government, and the area has been recognized by the Ministry of Culture and Tourism of the People's Republic of China as the first national demonstration area of regional tourism. With the development of the city, the Hailuo area is planned to be an international business headquarters park and a super-exclusive supporting service community with integrated industry and city, blending mountains and water and forming a functional structure of "one axis and one belt, one chain and multiple corridors, and three districts and multiple clusters" (Figure 5). Among them, "one axis" refers to the central business green axis; "one belt" refers to the waterfront cityscape belt of Linchun River; "one chain" refers to the green lifestyle service chain; "Multiple corridors" refers to the formation of multiple green corridor spaces relying on the Linchun River, the central business green axis, and the water catchment corridors; "three districts" refers to the Hailo International Business Service area, the International Talent Residential area, and the All-age Friendly Residential area; and "multiple clusters" refers to the central business green axis [24].

Society keeps developing, and the city needs to be renewed. In order to promote the development of the city and improve the quality of life of the residents, the Sanya government has decided to renovate and upgrade the Hailo area. Nowadays, part of the land grant and demolition work has been completed in this area, and the related municipal infrastructure and supporting facilities construction projects are also being planned and implemented.

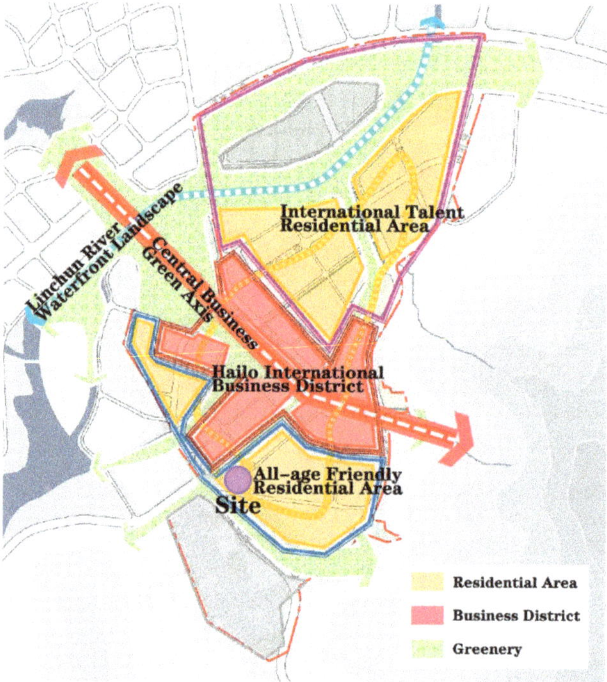

Figure 5. Hailo area planning (Data source: Detailed Control Planning for Hailuo area in Sanya City Center (Revision)).

The project site (Figure 6) is located at the intersection of the planned Xueyuan Road and Wangshan Road, with Hailuo Ridge to the southeast and the Linchun River winding through the northwest. These two planned roads are directly connected to Sanya's main arterial roads, Yingbin Avenue and Fenghuang Road. The site enjoys a prime geographical location with convenient transportation and is surrounded by a beautiful natural ecological environment. According to the land use planning layout, the surrounding area of the site is diverse in terms of land functions. To the northwest, there is commercial land with high foot traffic; to the east, there is residential land with target groups served by wellness centers; to the south, there are urban green spaces and financial zones, which can enhance greenery. This environment plays a positive role in improving the quality of life for residents and attracting people. Additionally, in the future, the surrounding area will feature well-developed public service facilities, providing a convenient and comfortable living environment for the residents of the district.

2.2. Building Codes Analysis

The first step is to review the code and understand the requirements related to evacuation distances, safety assistance facilities, and activity spaces of health care buildings for the elderly. The building codes relevant to this study include the "Standards for design of care facilities for the aged" JGJ450-2018 [25], the "General code for fire protection of buildings and constructions" GB 55037-2022 [26] and the "Codes for accessibility design" GB 50763-2012 [27] (Table 1).

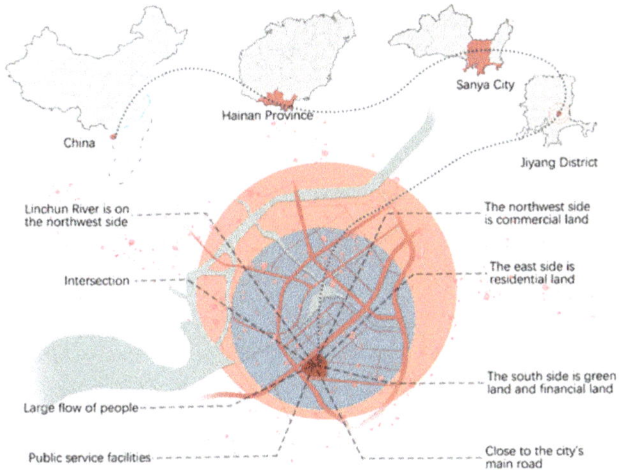

Figure 6. Location analysis.

Table 1. Relevant regulations in the codes.

Codes	Regulations		
"Standards for design of care facilities for the aged"	There should be entertainment and fitness rooms in the senior building	Seats in dining room: nursing care: $\geq$40% of beds; 4 m^2/seats non-nursing care: $\geq$70% of beds; $\geq$2.5 m^2/seats	Area of entertainment and fitness rooms $\geq$ 2 m^2/bed (person)
"General code for fire protection of buildings and constructions"	"The evacuation distance from any point in a room to the room's evacuation door should not exceed the maximum allowed evacuation distance from the evacuation door of rooms located on either side or end of a pocket corridor to the nearest safe exit".		
"Codes for accessibility design"	The size of cabin: 1500 mm (width) × 2700 mm (length) Handrails 850~900 mm high Mirrors: 900 mm high to the top	Single-layer handrails: 850~900 mm Double-layer handrails: upper 850~900 mm lower 650~700 mm	

Regulation 5.1.5 of the "Standards for design of care facilities for the aged" stipulates the following: "The entertainment and fitness rooms of 24 h care facilities for the aged shall be set up to meet the corresponding activity needs of the elderly, and may have rooms or spaces for reading, internet, chess and cards, calligraphy and painting, classrooms, fitness, and multifunctional activities". And regulation 5.3.1 stipulates the following: "The total usable area of the entertainment and fitness rooms in care facilities for the aged shall not be less than 2.00 m^2/bed (person)". Additionally, the number of seats in the dining room of 24 h care facilities for the aged shall be configured according to the number of people served; the number of seats in the dining room of care units for nursing care beds shall be configured according to not less than 40% of the number of beds served, and the usable area of each seat shall not be less than 4.0 m^2; the number of seats in the dining room of

non-nursing care beds shall be configured according to not less than 70% of the number of beds served, and the usable area of each seat shall not be less than 2.50 m^2 [25].

Regarding evacuation distance in the building for the elderly, the "General code for fire protection of buildings and constructions" stipulates the following: "The evacuation distance from any point in a room to the room's evacuation door should not exceed the maximum allowed evacuation distance from the evacuation door of rooms located on either side or end of a pocket corridor to the nearest safe exit". Furthermore, the fire resistance rating of elderly care buildings should not be lower than Class 3. In the new fire code, there is no clear data on the evacuation distance, so the evacuation distance is still calculated and discussed according to the original fire code. In the original code, if an automatic sprinkler system is installed, the evacuation distance can be increased by 25%. And the maximum allowed evacuation distance from the room's evacuation doors located on both sides or ends of the pocket corridor to the nearest safe exit in an elderly building with a fire-resistance rating of Class 3 is 15 m. Therefore, the maximum evacuation distance of elderly care buildings is 18.75 m [26].

We need to do more than think about the passing distance of the elderly; we also need to think about the safety of the elderly when they are walking. In the "Codes for accessibility design", there also are stipulations about accessible stairways and wheelchair-accessible elevators. The height of the handrail of the accessible stairway should be 850~900 mm, if it adopts a single-layer handrail; and if it adopts a double-layer handrail, the height of the upper handrail should be 850~900 mm, and the height of the lower handrail should be 650~700 mm. About the wheelchair-accessible elevator, the elevator should be specially chosen for hospital beds, and the size of its cabin is generally 1500 mm (width) $\times$ 2700 mm (length), and the cabin's three walls should be set up with handrails 850~900 mm high, on the front side of the cabin, a mirror or mirrored material should be installed from 900mm to the top of the cabin, which is convenient for the elderly in wheelchairs [27].

2.3. *Existing Problem*

As the physical function of the elderly declines with age, the speed, frequency, stride length, etc., with respect to walking, undergo changes; body regulation systems can show signs of aging or even lesions, and while walking, elderly people will appear to lift their feet slowly, dragging them on the ground, and they will show signs of stride shortening and so on. Studies have shown that the average step width for older men is 0.105 $\pm$ 0.016 m, for younger men is 0.122 $\pm$ 0.026 m, for older women is 0.090 $\pm$ 0.023 m, and for younger women is 0.103 $\pm$ 0.021 m [28]. In order to facilitate the calculation, we use the average data to calculate the number of walking steps; the average step width of the elderly is calculated as 0.098 m, and the step width of young people is calculated as 0.106 m.

In the original design, according to measurement, the number of service beds is 28 people, the total area of activity space is 842.18 m^2, and the average activity space is 30.08 m^2/bed(person). According to codes, the minimum area of activity space in this building is calculated as shown in Equations (1)–(3):

$$S_{min} = S_{entertainment} + S_{canteen} \tag{1}$$

$$S_{entertainment} = P \cdot c_1 \tag{2}$$

$$S_{canteen} = P_{nursing} \cdot \alpha \cdot c_2 + P_{non\text{-}nursing} \cdot \beta \cdot c_3 \tag{3}$$

where S_{min} is the minimum area of activity space, m^2; $S_{entertainmnet}$ is the minimum area of the entertainment and fitness rooms, m^2; $S_{canteen}$ is the minimum area of the dining room, m^2; P is the total number of people (person); $P_{nursing}$ is the number of nursing care beds (person); $P_{non\text{-}nursing}$ is the number of non-nursing care beds(person); the value of c_1 is 2 m^2/bed (person); the value of c_2 is 4 m^2/bed (person); the value of c_3 is 2.5 m^2/bed (person); the value of α is 40%; the value of β is 70%. After calculations are completed, the minimum area of activity space in this building is 102.6 m^2; this building is conformed to coeds, but during design, there was little consideration for the spiritual and entertainment needs of the elderly. The activity spaces are often large and empty without thinking about the diversity of activities, and only a few activity rooms are set up, which is far from meeting the needs of the elderly in their daily activities. Additionally, these rooms are far away from the living space of the elderly, resulting in excessive walking distances, and as the physical strength of the elderly declines, it is necessary to further shorten the elderly's walking distance in the building in order to better meet their needs. We took four representative points (Figure 3) in the living space section for measurement (see Table 2), and the average distance from the living space to the activity space is calculated as shown in Equations (4)–(6):

$$\overline{d_i} = \frac{d_1 + d_2 + d_3 + d_4}{4} \tag{4}$$

$$\overline{D} = \sum_{i=1}^{4} \overline{d_i} = \overline{d_1} + \overline{d_2} + \overline{d_3} + \overline{d_4} \tag{5}$$

$$F = \frac{\overline{D}}{x} \tag{6}$$

where $\overline{d_i}$ is the average distance from every point to one of the activity spaces, m; d_i is the walking distance from four points to one of the activity spaces, m; $\overline{D}$ is the average distance between the selected four points and every activity space, m; F is the average number of steps; x is the average step width, m, after the calculation can be obtained, the average distance from the living space to the activity space is 42.84 m, and the average number of steps for young people is 404, but the average number of steps for an elderly person is 437, which is 8.2% more than a young person. And the walking distance from the living space to the nearest safety exit is 28.25 m, and the average number of steps for a young person is 267, but the average number of steps for an elderly person is 288, which is 7.9% more than a young person, and the distance to the nearest exit has far exceeded the requirement of codes, and it is urgent to correct this. In addition, regarding the safety of walking, the handrail height of the original design is set unreasonably; the height of the height is 1100 mm, which is too high for the elderly, and there is only a single handrail; and the elevator design did not take into account the use of hospital beds, and the size of the cabin is relatively small. Furthermore, the placement and quantity of stairs and elevators were not adequately considered, resulting in some rooms at a considerable distance from both the elevator and staircase.

Table 2. Walking distance in the original health care building for the elderly.

	Point 1	Point 2	Point 3	Point 4	Average
To the nearest exit	28.82 m	17.82 m	33.75 m	32.62 m	28.25 m
To activity room on the first floor	48.69 m	37.69 m	53.63 m	52.49 m	48.12 m
To outdoor activity room	59.47 m	48.47 m	64.40 m	63.27 m	58.90 m
To communication space	58.33 m	6.12 m	5.88 m	7.81 m	19.54 m
To psychological counseling room	47.53 m	47.18 m	50.76 m	33.74 m	44.80 m

2.4. Optimization Strategies

The current situation and the problems of this health care center for the aged are integrated, and the optimization idea mainly starts from the perspective of improving the walking situation for the elderly and enhancing the public activity space (Figure 7). Through the analysis of the current situation and the study of the problems, it can be known that this health care building for the elderly has excessive walking distances, inadequate safety assistance facilities, and limited public activity spaces that lack a comprehensive design, which need to be optimized and renovated. Two optimization strategies are proposed to address the above problems: (1) Increasing activity spaces. Expanding the area of the public activity space, breaking up the large and empty activity space, adding different types of activity spaces, and integrating the activity space into other functional spaces to reduce the walking distance from the living room to the activity space. (2) Improving accessibility and safety. Increasing the number of accessible stairways and wheelchair-accessible elevators and shortening the average distance from each room to the safety exit. In addition, we should change the handrails of the staircases to a double-layer design and lower the height of the handrails to adapt them to the various needs of the elderly. At the same time, we should enlarge the size of the cabin to ensure that the wheelchair and the hospital bed can have sufficient space and add appropriate handrails on three sides of the elevator, along with a mirror on the front side.

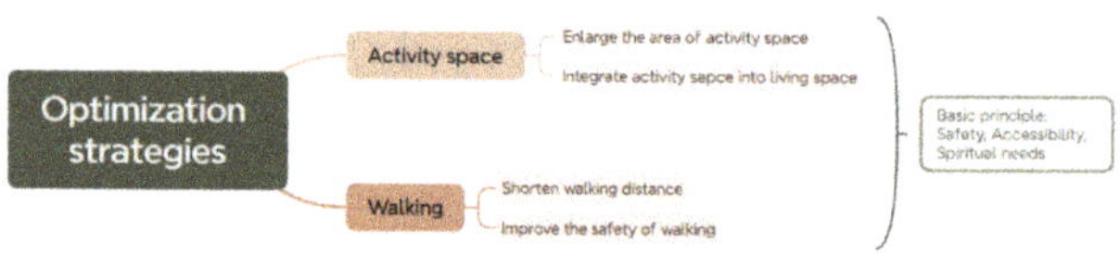

Figure 7. Ideas and renovation strategies for the health care center for the elderly.

After optimization, the building is still able to provide 24 h care for the elderly, and the area of medical consultations and check-ups has been reduced. In addition, increasing the number of people that can be served and improving the living environment for the elderly are important. After optimization, the total number of beds is 30, of which 18 are nursing care beds and 12 are non-nursing care beds (Figure 8).

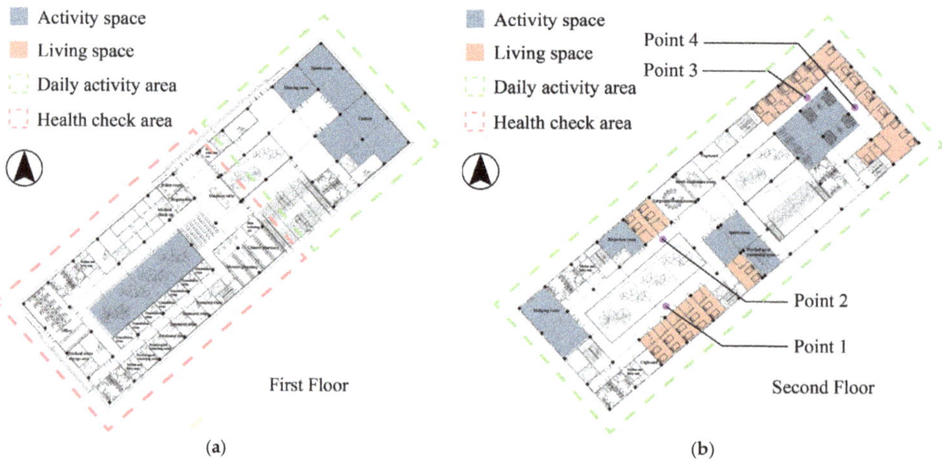

Figure 8. (**a**) First-floor plan of the optimal design; (**b**) second-floor plan of the optimal design.

3. Results and Discussions

3.1. Activity Space

In order to meet the spiritual and entertainment needs of the elderly, based on the original design, we undertook optimization and renovation projects. In the original design, the activity spaces were limited to basic functions, such as sporting, dining, and communication, but they lacked more targeted entertainment and social features. To address this shortcoming, we added a variety of activity spaces designed for seniors during the renovation process, such as a mahjong room, a projection room, and a dance area. The addition of these features not only enriches the daily activities of the elderly but also increases the possibility of social interaction, thus improving the overall quality of life. And the area of each activity room is more compact than the original [29], but the total area of the activity space as well as the per capita area of the activity has increased. After the renovation, the number of beds (people) served by the health building is 30, the total activity area is 1114.18 m^2, and the per capita area of the activity is 37.14 m^2/bed (person), a 23.5% increase compared with the original design (Figure 9). So, this redesign not only provides the elderly with more diverse options for their daily activities but also ensures a more comfortable and spacious environment. By offering a variety of engaging and well-structured spaces, the health care building promotes a better quality of life for its residents, fostering both their physical and mental well-being.

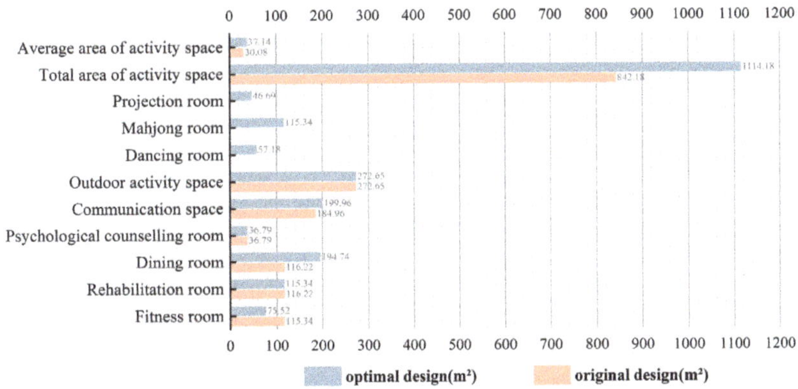

Figure 9. Comparison of activity space.

3.2. Accessibility of Walking

To enhance the accessibility of the building and reduce the walking distance for elderly residents, a series of optimizations were made to the original design. These improvements focus on making it easier and safer for elderly individuals to move in the building. Key modifications include increasing the number of stairways and elevators, which significantly shortens the distance to safety exits. Additionally, activity spaces have been strategically integrated into functional areas of the living spaces, minimizing the distance elderly residents need to travel to reach public activity areas. Taking four representative points within the living space for analysis (Figure 8), we obtained the following results (Figure 10): After the renovation, the average walking distance from living areas to the nearest safety exit has been reduced to 17.83 m, with an average of 182 steps, representing a 36.8% reduction compared to the original design. Similarly, the average walking distance to activity spaces has decreased to 39.25 m, with an average of 401 steps, marking an 8.4% improvement over the original design. So, these renovations ensure that elderly residents can move more conveniently and quickly within the building, whether heading to safety exits or accessing activity spaces.

3.3. Safety of Walking

To enhance the safety and accessibility of the building for elderly residents, key modifications were made to the staircases and elevators. The single handrail was optimized by replacing it with a double handrail of a reasonable height. This adjustment ensures that elderly individuals of varying heights can easily grip the handrails, providing greater stability and reducing the risk of falls [30]. Similarly, the elevators were upgraded to improve both safety and comfort. The cabin size was increased to accommodate mobility aids such as wheelchairs and walkers, ensuring ease of use. Handrails were installed inside the elevator to offer additional support and stability during movement, and mirrors were also added to the elevator interiors. These improvements create a safer and more user-friendly environment for elderly residents.

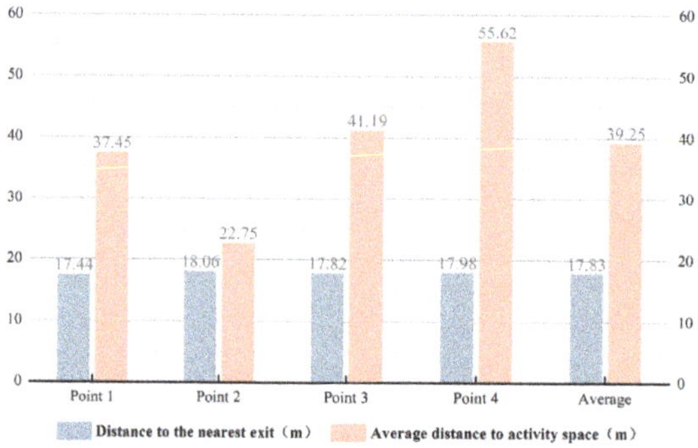

Figure 10. Walking distance in the optimal health care building for the elderly.

3.4. Daylight Factor and Ventilation

Lighting and ventilation are important criteria to judge whether the building is suitable for the elderly to live in or not; this paper does not specialize in the light environment and ventilation of the building, but here we can use these two indicators to evaluate whether the transformed building is suitable for the elderly to live in or not.

According to the "Standard for daylighting design of buildings" (GB 50033-2013 [31]) and the "General code for building environment" (GB 55016-2021 [32]), we used the Dali2023 of the Thsware to calculate the lighting coefficient of the building. This project adopts the simulation method of Radiance to simulate lighting and calculate the daylight factor. In the simulation analysis, the CIE full cloudy sky is used to simulate the lighting conditions, and the light reflection is set to be three times, the functional rooms take a height of 0.75 m from the ground as the reference plane, while the public space takes the ground as the reference plane directly. The calculation grid is divided according to the size of the room to ensure the accuracy of the simulation: for rooms with an area of less than or equal to 10 square meters, the grid size is set to 0.25 m; for rooms with an area between 10 and 100 square meters, the grid size is 0.50 m; and for rooms with an area of more than or equal to 100 square meters, the grid size is 1.00 m. Outdoors, the shading effect between neighboring buildings was considered, while indoors, the effect of indoor furniture and other facilities was ignored, and only indoor permanently fixed structures, such as ceilings, floors, and walls, were considered.

According to research findings, the elderly not only have a greater need for daylight due to a decline in the ability of the physiological perception system but also have a psychological need; adequate sunlight can reduce the emergence of psychological problems in the elderly [33]. In this building (Figure 11), on the first floor, the highest daylight factor is 17.8; it is mainly found in the corner area, where windows are opened on both sides for lighting. This area receives sufficient natural light and has the best lighting effect. The lowest daylight factor is 0, which mainly occurs in the hallway section. Due to the lack of direct window lighting, the aisles rely on artificial lighting to meet the light demand. And the main activity spaces are well-lit and meet the requirements of the relevant codes, ensuring that these spaces are in good condition for use. However, some of the transportation spaces lack windows with direct access to the outside, so natural lighting in these areas is limited and relies on artificial lighting as a supplement, and the light factor of the consulting rooms on the first floor near the atrium does not meet the light requirements (Red box in Figure 11). On the second floor, the highest daylight factor is 14.9, which is mainly found in the connecting corridors, where more natural light is available due to the high number of windows. The lowest daylight factor is 0, which occurs in the interior of the communication hall and part of the transportation space. Due to the limitations of the architectural layout and the location of the windows, these areas are unable to obtain natural light directly, and they can only satisfy basic lighting needs through artificial lighting. The main living space for the elderly on the second floor has sufficient lighting; it meets the requirements of the relevant codes and takes into full consideration the needs of the elderly to ensure their comfort and safety in daily life. Overall, the building design has fully considered the lighting needs of each functional area, and in the main activity space and living area, the daylight factors are good and meet the standards of the design specifications, while in part of the transportation space, the insufficient light is supplemented by artificial lighting to ensure the lighting needs in the overall environment.

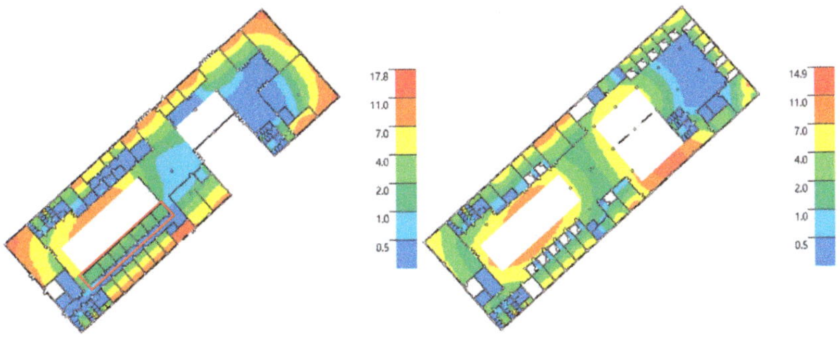

Figure 11. Building lighting (from left to right: first floor, second floor).

Elderly people in health care buildings spend a very long time indoors, and air quality is an important factor in the health status of older people; if effective measures are not taken, elderly people will be exposed to poor air most of the time, and ventilation is a very effective measure to improve indoor air quality [34]. According to the "Assessment standard for green building", the indoor and outdoor airflow distribution and flow rate of the building were calculated using VENT2023 of Thsware and CFD calculation method.

Based on the simulation results, we obtained the following results: the ventilation of the exterior of this building is good, and the maximum wind speed can reach 1.9 m/s (Figure 12). It is mainly concentrated in the corners of the building, and these locations usually have high wind speeds due to the geometric shape of the building and the influence of the surrounding environment. The overall design of the building takes into account the need for natural ventilation, ensuring that the wind speed in the external environment is appropriate and conducive to air circulation. Inside the building (Figure 13), the highest wind speed on the first floor is 1.59 m/s, and the highest wind speed on the second floor is 1.67 m/s. This indicates that the air circulation inside the building is good, especially in the activity space and the living space for the elderly, it can achieve effective air exchange and keep the indoor air fresh and smooth. Overall, the ventilation effect of the building is satisfactory both externally and internally, providing a healthy and comfortable living environment for the occupants and giving full consideration to the needs of the elderly.

The lighting factor and wind speed inside the building should only be increased to a certain extent; they have ranges that make the residents feel the most suitable; the lighting and ventilation of the building basically meet the requirements of the code, but the question of how to make the building more suitable for the elderly to live in should be further studied.

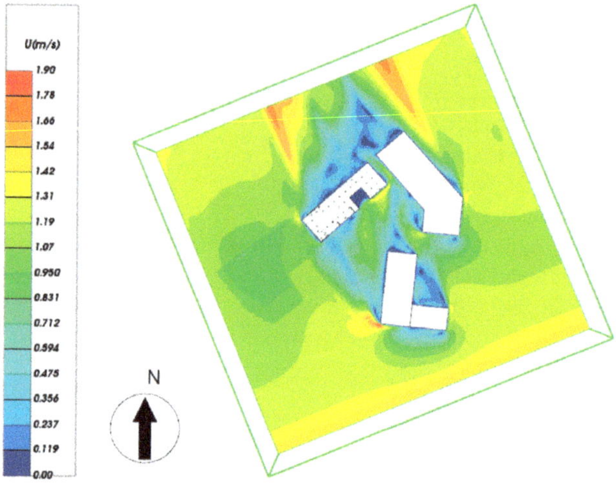

Figure 12. Outdoor ventilation.

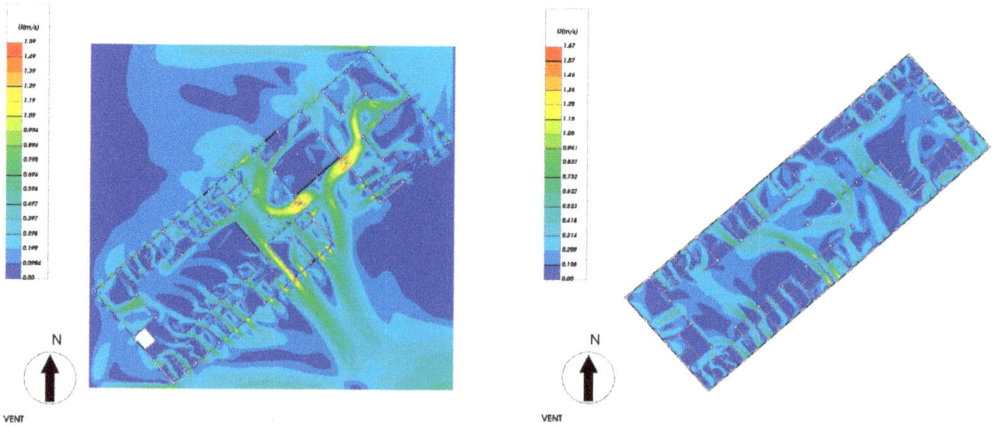

(a) (b)

Figure 13. (**a**) Building ventilation of first floor; (**b**) building ventilation of second floor.

3.5. Discussions

This paper is in line with the direction of most previous research on age-friendly buildings, with the goal of improving the comfort and convenience of the elderly in buildings. However, most of the previous studies on public activity space and walking distance for the elderly are set in outdoor environments, while this paper focuses on the optimization of public activity space and walking distances for the elderly in buildings.

Shortening the walking distances of the elderly in the building can reduce the burden of the elderly when walking and make the elderly more willing to go out of their rooms to go out for activities. And the optimization of the public activity space in the recreational building for the elderly can provide a richer choice of activities for the elderly, encourage the elderly to carry out activities, and provide a space for the elderly to socialize and maintain social relationships, reducing the loneliness of the elderly [35]. According to relevant research, elderly people who carry out daily activities and maintain a healthy lifestyle will have a favorable mental health impact, increase their positive emotions, and improve their subjective sense of well-being and life satisfaction [36].

Thanks to the results of this study, it is hoped that future aging-friendly buildings can be designed with reasonable spatial layouts, with core functional areas arranged in easily accessible locations to effectively shorten the walking distance of the elderly in the building, and with resting nodes and auxiliary facilities set up on the walking paths to improve the safety of the elderly walking. And, in the design of public activity space, a moderate size is adopted, and a variety of functional activity spaces are designed to meet different activity needs and enhance the flexibility and practicability of space.

This optimization is expected to provide long-term comfort and convenient lives for the elderly. In order to evaluate the effectiveness of this optimization over time, in the future, regular return visits and questionnaire surveys can be planned at certain intervals to collect the opinions of the elderly, check the frequency of the use of each flow line inside the building, and test the use of the accessible elevator and the staircase, to evaluate whether their safety and convenience can be maintained effectively over time.

4. Conclusions and Prospects

This paper focuses on the age-friendly renovation of health care buildings, primarily addressing two aspects: elderly accessibility and public activity spaces. The goal is to create a more suitable environment for the daily lives of the elderly. Through the improvement of the existing building layout and the optimization of the existing facilities, the walking distance to the nearest exit of the elderly in the building has been reduced by 36.8%, the walking distance to activity spaces for the elderly has been reduced by 8.4%, and the average public activity space of each elderly person has been increased by about 23.5%. Optimization enables the elderly to walk more safely and go to the targeted room more quickly in the building, while also addressing their entertainment needs and enriching their mental and emotional well-being.

However, there are limitations to this paper. This paper only analyses the retrofitting of aging buildings from the perspective of the public activity space in the buildings for the elderly and the walking distance inside the buildings; however, in practice, those retrofitting aging buildings have to consider more influencing factors, such as the existing building design, the physical conditions of the elderly, mental health problems, socialization, safety, the sense of belonging and identity, and so on. To address the shortcomings of this study, the following prospects are proposed for the future:

(1) A lot of research on age-friendly building renovation focuses on the physical factors of the elderly, but the mental health of the elderly is also an important factor that affects

their happiness in daily life. Future research on age-friendly buildings for the elderly could consider and address the psychological needs of the elderly.

(2) In addition to the per capita area of public activity spaces, future research could further explore the association between the area of public spaces and the physical health of the elderly, in addition to how the planning of public spaces can help the elderly to be more active in various activities and help improve their health.

(3) Future research could also integrate age-friendly building design with sustainable development by conducting in-depth studies on lighting and ventilation in buildings for the elderly and using green and energy-efficient designs to reduce energy consumption, while making the built environment more suitable for the elderly to live in, thereby enhancing the quality of life and happiness of the elderly.

Although this research on the aging adaptation transition of the walking distance and average area of activity spaces cannot comprehensively examine the strategies for the aging adaptation transition of senior buildings, this renovation project hopes to serve as a reference and guide for future relevant age-friendly renovations and relevant studies of other aging buildings, contributing to the development of aged-friendly buildings' design.

Author Contributions: Conceptualization, C.Y. and Y.Q.; methodology, W.H.; software, Y.L. (Yiheng Liu); validation, C.Y., W.H. and Y.Q.; formal analysis, Y.L. (Yijie Lin); investigation, Y.L. (Yiheng Liu); resources, Y.Q.; data curation, C.Y.; writing—original draft preparation, C.Y.; writing—review and editing, W.H.; visualization, Y.L. (Yijie Lin); supervision, Y.Q.; project administration, C.Y.; funding acquisition, Y.Q. All authors have read and agreed to the published version of the manuscript.

Funding: This research was funded by the Central University Basic Research Business Fee Special Fund Project.

Data Availability Statement: Data, original images, and detailed design model materials will be available on request.

Acknowledgments: Thanks to everyone for their help and support during the writing process.

Conflicts of Interest: The authors declare no conflicts of interest.

References

1. World Health Organization. Available online: https://www.who.int/china/health-topics/ageing (accessed on 23 January 2025).
2. Langenbrunner, J.C.; Marquez, P.V.; Wang, S. *Toward a Healthy and Harmonious Life in China: Stemming the Rising Tide of Non-Communicable Diseases*; The World Bank: Washington, DC, USA, 2011.
3. Wang, X.-Q.; Chen, P.-J. Population ageing challenges health care in China. *Lancet* **2014**, *383*, 870. [CrossRef] [PubMed]
4. Harwood, R.H.; Sayer, A.A.; Hirschfeld, M. Current and future worldwide prevalence of dependency, its relationship to total population, and dependency ratios. *Bull. World Health Organ.* **2004**, *82*, 251–258. [PubMed]
5. Wong, Y.C.; Leung, J. Long-term care in China: Issues and prospects. *J. Gerontol. Soc. Work.* **2012**, *55*, 570–586. [CrossRef] [PubMed]
6. Trivedi, A.J.; Mehta, A. Maslow's hierarchy of needs-theory of human motivation. *Int. J. Res. All Subj. Multi Lang.* **2019**, *7*, 38–41.
7. Nie, H.; Xu, Y.; Liu, B.; Zhang, Y.; Lei, T.; Hui, X.; Zhang, L.; Wu, Y. The prevalence of mild cognitive impairment about elderly population in China: A meta-analysis. *Int. J. Geriatr. Psychiatry* **2011**, *26*, 558–563. [CrossRef]
8. Chen, Y.; Hicks, A.; While, A.E. Depression and related factors in older people in China: A systematic review. *Rev. Clin. Gerontol.* **2012**, *22*, 52–67. [CrossRef]
9. Han, J.; Chan, E.H.; Yung, E.H.; Qian, Q.K.; Lam, P.T. A policy framework for producing age-friendly communities from the perspective of production of space. *Int. J. Environ. Res. Public Health* **2022**, *19*, 2031. [CrossRef]
10. Xu, Y.; Liu, H.; Su, S.; Mao, P. Ageing suitability evaluation of residential districts based on Active Ageing Theory. *Buildings* **2023**, *13*, 1041. [CrossRef]
11. Fallah, H.; Nazari, J.; Choobineh, A.; Morowatisharifabad, M.A.; Jafarabadi, M.A. Inclusive Design: An Approach to Adapt Homes for the Elderly. *Pak. J. Med. Health Sci.* **2021**, *15*, 1599–1603. [CrossRef]

12. Liu, J.; Wei, Q. Discussion on the spatial design of elderly care buildings under the concept of ageing-friendly concept. In Proceedings of the SHS Web of Conferences, Harbin, China, 6–8 December 2024; p. 01015.
13. Han, J.; Chan, E.H.W.; Qian, Q.K.; Yung, E.H.K. Achieving sustainable urban development with an ageing population: An "age-friendly city and community" approach. *Sustainability* **2021**, *13*, 8614. [CrossRef]
14. Han, L.; Wang, B. Research on the Community Renewal of "Active Adaptation to the Elderly" in the Unit Compound. In Proceedings of the ECSF 2021: Engineering, Construction, and Infrastructure Solutions for Innovative Medicine Facilities, St. Petersburg, Russia, 19–21 May 2021; Springer: Cham, Switzerland, 2022; pp. 1–14.
15. Tao, Y.; Zhang, W.; Gou, Z.; Jiang, B.; Qi, Y. Planning walkable neighborhoods for "aging in place": Lessons from five aging-friendly districts in Singapore. *Sustainability* **2021**, *13*, 1742. [CrossRef]
16. Tort Ayala, L.S. Principles and Recommendations to Design Aging-Friendly Homes. Doctoral Dissertation, Massachusetts Institute of Technology, Cambridge, MA, USA, 2019.
17. Liu, J. Data Mining and Retrofit Design for Age-friendly Spaces. In Proceedings of the International Conference on Green Building, Civil Engineering and Smart City, Guiyang, China, 9–12 June 2023; pp. 835–843.
18. Fang, M.L.; Sixsmith, J.; Hamilton-Pryde, A.; Rogowsky, R.; Scrutton, P.; Pengelly, R.; Woolrych, R.; Creaney, R. Co-creating inclusive spaces and places: Towards an intergenerational and age-friendly living ecosystem. *Front. Public Health* **2023**, *10*, 996520. [CrossRef] [PubMed]
19. Luciano, A.; Pascale, F.; Polverino, F.; Pooley, A. Measuring age-friendly housing: A framework. *Sustainability* **2020**, *12*, 848. [CrossRef]
20. Tao, Y.; Gou, Z.; Yu, Z.; Fu, J.; Chen, X. The challenge of creating age-friendly indoor environments in a high-density city: Case study of Hong Kong's care and attention homes. *J. Build. Eng.* **2020**, *30*, 101280. [CrossRef]
21. Jiao, Y.; Yu, Y.; Yu, H.; Wang, F. The impact of thermal environment of transition spaces in elderly-care buildings on thermal adaptation and thermal behavior of the elderly. *Build. Environ.* **2023**, *228*, 109871. [CrossRef]
22. Hickerson, B.; Moore, A.; Oakleaf, L.; Edwards, M.; James, P.A.; Swanson, J.; Henderson, K.A. The Role of a Senior Center in Promoting Physical Activity for Older Adults. *J. Park Recreat. Adm.* **2008**, *26*, 22–39.
23. The People's Government of Sanya City. Available online: https://www.sanya.gov.cn/sanyasite/tjgb/202105/01681d7550e343c69c2b892709e504dd.shtml (accessed on 23 January 2025).
24. The People's Government of Sanya City. Available online: https://www.sanya.gov.cn/sanyasite/syyw/202311/1edf5e69fcf949a89fbda0fab5b52461.shtml (accessed on 23 January 2025).
25. *JGJ450-2018*; Standards for Design of Care Facilities for the Aged. Architecture & Building Press: Beijing, China, 2018.
26. *GB 55037-2022*; General Code for Fire Protection of Buildings and Constructions. China Planning Press: Beijing, China, 2022.
27. *GB 50763-2012*; Codes for Accessibility Design. China Architecture & Building Press: Beijing, China, 2012.
28. Shao, Q.; Wang, H.; Hu, T.; Gu, D. Effects of age and gender on biomechanical characteristics of human gait. *Chin. J. Appl. Mech.* **2022**, *39*, 1193–1202.
29. Lee, S.; Dilani, A.; Morelli, A.; Byun, H. Health supportive design in elderly care homes: Swedish examples and their implication to Korean counterparts. *Archit. Res.* **2007**, *9*, 9–18.
30. Tural, E.; Lu, D.; Cole, D.A. Factors predicting older Adults' attitudes toward and intentions to use stair mobility assistive designs at home. *Prev. Med. Rep.* **2020**, *18*, 101082. [CrossRef]
31. *GB 50033-2013*; Standard for daylighting design of buildings. China Architecture & Building Press: Beijing, China, 2013.
32. *GB 55016-2021*; General code for building environment. China Architecture & Building Press: Beijing, China, 2021.
33. Yang, H.; Guo, B.; Shi, Y.; Jia, C.; Li, X.; Liu, F. Interior daylight environment of an elderly nursing home in Beijing. *Build. Environ.* **2021**, *200*, 107915. [CrossRef]
34. Cerqueira, F.; Nogueira, S.; Aelenei, D.; Viegas, J. Assessment of ventilation in elderly care centres. In Proceedings of the 40th IAHS World Congress on Housing Sustainable Housing Construction, Funchal, Portugal, 16–19 December 2014.
35. Abbas, N.; Abrar ul Haq, M.; Ashiq, U.; Ubaid, S. Loneliness among elderly widows and its effect on social and mental well-being. *Glob. Soc. Welf.* **2020**, *7*, 215–229. [CrossRef]
36. Lara, R.; Vázquez, M.L.; Ogallar, A.; Godoy-Izquierdo, D. Psychosocial resources for hedonic balance, life satisfaction and happiness in the elderly: A path analysis. *Int. J. Environ. Res. Public Health* **2020**, *17*, 5684. [CrossRef]

buildings

MDPI

Article

Energy-Efficient Architectural Design of a Banquet Hall with Integrated Tunnel Ventilation: Monitoring Performance During the Transitional Season in China

Jianwu Xiong [1], Jing Wu [1], Jifan Cao [2], Zexuan Tian [3,*] and Qianru Yang [1,*]

[1] School of Architecture, Southwest Minzu University, Chengdu 610225, China; 80300151@swun.edu.cn (J.X.); xwykxa@163.com (J.W.)
[2] College of Design and Engineering, National University of Singapore, Singapore 119077, Singapore; caojifan2002@163.com
[3] School of Architecture and Urban Planning, Chongqing University, Chongqing 400030, China
* Correspondence: 202415021089t@stu.cqu.edu.cn (Z.T.); yangqianru@swun.edu.cn (Q.Y.); Tel.: +86-18131202606 (Z.T.); +86-15680016301 (Q.Y.)

Abstract: The construction industry, a significant contributor to global energy consumption and greenhouse gas emissions, is under considerable pressure to adopt transformative approaches. Public buildings, which account for a substantial portion of total energy usage, must balance high standards of thermal comfort with ventilation efficiency. In China, many public buildings are part of urban landscapes, where façade designs often limit natural ventilation. Consequently, technologies like earth-to-air heat exchangers and wind towers are increasingly essential for enhancing natural ventilation. However, research on the efficacy of these systems remains sparse. This study examines the transitional seasonal environment by evaluating the thermal-humidity index of a banquet hall equipped with an earth-to-air heat exchanger system. Using DeST software [DeST 2.0], the study simulates indoor natural ventilation, calculates ventilation rates, and assesses residual heat removal efficiency. The system's performance is also modeled under various thermal design zones. Results demonstrate that under natural ventilation, the system can achieve a residual heat removal efficiency of up to 490%. Simulations across different climate zones indicate that the system performs best in regions with extreme temperature fluctuations, particularly those with hot summers and warm winters. In these areas, the system reduces the annual temperature difference by up to 56.7%, significantly improving thermal comfort and reducing dependency on air conditioning. In contrast, performance in milder regions like Kunming achieves only a 37.5% reduction in temperature difference. Overall, this study provides valuable insights into energy-efficient design strategies and thermal optimization for banquet halls, with significant potential for energy savings and enhanced occupant comfort.

Keywords: building space; heat pressure regulation; room temperature measurement; energy efficiency optimization; indoor space; natural ventilation; operation optimization; experiment

1. Introduction

The world has entered an era of global climate change, becoming the greatest non-traditional security challenge facing human development [1,2]. As the most populous nation and the world's largest carbon emitter, China faces unique environmental challenges. [1,3]. In response, the Chinese government has introduced several key policy documents, including the "Opinions on Fully and Accurately Implementing the New Development Concept to Achieve Carbon Peak and Neutrality" and the "2024–2025 Energy Conservation and Carbon Reduction Action Plan". These policies set clear goals: to reach peak carbon emissions by 2030 and achieve carbon neutrality by 2060 [4–7]. The construction industry is one of the main sectors contributing to China's carbon emissions (alongside industry and transportation) and is a major area of responsibility for both direct and indirect carbon emissions [8].

Public buildings play a critical role not only as a fundamental part of urban architecture but also as key drivers of urban economic, cultural, and social development. However, due to their large spaces and complex operational systems, public buildings typically consume a significant amount of energy and contribute to high levels of carbon emissions [9]. According to the "China Building Energy Consumption and Carbon Emission Research Report (2023)" [7], public buildings occupied 14 billion m^2 in 2021, accounting for 20% of the total building area in the country. However, their energy consumption represented 37% of the total operational energy consumption of all buildings. Furthermore, operational carbon emissions from public buildings increased by 0.95 billion tons of CO_2, contributing 63% of the total rise in building-related operational carbon emissions. This highlights the need for sustainable and green transformation within the sector.

Numerous studies indicate that the frequency of natural disasters such as earthquakes, typhoons, and floods has increased in recent years. In response, large open spaces, including green areas, have become vital for accommodating evacuees [10,11]. However, as urban populations continue to grow and land becomes increasingly scarce, green spaces are limited. Consequently, public buildings, with their large interiors, durable structures, and advanced facilities, are often used as temporary shelters during emergencies [11,12]. This dual function of public buildings necessitates careful consideration of not only their everyday operational efficiency but also their adaptability for use as emergency shelters. In particular, when these buildings are utilized as shelters, it is crucial to evaluate their ventilation efficiency and ability to maintain thermal comfort for large numbers of occupants [12].

Existing research demonstrates that human comfort levels are generally higher in naturally ventilated environments compared to air-conditioned spaces [13]. Moreover, natural ventilation is a primary passive strategy for reducing air conditioning energy consumption [14]. During transitional seasons in China, natural ventilation plays a crucial role in energy saving, emission reduction, and thermal environment optimization in large public buildings [13,14]. Sui Xuemin, Tian Zhongjie, Yu Senfeng, and Yan Bo conducted an empirical study in Xi'an, analyzing the ventilation rates and indoor air quality of different ventilation methods in residential buildings [15]; Liu Tao proposed a technical solution for reducing energy consumption in railway passenger stations using natural ventilation and conducted simulation analyses [1]; Zhang Bing studied the impact of building orientation on natural ventilation energy consumption and found that natural ventilation can significantly reduce building energy consumption [16]; Hu Yuting researched the energy-saving potential of natural ventilation in residential buildings in regions with hot summers and cold winters, based on thermal balance theory [17]; Xu Jun, Wang Qi, and Yin Ziwen discovered through simulation analysis that the rational use of natural ventilation can significantly reduce the operational energy consumption of public buildings [18]; Jing Chen and Xin Ye studied the strategies and designs for natural ventilation in public buildings in Hong Kong, noting that indoor ventilation design measures are gradually being replaced by air conditioning systems [19]. However, when outdoor temperatures are too high, natural ventilation cannot meet people's needs [20]. The earth duct system, a building energy-saving ventilation technology that has been in use for decades, was initially applied mainly in large spaces such as auditoriums, cinemas, and factories. In recent years, through extensive theoretical and practical research, it has gradually been widely applied in various types of public buildings [21]. For example, in the design of the air conditioning system for the visitor service center at the Mogao Caves, a combination of the earth duct system and natural ventilation was used to minimize the energy consumption of air conditioning equipment [22]. Thus, combining the earth duct system with natural ventilation can both improve the indoor thermal environment and reduce energy consumption [23–25]. Nowadays, many public buildings in China are considered integral components of urban landscapes. Their facade designs are influenced by urban aesthetics and appearance requirements, which often restrict window openings and thus limit opportunities for natural ventilation. As a result, technologies such as earth duct systems and ventilation shafts are increasingly important for enhancing natural ventilation in public buildings. The subject of this study, the Banyan Tree Garden Banquet Hall, also applies this technology, utilizing a "bottom earth duct and top ventilation shaft" natural ventilation system.

This study focuses on a large banquet hall located in Banyan Tree Garden Park in Nanlin Village, Xiwai Town, Guanghan City, Sichuan Province. The hall is typically used for hosting large events and gatherings, with high attendee numbers but relatively short durations and frequent movement of people. Such events demand a ventilation system capable of efficiently managing large volumes of respiratory gases and pollutants. Additionally, the thermal load in the banquet hall is often significant due to the high occupancy, combined with the extensive use of lighting and sound equipment. This poses substantial challenges for ventilation systems, especially in terms of thermal load management and temperature control [26]. The nature of banquet events, such as weddings and longevity celebrations, often requires inward-focused spaces for privacy, resulting in relatively enclosed designs. To maintain privacy and sound insulation, the banquet hall is built with high airtightness, which complicates natural ventilation efforts. Thus, balancing thermal comfort and ventilation energy consumption becomes critical [27,28].

In this study, only two typical field tests were conducted. To validate the results and extrapolate conclusions, further simulation calculations will be performed using DeST (Designer's Simulation Toolkit), developed by Tsinghua University's School of Architecture, which is designed for China's specific climate and building conditions [29–31]. While its performance in simulating HVAC systems is comparable to international tools like DOE-2, EnergyPlus, and TRNSYS, DeST provides more accurate data for energy simulations of Chinese buildings, with deviations in calculations generally within 5% [29]. Additionally, DeST is more suitable for energy simulations of Chinese building types, accurately reflecting their energy consumption characteristics and providing more precise data support for energy-efficient design.

Therefore, this study intends to combine field tests and DeST simulation calculations to explore the impact of natural ventilation on the indoor thermal environment of the banquet hall during transitional seasons. The goal is to ensure occupant comfort while reducing energy consumption and exploring potential ventilation strategies. Additionally, key indicators such as residual heat variation and heat removal efficiency will be analyzed to assess the impact of different adjustment strategies to influence the hall's indoor temperature. This will help evaluate the role of earth duct ventilation in regulating indoor temperature, highlighting its potential as an energy-saving cooling method. The findings are expected to offer scientific evidence for decision-makers, significantly contributing to building energy conservation, emission reduction, and occupant comfort, key elements in China's sustainable development trajectory.

2. Materials and Methods

2.1. Experimental Principles and Framework

2.1.1. Principles of Stack-Driven Natural Ventilation and Associated Formula

The principle of stack-driven natural ventilation is based on the temperature difference between indoor and outdoor air, which creates a pressure differential and causes changes in air density [32]. This process induces warm air to exit through openings at the top of the building, while cooler air enters through openings at the lower part of the building, thereby achieving natural ventilation [33,34]. This upward and downward movement forms a natural convection loop, resulting in air circulation that facilitates stack-driven natural ventilation. The theoretical relationship between indoor thermal pressure and ventilation can be described using the following formula:

$$Q = C_P \rho A \Delta T h \tag{1}$$

where,

Q is the ventilation volume (m^3/s);

C_P is the specific heat capacity of air (J/kg·°C);
ρ is the density of air (kg/m^3);
A is the effective area of the vent or ventilator (m^2);
ΔT is the difference between indoor and outdoor temperatures (°C);
h is the height difference of ventilation (m).

The formula indicates that a greater temperature difference results in a higher ventilation rate. Similarly, larger areas of openings or ventilation devices also lead to an increased ventilation rate. Additionally, the height difference between the openings or ventilation devices influences the magnitude of the ventilation rate.

2.1.2. Experimental Framework Design

The objective of this experiment is to precisely measure the natural ventilation efficiency of a large banquet hall. Using the YOWEXA air quality and environmental tester, ventilation data was gathered from the bottom inlet, the top outlet, and the underground duct inlet of the banquet hall. The experiment aimed to measure how effectively the natural ventilation system removes excess heat and to assess the performance and energy-saving potential of the earth-air tunnel system as a natural cooling solution. By analyzing these data, the study identified key factors influencing the system's performance, particularly its impact on indoor temperature regulation. The results provide insight into the effectiveness of the earth-air tunnel in maintaining a comfortable thermal environment while minimizing energy consumption, demonstrating its potential as a sustainable, energy-efficient cooling method for large spaces.

2.2. Study Area

The subject of this study is Unit 2 of the Banyan Tree Garden Banquet Hall (Hall No.2) in Nanlin Village, Xiwai Township, Guanghan City, Sichuan Province. This hall is primarily used for significant local events such as weddings, birthday parties, and longevity celebrations (Figure 1a). It has a longitudinal north-south spatial configuration, with the northern side serving as a central outdoor parking lot for the Banyan Tree Garden. The eastern side borders a rural road and includes public service facilities such as the Nanlin Village committee office and a senior activity center. The western and southern sides are landscaped and recreational areas within the Banyan Tree Garden complex (Figure 1b). Construction of the banquet hall began in May 2015 and was completed in October 2015. The main structure is composed of frame steel construction, with a total building area of 4975 m^2.

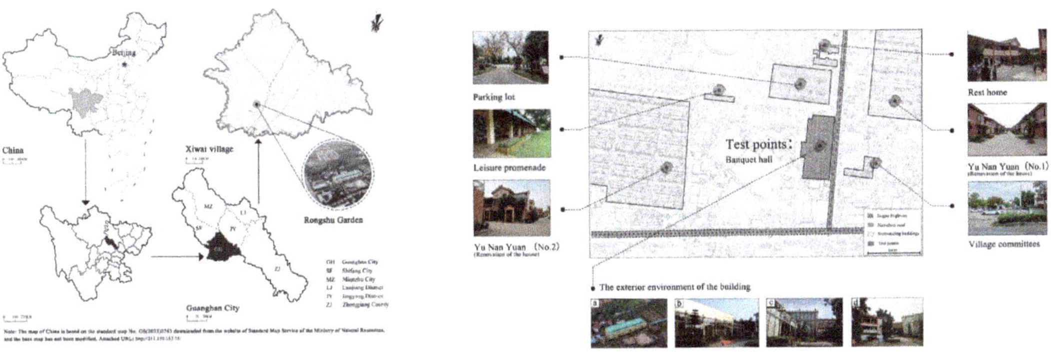

(a) (b)

Figure 1. Location map of the banquet hall. (**a**) The banquet hall is located in the Banyan Tree Garden area of Nanlin Village, Xiwai Township, Guanghan City, Sichuan Province, China [35]; (**b**) Schematic diagram of the project's surrounding functional areas, including parking lots, recreational zones, and residential points.

The functional spaces of the building are primarily divided into three main parts: the kitchen service area, the ceremony waiting and rest area, and the banquet units (Figure 2). The banquet units, which are primarily used for dining and wedding ceremonies, form the core functional space of the building. The design accommodates the simultaneous hosting of multiple events, with the capacity to serve over 1000 guests. To address the variability in the number of attendees for different banquets, the design was based on statistical conclusions drawn from over 100 previously hosted events in the park. The approach used was to design the banquet hall units based on "high-frequency attendance (Hall No.1, No.2, and No.3) + flexible space design (Hall No.4)", resulting in a layout of three high-frequency banquet units (each with a standard capacity of 240 people) joined end-to-end from north to south, along with one medium-frequency banquet unit (with a standard capacity of 360 people). All four banquet units utilize a large-span frame structure with a depth of 24 m. The high-frequency units have a width of 21 m, while the medium-frequency unit has a width of 32 m. Each banquet unit features high single-story vertical spaces, with interior heights gradually increasing from 8.8 m at the sides to 10.7 m at the center. A narrow skylight at the highest central point of each unit allows sunlight to stream in during ceremonies, illuminating the red carpet area and creating a solemn and sacred atmosphere. To accommodate the less frequent demand for super large-scale banquets with over 500 guests, flexible partitions are used between each banquet unit along the north-south axis, allowing the units to be combined as needed. Additionally, to meet the visual requirements of the facade, the need for guest waiting areas before banquets, and the service requirements for large-scale banquet catering, the east and west sides of the banquet units are designed as kitchen service corridors and waiting spaces. This arrangement of multi-hall connections along the north-south axis and functional zones along the east-west axis constitutes the banquet units (Figure 3).To enhance indoor-outdoor air exchange and ventilation, and to address the conflict between thermal comfort demands and ventilation energy consumption, the banquet hall employs a ventilation system where air enters from the northern end of the building and is distributed through a ground-level tunnel system, with exhaust vents located at the top of each hall. A subterranean ventilation system is installed at specific locations on the lower basement level to serve the banquet halls. This system includes four ducts of varying lengths that connect to the interior of the banquet hall, with each of the first three units having four ground-level air inlets, while the fourth unit has six. Each unit also features eight exhaust vents integrated with lighting at the top, making a total of 32 exhaust vents across all four units. This experiment aims to determine, through empirical data and calculations, whether this natural ventilation system of "ground-level tunnel air supply and top-level exhaust" can meet the daily ventilation requirements of a large, windowless public building with high occupancy.

2.3. Instrument Commissioning

To ensure instrument precision and test result reliability, a thorough calibration process was conducted before starting the official experiments. The procedure involved several key steps (Figure 4): First, we checked the battery levels and basic functions of all YOWEXA air quality and environmental testers to ensure they were in good working condition. Then, the six experimental devices were set with standard temperature, humidity, and other relevant parameters. They were placed in a temperature and humidity-controlled empty classroom. The readings displayed by the instruments were recorded and compared with the standard values. If discrepancies were found between the readings and the standard values, the instrument settings were adjusted until the displayed values matched the actual parameters of the standard environment. Once the instrument data matched the actual parameters, the six devices were relocated to different positions within the temperature and humidity-controlled classroom, and their readings were recorded again. Finally, the results of each calibration were documented and saved as part of the subsequent analysis, ensuring that the instruments could maintain optimal working conditions throughout the entire experiment. The following are the specific calibration steps and results:

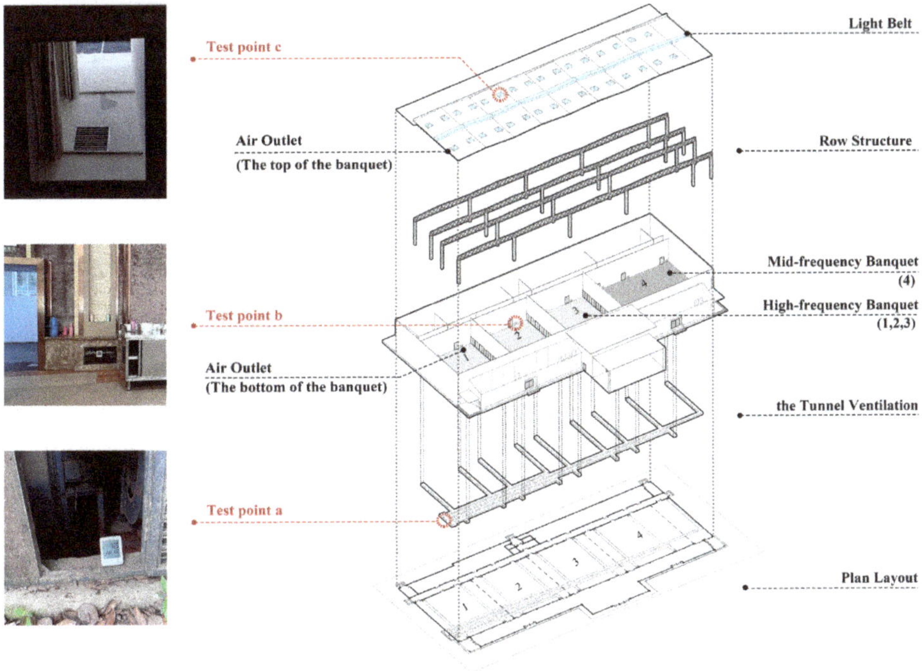

Figure 2. Schematic diagram of the internal structure of the banquet hall.

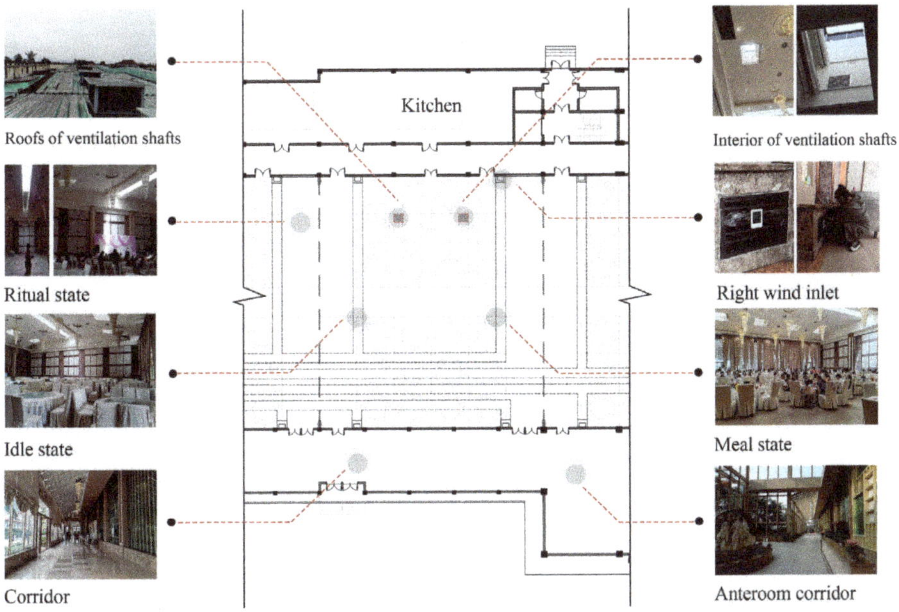

Figure 3. The functional layout of the banquet hall interior and schematic diagrams of its different usage states.

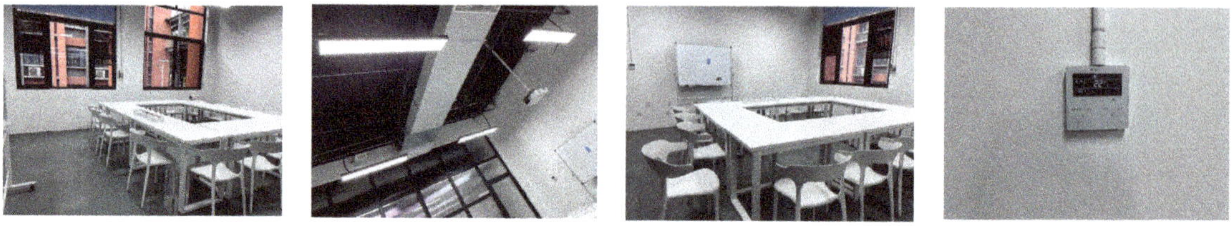

(**a**) (**b**) (**c**) (**d**)

Figure 4. Real pictures of the instrument debugging process. (**a**) Six devices together in a temperature and humidity-controlled empty classroom; (**b–d**) six instruments in different locations within the controlled classroom.

Test Location: Smart Classroom 1005, Engineering Laboratory Building, Southwest Minzu University, Chengdu, Sichuan Province

Test Time: 19 April 2023, 18:40

Air Conditioning Temperature: 22 °C

Note: The test recording interval was 5 min, with a total test duration of 1 h. The test environment was enclosed, and temporary air conditioning adjustments were made to maintain the stability of the test environment.

The data from the calibration tests showed that the six YOWEXA air quality and environmental testers displayed consistent results throughout the experiment, without any significant fluctuations (Figures 5 and 6). Importantly, there were no instances where one or more instruments consistently gave higher or lower readings. This consistency in performance indicates that the instruments exhibited high levels of accuracy and precision, free from any major malfunctions. Consequently, it was concluded that all six instruments could be reliably utilized in the subsequent testing phase, ensuring dependable data collection for the natural ventilation efficiency experiments.

YOWEXA Air Quality and Environment Tester Calibration Experiment Record Form（2023.4.18）

Measurement type/Instrument code	Temperature(−10～65C°)	Humidity(0～100RH)	Carbon dioxide(0～10,000PPM)
Instrument No.1	26.2	52.5	402
Instrument No.2	26.2	52.9	407
Instrument No.3	26.4	52.9	406
Instrument No.4	25.8	54.3	409
Instrument No.5	26.2	53.5	406
Instrument No.6	25.9	53.4	410

Note: The instrument calibration began at 17:50, with a CO_2 measurement interval set to five seconds. The calibration duration was five minutes, conducted in Room 1005 of the Engineering Building Smart Classroom. The experimental environment was comfortable, with ventilation meeting calibration requirements and excellent air quality.

Figure 5. YOWEXA air quality and environmental tester calibration experiment records.

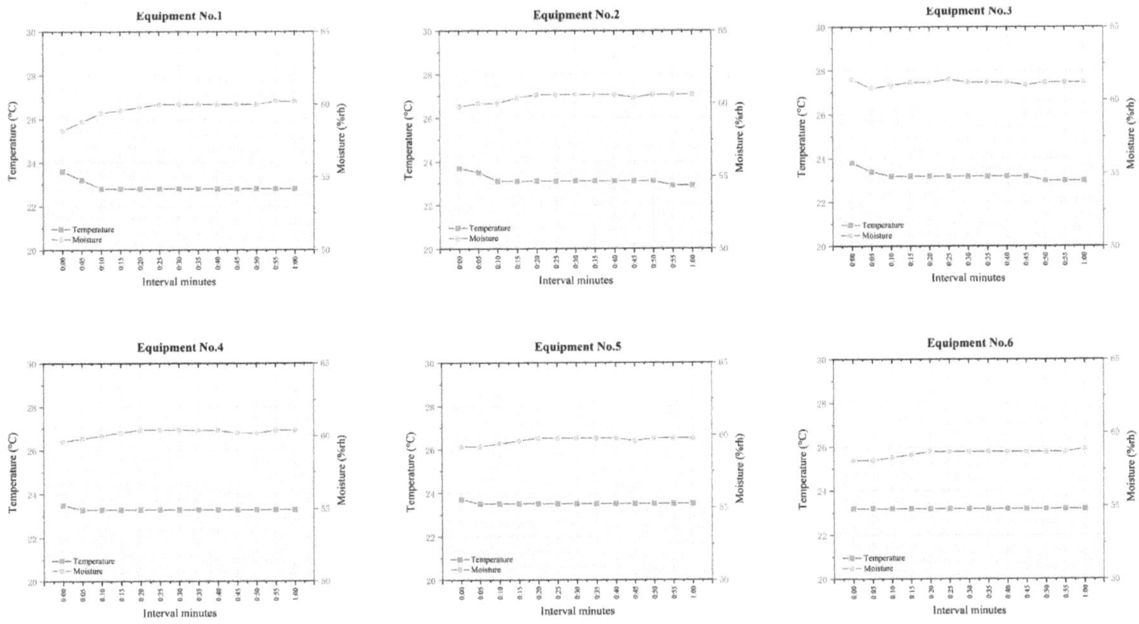

Figure 6. Calibration experiment data fluctuation chart.

3. Results

On 22 April 2023 and 2 May 2023, the team members conducted indoor and outdoor temperature tests in Banquet Hall No.2 under conditions of both non-use and dining activities. On 22 April, the gathering had 100 attendees, while on 2 May, the gathering had 200 attendees. In Hall No.2, there were three testing points (Figure 2): testing point 'a' was located at the inlet of the underground ventilation duct of the banquet hall, testing point 'b' was located at the bottom inlet of Banquet Hall Unit 2, and testing point 'c' was located at the top exhaust outlet of Banquet Hall No.2. The primary parameter tested was the air dry-bulb temperature. The specific instruments used for testing, their accuracy, and the testing frequency are detailed in Table 1.

Table 1. Test Parameters, Instruments, and Accuracy.

Test Parameters	Test Devices	Measurement Accuracy	Sampling Frequency
indoor temperature	YOWEXA Air Quality and Environment Tester	±0.2 °C	for 5 min each time
air inlet temperature	YOWEXA Air Quality and Environment Tester	±0.2 °C	for 5 min each time
air outlet temperature	YOWEXA Air Quality and Environment Tester	±0.2 °C	for 5 min each time

Based on the data obtained from various instruments in Banquet Hall No.2, this study plotted the temperature variation curves for the three air vents (Figure 7). The temperature at testing point 'a', located at the inlet of the underground ventilation duct of the banquet hall, exhibited a relatively stable trend over the two days of testing. Regardless of whether the hall was unoccupied or during a banquet, the inlet temperature remained within a stable range. As shown by the curve variations in Figure 7, the temperature fluctuation at the inlet of the underground ventilation duct was minimal, with a maximum temperature of 20.2 °C and a minimum temperature of 20.05 °C, resulting in an overall variation of no more than 0.15 °C. This stability demonstrates the effectiveness of the "bottom underground ventilation + top exhaust" natural ventilation system in maintaining indoor environmental control despite external temperature fluctuations. This result indicates that, regardless of the number of banquet participants, the temperature control of the inlet ventilation system remains stable, providing a constant air temperature for the indoor environment.

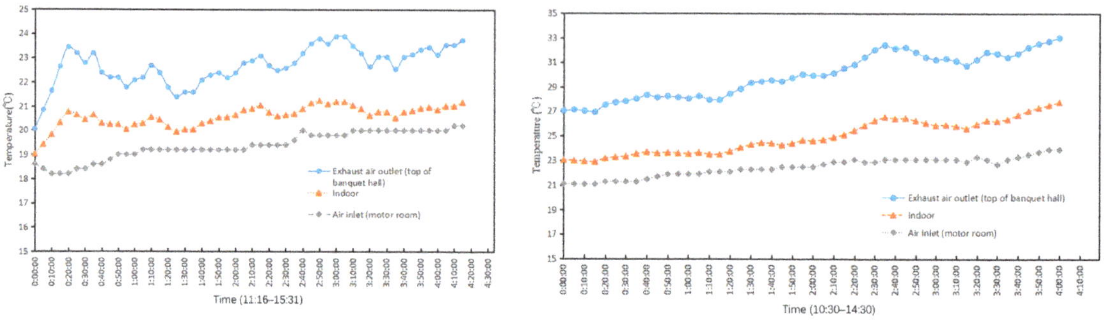

(a) (b)

Figure 7. Temperature data statistics at various points in the banquet hall utilizing the "bottom underground ventilation + top exhaust" natural ventilation system: (**a**) Measured temperature variation data at each point on 22 April with 100 participants; (**b**) Measured temperature variation data at each point on 2 May with 200 participants.

Additionally, the data from testing point 'b', located at the bottom inlet of Banquet Hall No.2, indicated that indoor temperature variations are closely related to the number of occupants during the banquet. As shown in Figure 7, on 22 April with 100 participants, the indoor temperature gradually increased with the rise in the number of people and their activities. When no activities were occurring in the banquet hall, the indoor temperature was at its lowest, at 19.03 °C. As the attendees began to gather, the temperature steadily increased, eventually reaching a peak of 21.20 °C, yielding a temperature difference of approximately 2.17 °C. In contrast, on 2 May with 200 participants, the lowest temperature in the unoccupied banquet hall was 22.63 °C, while the highest temperature, after the attendees had gathered, was 27.78 °C, resulting in a temperature difference of about 5.15 °C. The data clearly show a significant increase in the rate of indoor temperature rise. This indicates that as the number of people in the banquet hall increases, the accumulation of indoor heat occurs rapidly, especially during the afternoon gathering period when the heat reaches its peak.

As shown by the curve variations in Figure 7, the temperature data at testing point 'c', located at the exhaust vent at the top of Banquet Hall No.2, exhibited irregular fluctuations. Compared to testing points 'a' and 'b', its curve is more complex. On 22 April, with 100 participants, the test data show that the exhaust vent temperature fluctuated between 23.9 °C and 24 °C. Although the overall fluctuation range was small, it consistently remained higher than both the inlet and indoor temperatures. On 2 May, with 200 participants, the exhaust vent temperature gradually increased from 26.15 °C to 33.05 °C, significantly higher than both the indoor and inlet temperatures. The fluctuation range increased by 6.9 °C compared to the 22 April scenario with 100 participants. This suggests that during periods of high occupancy and intense activity, the exhaust vent temperature experiences significant fluctuations. These irregular fluctuations are directly related to the density of people, the intensity of their activities, and the accumulation of heat processed by the exhaust system during the banquet.

Overall, the inlet temperature remains stable, with minimal fluctuations before and after the banquet, indicating that the "bottom underground ventilation + top exhaust" natural ventilation system performs well in maintaining steady inlet conditions. Indoor temperature, however, is significantly affected by the number of occupants, showing a gradual increase with the activities and progression of the banquet. The temperature variation on 2 May with 200 participants is notably greater than that on 22 April with

100 participants. Among the three temperature indicators, the exhaust vent temperature exhibits the greatest fluctuation, particularly during peak banquet periods, where it is significantly higher than both indoor and inlet temperatures. This reflects the system's sensitivity to indoor heat discharge.

4. Discussion

4.1. Analysis of Residual Heat Removal Efficiency and Its Key Influencing Factors

Based on the data from various instruments in Banquet Hall No.2, this study plotted the curves of room residual heat and residual heat removal efficiency for the two days (Figure 8). The variations in residual heat removal efficiency shown in Figure 8 indicate that the "bottom underground ventilation + top exhaust" natural ventilation system can effectively expel indoor heat, particularly when there are fewer occupants and lower activity intensity in the hall, resulting in higher heat removal efficiency. The data reveal a clear inverse relationship between indoor residual heat and heat removal efficiency. In the test on 22 April with 100 participants, indoor residual heat decreased from a peak of 28 kJ to 12 kJ, while the removal efficiency increased from 200% to 490%. In contrast, during the peak period of the banquet on 2 May with 200 participants, although the initial heat removal efficiency was high at 420%, as indoor heat increased to 50 kJ, the efficiency gradually decreased to 240%. These data suggest that the "bottom underground ventilation + top exhaust" natural ventilation system is more effective in expelling lower heat loads.

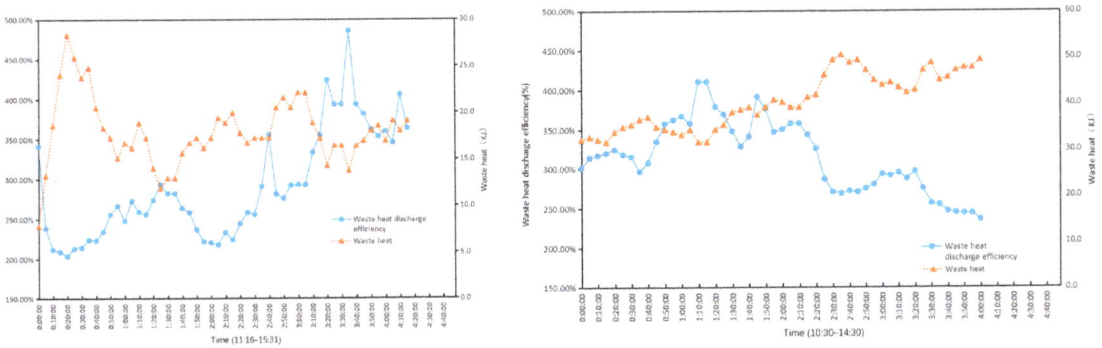

(**a**) (**b**)

Figure 8. Residual heat and residual heat removal efficiency in Banquet Hall No.2: (**a**) Residual heat and heat removal efficiency in the indoor space during the test on 22 April with 100 participants; (**b**) Residual heat and heat removal efficiency in the indoor space during the test on 2 May with 200 participants.

Thus, it is evident that the current "bottom underground ventilation + top exhaust" natural ventilation system used in the banquet hall is capable of effectively removing residual heat. However, during hotter external conditions and high-density banquet peak periods, indoor residual heat accumulates, leading to a decrease in removal efficiency. This highlights the significant impact of external climatic conditions and indoor occupancy density on the performance of the ventilation system, with higher density and activity levels leading to reduced effectiveness.

4.2. Discussion on the Dynamic Environmental Response Capability of the "Bottom Inlet + Top Outlet" Ventilation System

To accurately assess the impact of the "bottom underground ventilation + top exhaust" natural ventilation system on the indoor thermal environment, this study conducted ventilation simulations of the Banyan Tree Garden Banquet Hall using the DeST software based on field measurement data [36,37]. Additionally, to enhance the generalizability and broader applicability of the study results, the scope of discussion was further expanded to include simulation experiments in five representative cities within China's architectural climate zones. As shown in Figure 9, the selected typical cities are Harbin, representing the severe cold region; Beijing, representing the cold region; Chengdu, representing the hot summer and cold winter region; Guangzhou, representing the hot summer and warm winter region; and Kunming, representing the temperate region. The main parameters and model overview of this simulation experiment are presented in Table 2 and Figure 10. The architectural structure and construction parameters are based on measured data, while potential heat-generating equipment in the indoor environment, including lighting, computers, and other appliances, are treated as control variables in this study and are not the focus of discussion. Therefore, simulations were conducted using default values in the DeST software that conform to existing standards.

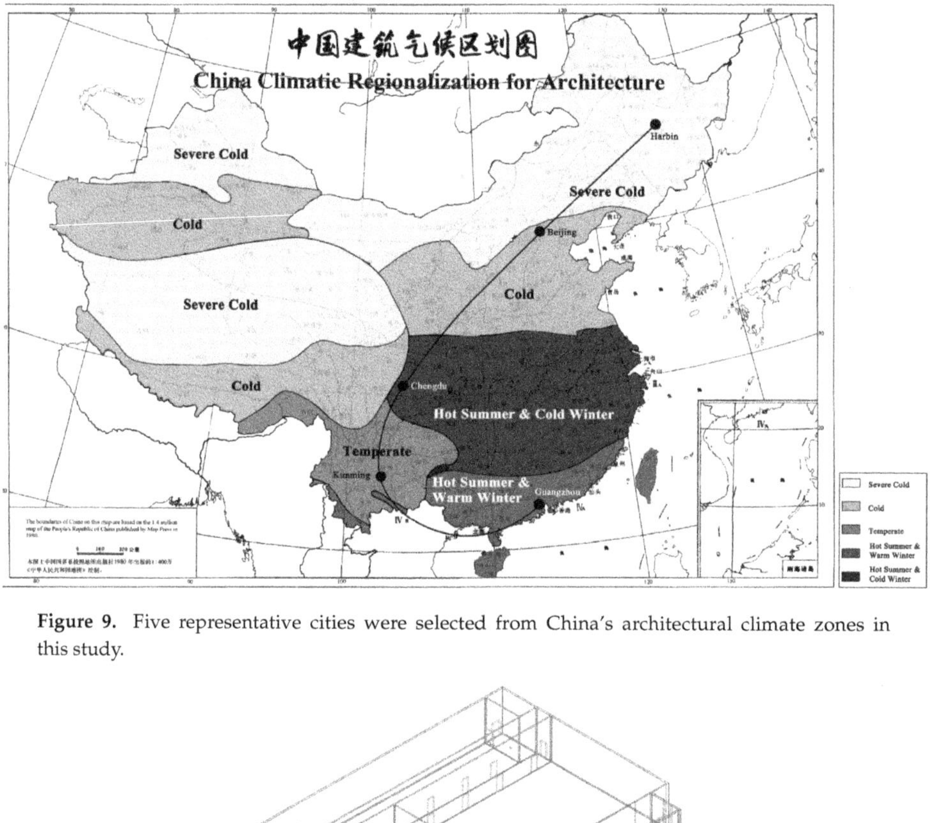

Figure 9. Five representative cities were selected from China's architectural climate zones in this study.

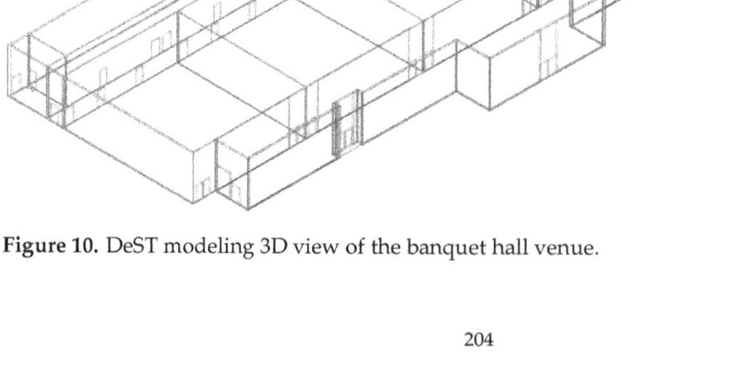

Figure 10. DeST modeling 3D view of the banquet hall venue.

Table 2. Model setting parameters.

Category	Name	Heat Transfer Coefficient W/(m²·k)	Notes
Exterior Wall	24 Brick walls + polystyrene board internal insulation	0.564	/
Interior Wall	Ceramic concrete interior wall	1.515	/
Exterior Window	Inert gas filled + low-e film coated hollow	2.000	/
Outer Door	Single solid wood exterior doors	0.35	/
Air Ventilation	Room interoperability wind	/	3 times/h
	Room and exterior ventilation	/	3 times/h

Before conducting simulation experiments for different representative cities, this study first performed a simulation experiment on the indoor temperature variations of Banquet Hall No.2 under natural ventilation conditions, based on the existing site conditions of the banquet hall. The experimental results, as shown in Figure 11, indicate that the lowest indoor temperature throughout the year is 15 °C, while the highest is 33 °C. The annual temperature trend shows an initial rise followed by a decline, closely mirroring the fluctuations in outdoor temperature. This indicates that the thermal insulation performance of the building envelope in the model plays a crucial role in maintaining indoor temperature stability, thus avoiding potential errors in the simulation process.

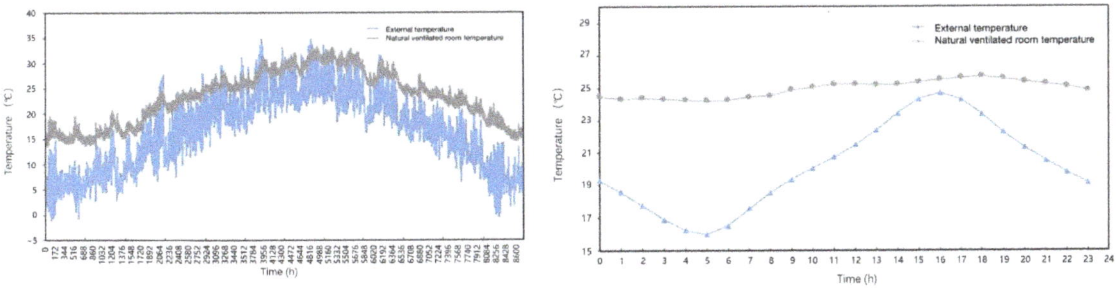

(**a**) (**b**)

Figure 11. Schematic diagram of test data for Banquet Hall No.2 based on real environmental conditions: (**a**) Comparison of outdoor and indoor temperatures throughout the year in the simulation experiment; (**b**) Comparison of outdoor and indoor temperatures throughout the day.

In the simulations for representative cities, the aim was to evaluate the performance of the "bottom ventilation duct + top exhaust" natural ventilation system by comparing indoor and outdoor temperature variations under natural ventilation conditions. The simulation data of indoor and outdoor temperatures for each city indicate that the overall trend of indoor temperatures closely follows the outdoor temperatures in all regions. However, applying the "bottom ventilation duct + top exhaust" natural ventilation system effectively reduced the amplitude of indoor temperature fluctuations. As shown in Figure 12, the effectiveness in reducing the amplitude of indoor-outdoor temperature differences across the five cities is ranked as follows: Guangzhou > Harbin > Chengdu > Beijing > Kunming. In Guangzhou, located in a hot summer and warm winter region, the annual amplitude of indoor-outdoor temperature fluctuations reaches a maximum of 56.7%, the best performance among all cities. This indicates that the system effectively mitigates the impact of high summer temperatures on the indoor environment, reducing the reliance on air conditioning. In Harbin, located in a severely cold region, the system reduces the amplitude of temperature fluctuations by 50.1%, demonstrating its capability to significantly lower heating demand even in winter. In Chengdu, characterized by hot summers and cold winters, the temperature fluctuation amplitude is 48.5%, indicating that the system ensures good stability and comfort of indoor temperatures during both hot and cold seasons. For Beijing, representing cold regions, and Kunming, representing temperate regions, the temperature fluctuation amplitudes are 40.4% and 37.5%, respectively. These lower amplitudes suggest relatively lower performance of the system in these two regions compared to the aforementioned cities. Through the performance evaluation of the "bottom ventilation duct + top exhaust" natural ventilation system in different cities, it can be observed that the system effectively alleviates discomfort caused by high temperatures and reduces air conditioning dependence. In severely cold regions, it also effectively lowers heating demand. Although the amplitude of temperature fluctuations varies across different climate zones, the system consistently demonstrates positive effects in maintaining indoor temperature stability and optimizing the indoor thermal environment.

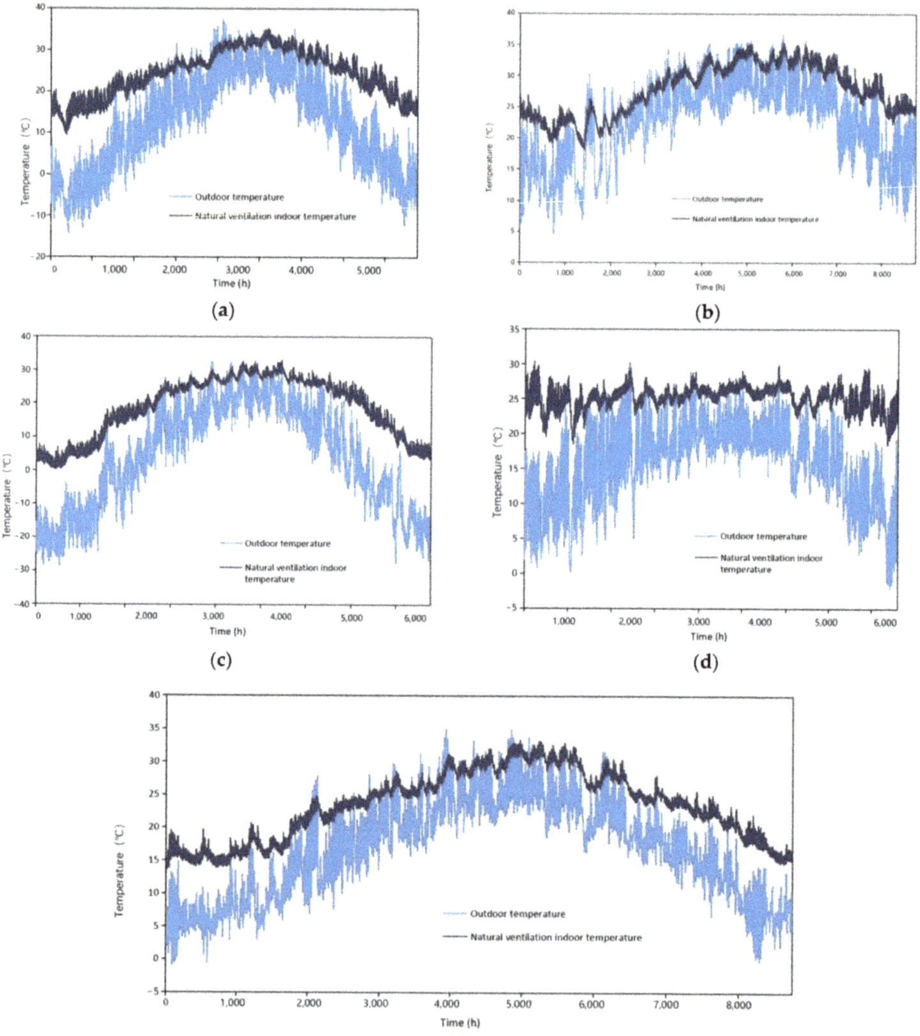

Figure 12. Schematic diagram of the test data of the banquet hall under different working conditions: (**a**) Beijing; (**b**) Guangzhou; (**c**) Harbin; (**d**) Kunming; (**e**) Chengdu.

5. Conclusions

This study highlights the effectiveness of the "bottom ventilation duct + top exhaust" natural ventilation system in managing the indoor thermal environment of the Banyan Tree Garden Banquet Hall during transitional seasons. Here are the key findings and limitations.

5.1. Key Findings

- Heat Removal Efficiency: The system effectively removes residual indoor heat, maintaining stable temperatures even when windows cannot be opened. At lower occupancy levels (up to 100 participants), the heat removal efficiency exceeds 490%. However, as occupancy increases to 200 participants, this efficiency drops to 240%, indicating that higher occupant density and external temperatures significantly affect the system's performance.
- Adaptability Across Climate Zones: The system's effectiveness varies by region, with temperature fluctuation reductions ranking from Guangzhou (56.7%), Chengdu, Harbin (50.1%), and Beijing, to Kunming. This demonstrates the system's potential to reduce reliance on air conditioning or heating across different climatic conditions.

5.2. Limitations

- Influencing Variables: The study primarily focused on occupant density, neglecting wind speed as a factor. The high frequency of calm winds in the Guanghan area (around 40%) contributed to this decision [38]. Future studies should include additional variables for a more comprehensive assessment.
- Seasonal Analysis: The current research only examines transitional seasons, lacking insights into ventilation performance during extreme summer and winter conditions. Future investigations will analyze data throughout the year.
- Occupant Density Variability: The study considered only two occupancy scenarios (100 and 200 participants), which may not capture the full range of occupant densities in actual use. Future research should include simulations for a wider variety of densities.
- Applicability to Other Building Types: While the findings indicate good performance for banquet halls, the research does not include a cost-benefit analysis for different types of public or residential buildings. Future studies should broaden the scope to evaluate system performance across various building types.

This study contributes valuable insights into the use of natural ventilation systems in public buildings, aligning with dual carbon goals and disaster mitigation efforts. Further research addressing the identified limitations will enhance the understanding of these systems' effectiveness in diverse contexts.

Author Contributions: Conceptualization, J.X.; methodology, J.X. and Z.T.; validation, Q.Y.; formal analysis, J.W. and J.C.; investigation, J.W. and Z.T.; data curation, Z.T. and J.C.; writing—original draft preparation, J.X. and Z.T.; writing—review and editing, J.X. and Q.Y.; visualization, J.W. and Z.T.; supervision, Q.Y.; project administration, J.X.; funding acquisition, J.X. All authors have read and agreed to the published version of the manuscript.

Funding: This research was funded by the National Natural Science Foundation of China (No. 52108032), the Sichuan Science and Technology Research Program (No. 2022NSFSC1148 and No. 25NSFSC1102), and Chengdu Social Science Research Program (No. YY2920240088).

Data Availability Statement: All data generated or analyzed during this study are included in this published article.

Conflicts of Interest: The authors declare no conflicts of interest. The funders had no role in the design of the study; in the collection, analyses, or interpretation of data; in the writing of the manuscript; or in the decision to publish the results.

References

1. Liu, Z.; Pan, X.; He, W.; Li, Y. Simulation Study on Natural Ventilation Performance in a Low-Carbon Large-Space Public Building in Hot-Summer and Cold-Winter Region of China. *Buildings* **2023**, *13*, 2263. [CrossRef]
2. Zhang, C.; Luo, H. Research on Carbon Emission Peak Prediction and Path of China's Public Buildings: Scenario Analysis Based on LEAP Model. *Energy Build* **2023**, *289*, 113053. [CrossRef]
3. Liu, K.; Tian, J.; Chen, J.; Wen, Y. Low-Carbon Retrofitting Path of Existing Public Buildings: A Comparative Study Based on Green Building Rating Systems. *Energies* **2022**, *15*, 8724. [CrossRef]
4. The State Council of the People's Republic of China. Notice of the State Council on Issuing the "2024–2025 Action Plan for Energy Conservation and Carbon Reduction. 2024. Available online: https://www.gov.cn/zhengce/content/202405/content_6954322.htm (accessed on 29 September 2024).
5. Xinhua News Agency. Opinions of the Central Committee of the Communist Party of China and the State Council on Fully, Accurately, and Comprehensively Implementing the New Development Philosophy to Achieve Carbon Peak and Carbon Neutrality. 2021. Available online: https://www.gov.cn/zhengce/2021-10/24/content_5644613.htm (accessed on 29 September 2024).
6. Shen, Z. Exploration on energy conservation of near zero energy consumption buildings under the goal of "carbon peaking and carbon neutrality. *New Build. Mater.* **2022**, *49*, 14–18.
7. China Building Energy Efficiency Association; Chongqing University Urban and Rural Construction; Development Research Institute. China Building Energy Consumption and Carbon Emissions Research Report (2023). *Constr. Archit.* **2024**, *2*, 46–59. (In Chinese)
8. Huang, H.; Cheng, F.; Su, Y.; Yao, L.; Hu, J. Energy-Saving Potential Analysis and Countermeasures for Carbon Peaking in China. *Strateg. Study CAE* **2021**, *23*, 81–91. [CrossRef]
9. Zhang, S.; Wang, K.; Xu, W. Research on Carbon Emission Control Index of Low-carbon, Nearly Zero-carbon and Zero-carbon Public Buildings. *Build. Sci.* **2023**, *39*, 35. [CrossRef]
10. Chen, X.; Kwan, M.-P.; Li, Q.; Chen, J. A Model for Evacuation Risk Assessment with Consideration of Pre- and Post-Disaster Factors. *Comput. Environ. Urban Syst.* **2012**, *36*, 207–217. [CrossRef]
11. Huang, L.; Xu, X.; Luo, K. Design of Cultural Architecture Refuge Space from the Combination of Normal Conditions and Calamity: Taking Case of a Northeastern University Libraries. *Archit. Cult.* **2022**, *2*, 30–31. [CrossRef]
12. Group Discussions: Contemporary City, New Human Habitat, and Architectural Design. *Archit. J.* **2020**, *Z1*, 2–27. [CrossRef]
13. Wang, Y.; Hua, J.; Ouyang, Q.; Li, X.; Zhu, Y. Human thermal comfort under simulated natural wind with different turbulence intensities. *Heat. Vent. Air Cond.* **2013**, *43*, 91–96.
14. Xu, J.; Chen, H.; Xiong, Y. Comparison of Natural Ventilation of Buildings in China and Abroad Based on Bibliometric Analyses. *South Archit.* **2021**, *6*, 114–123. (In Chinese)
15. Sui, X.; Tian, Z.; Yu, S.; Yan, B. Measurement and Analysis of Ventilation Rate and Indoor Air Quality of Two Typical Newly-built Residential Buildings in Xi'An in Winter: Natural and Mechanical Ventilation Scenarios. *Build. Sci.* **2024**, *40*, 120–130. [CrossRef]
16. Zhang, B.; Yan, Z. Simulation study on influence of residential building orientation on building energy consumption based on natural ventilation: Taking Ankang as an example. *Heat. Vent. Air Cond.* **2023**, *53*, 78–84. [CrossRef]
17. Ding, Y.; Hu, Y. Analysis of Key Influencing Factors of Indoor Natural Ventilation Effect in Residential Buildings: Taking Chongqing as an Example. *Archit. Tech.* **2021**, *27*, 106–108. [CrossRef]
18. Xu, J.; Wang, Q.; Yin, Z. Analysis of Energy Saving Potential of Natural Ventilation for Public: Buildings in Chaobei Area. *J. Lanzhou Inst. Technol.* **2021**, *28*, 46–51.
19. Chen, J.; Ye, X. A Study on Natural Ventilation Patterns of Public Buildings in Hongkong. *Inter. Des. Constr.* **2021**, *7*, 112–113. (In Chinese)
20. Guo, P.; Wang, S.; Xu, B.; Meng, Q.; Wang, Y. Reduced-Scale Experimental Model and Numerical Investigations to Buoyance-Driven Natural Ventilation in a Large Space Building. *Build. Environ* **2018**, *145*, 24–32. [CrossRef]
21. Weng, J.; Wang, H. Application of Tunnel Ventilation Cooling Technology in the Rural Residences for Hot-summer and Cold-winter Areas. *J. Hum. Settl. West China* **2013**, *2*, 114–117. [CrossRef]
22. Wang, J. Application of underground tunnel ventilation to air conditioning system for Mogao Grottoes' Tourist Service Center. *Heat. Vent. Air Cond.* **2011**, *41*, 15–17, 33.
23. Dai, X.; Su, H. Review on the influence of tunnel air system on indoor thermal environment. *Shanxi Archit.* **2022**, *48*, 156–159. [CrossRef]
24. Gui, L. Utilization of Air through Tunnel in Ventilation and Air Conditioning System of Buildings. Master's Thesis, Nanjing University of Science and Technology, Nanjing, China, 2010.
25. Chi, X. Study on the Developability of Hotel Ballroom Design. *Inter. Archit. China* **2022**, *2*, 175–176. (In Chinese)
26. Zhou, W.; Peng, C.; Liu, S.; Hao, B. Analysis on Factors Influencing the Energy Consumption of Hotel Building. *Build. Sci.* **2015**, *31*, 31–37. [CrossRef]
27. Wang, Y.; Tang, Y.; Liang, S. Influences of Different Ventilation Modes on Public Buildings' Base Room Temperature and Energy Consumption in Mianyang. *Build. Sci.* **2012**, *28*, 82–85, 95. [CrossRef]
28. Pei, F.; Liu, Y.; Wu, T.; Zhu, J. Bi-level Optimal Control Strategy for Heating, Ventilation, and Air Conditioning System with Collaborative Consideration of Air Quality and Thermal Comfort. *Autom. Electr. Power Syst.* **2024**, *48*, 151–160.

29. Tian, S.; Zhou, X.; An, J.; Yan, D.; Wu, R.; Sun, H.; Shi, X.; Jin, X.; Jiang, X. Comparing and Verifying the Capability of DeST in Calculating the Performance of Space—Cooling and Space—Heating Equipment Based on ASHARE Standard 140. *Build. Sci.* **2020**, *36*, 19–30. [CrossRef]
30. Yang, G.; Fu, Y.; Yan, M.; Zhang, J. Exploring the Distribution of Energy Consumption in a Northeast Chinese City Based on Local Climate Zone Scheme: Shenyang City as a Case Study. *Energy Explor. Exploit.* **2020**, *38*, 2079–2094. [CrossRef]
31. Shi, L.; Zhang, H.; Li, Z.; Luo, Z.; Liu, J. Optimizing the Thermal Performance of Building Envelopes for Energy Saving in Underground Office Buildings in Various Climates of China. *Tunn. Undergr. Space Technol.* **2018**, *77*, 26–35. [CrossRef]
32. Peng, B.; Hsieh, S.-J. Cyber-Enabled Optimization of HVAC System Control in Open Space of Office Building. *Sensor* **2023**, *23*, 4857. [CrossRef]
33. Elshafei, G.; Vilcekova, S.; Zelenakova, M.; Negm, A.M. Towards an Adaptation of Efficient Passive Design for Thermal Comfort Buildings. *Sustainability* **2021**, *13*, 9570. [CrossRef]
34. Li, Z.; Zhao, X. Principles of natural ventilation and its application in building design. *Sci. Technol. Innov.* **2009**, *3*, 256. (In Chinese)
35. Ministry of Natural Resources, People's Republic of China. Standard Map Service System. Available online: http://211.159.153.75/ (accessed on 18 October 2024).
36. Ye, K.; Li, Y.; Long, T.; Zheng, D.; Zhang, S. Experimental study on the natural ventilation of earth-to-air heat exchanger combined with solar chimney on the indoor thermal environment. *J. Chongqing Univ.* **2023**. (In Chinese)
37. Li, X.; Huang, K.; Feng, G.; Zhang, D. Analysis of Night Ventilation Energy Saving Potential for Ultra-Low Energy Consumption Buildings in Severe Cold Area. *J. Shenyang Jianzhu Univ. (Nat. Sci.)* **2022**, *38*, 520–527.
38. Chengdu Institute of Planning & Design. Interpretation of Chengdu's Wind Environment. Available online: https://www.cdipd.org.cn/index.php?m=content&c=index&a=show&catid=82&id=138 (accessed on 15 October 2024).

buildings MDPI

Article

Effects of Contractual and Relational Governance on Project Performance: The Role of BIM Application Level

Bing Yi [1,2] and Nina Lee See Nie [2,*]

[1] College of Civil Engineering, Sichuan University of Science & Engineering, Zigong 643000, China; yibing@suse.edu.cn
[2] Graduate School of Business, SEGi University, Petaling Jaya 47810, Selangor, Malaysia
[*] Correspondence: leeseenie@segi.edu.my; Tel.: +60-10-270-6171

Abstract: This study aims to explore the moderating role of Building Information Modeling (BIM) between project governance and project performance. The theoretical foundation of this research is rooted in transaction cost economics. The data come from the construction industry in China's Sichuan province. A dataset comprising 175 survey responses was subjected to analysis through the Partial Least Squares (PLS) method. The findings confirm that contract completeness and contract flexibility positively influence project performance, and trust in relational governance also has a positive impact on project performance. Additionally, the level of BIM application moderates the relationships between contract flexibility and trust with project performance. However, a significant positive relationship between contract completeness and project performance was not observed. These findings establish a groundwork for transitioning project governance research from a static to a dynamic viewpoint, thereby facilitating the practical implementation of BIM technology. As a result, this study enriches the academic comprehension of governance amidst digital transformation and provides actionable suggestions for fostering efficient governance practices within a technologically progressive landscape.

Keywords: contractual governance; relational governance; BIM application level; project performance

1. Introduction

As construction projects grow increasingly complex and expand in scale, a more diverse set of skills becomes essential for effectively managing project execution and delivery [1]. However, according to perspectives from construction firms on development and management, low efficiency and performance present significant challenges [2] that can impact the management approaches of researchers and workers involved in real projects [3]. The emergence of Industry 4.0 and advancements in technology are poised to introduce new criteria for evaluating project management [4]. This shift entails a departure from traditional management systems that prioritize the performance of projects over the conventional cost–schedule–quality "Iron Triangle" management paradigm [5]. Emphasizing the importance of fostering collaborative partnerships among project stakeholders has become a critical consideration [6]. Moreover, project governance has been utilized to address management system issues in recent years [7].

Numerous studies indicate that project management is often viewed as primarily focused on execution rather than as a comprehensive management system [1,8]. Given the increasing intensity of competition within the construction industry, it is imperative for firms to adopt effective management systems that encompass collaborative behaviors and information exchange among various stakeholders [9,10]. Extensive research consistently demonstrates that effective project governance plays a pivotal role in enhancing project performance [11,12]. This is because project governance not only addresses policy-making issues but also resolves inter-organizational challenges among project partners [13,14]. For example, Fernandes et al. [15] suggested that trust and interpersonal relationships are essential elements in the emergence of project governance in a study of project governance in organizations in Latin America and Europe. They also suggest that the professionalism, centralization, coordination and long-term control of partners have a positive effect on project governance, which contributes to the achievement of successful project management. Consequently, project governance has emerged as a prominent topic in management research. This growing interest is driven by mounting evidence of its substantial positive impact on project performance [16,17].

In China, research has shown that industries other than construction have higher profit rates [18]. Project performance is a critical factor influencing the profitability of construction projects. It is typically associated with meeting predetermined time, budget, and quality standards while fulfilling stakeholder expectations [19]. However, numerous factors beyond project performance can lead to cost overruns and delays in construction projects [20]. These include external factors such as adverse weather conditions, regulatory or market changes, supply chain disruptions, and even political instability, as well as internal factors like inadequate planning, insufficient resource allocation, communication failures, or technical challenges [21].

To address the challenges facing the industry, this study focuses on project performance, aiming to identify key performance indicators and management strategies to mitigate project management risks and prevent cost overruns and delays [20]. Moreover, well-defined project performance indicators can optimize resource allocation, ensure stakeholder satisfaction, enhance project partnerships, and improve social reputation [22]. These factors can strengthen a project's competitiveness and promote long-term sustainable development. In addition, the International Program in the Management of Engineering and Construction (IMEC) has reported that mega-construction projects with a value of one billion dollars often experience cost overruns of approximately 18% compared to the original cost plan [23]. Cost overruns and delays remain significant indicators of project performance [24,25]. Despite advancements in control tools and cost management technology, the incidence of cost overruns has not decreased over the past 70 years [26,27].

Currently, the majority of academic literature has concentrated on examining the positive influence on project performance, and has highlighted its potential role as a justice mechanism [28–31]. The contract governance and relationship governance approaches represent two primary research directions within the project governance mechanism framework [32]. The two forms of governance are not mutually exclusive; rather, they complement each other. While contractual governance provides the necessary structure and clarity for operationalizing agreements, relational governance enhances cooperation and facilitates the sharing of information necessary for adapting to changing circumstances [27,33]. It is worth noting that these studies have primarily been conducted within the manufacturing industry and other related sectors [34].

The construction industry, in comparison to other sectors, is characterized by its complexity and involvement of numerous stakeholders [35]. Projects within this industry are temporary in nature, concluding upon the achievement of specific objectives without a stable organizational structure [36]. Effective management methods are essential to address the unique challenges presented by project management in this context. For example, issues such as opportunism and communication efficiency can pose difficulties in project management [37]. Existing research has identified two key gaps in the literature. Firstly, prior studies have primarily focused on governance structures based on internal management policies within companies, overlooking the significance of contractual governance frameworks in enhancing performance outcomes. Mature contractual agreements can help mitigate opportunistic behaviors [38]. Secondly, while some research has explored the relationship between project governance and performance from a relational perspective, findings suggest that performance is influenced by managerial behavior [39]. However, these studies have not adequately addressed issues related to information transfer efficiency and the active role of managers in this process.

While contractual governance and trust can help establish cooperative mechanisms and relationship rules among stakeholders, communication and coordination platforms are still necessary to enhance information exchange and execution efficiency [39]. Borkowski [40] suggested a Building Information Model (BIM), referring to a digital information model used for design, construction, and operational decision-making in building projects. BIM fosters a collaborative environment by enabling real-time interactions among project stakeholders, leading to improved decision-making and problem-solving [38]. BIM enhances visualization through the construction of 3D models, allowing project managers and stakeholders to better understand design and construction aspects, thereby enabling more informed decision-making [41]. This approach improves collaboration, strengthens visualization, and addresses technical issues. The data management capabilities of BIM promote equity among stakeholders by allowing each member to access project data and make necessary corrections [42]. This ensures that changes are visible to all and can be collectively discussed before implementation. Furthermore, BIM provides real-time updates and facilitates centralized data management, creating a collaborative environment that enables more effective communication and coordination among project team members [43]. For example, during the foundation construction phase of engineering projects, the application of BIM technology can optimize construction strategies under both static and dynamic loading conditions, thereby significantly enhancing the efficiency of foundation construction [39].

Therefore, this collaborative approach helps mitigate risks, ensure alignment with strategic objectives, and increase the overall success rate of construction projects [44]. Despite the numerous benefits of BIM technology, its adoption is not without challenges. Significant barriers remain, including lack of skills and experience, access-related obstacles, and the need for continuous training and upskilling of professionals in the Architecture, Engineering, and Construction (AEC) sector [45].

This study addresses two significant gaps in the current research by examining the relationship between project governance and project performance. Specifically, we explore the following research questions:

1. How does project governance impact project performance in China?
2. Does the application of BIM technology in project governance have a positive impact on project performance?

Currently, quantitative research examining the relationship between project governance and project performance remains limited, despite the growing importance of this field, and most studies tend to explore it qualitatively [46]. To solve the situation, the study will use empirical analysis methods for the in-depth exploration of the mechanisms of action between project governance and project performance in the Chinese construction industry. Specifically, this paper will analyze how contract governance and relational governance affect project performance. It will also examine the relationship between project governance and project performance, and whether this will be influenced by the level of BIM application.

This study has the following components. In Section 2, we will elaborate on the theoretical background of concepts such as project performance, contract management, and relationship management, as well as relevant knowledge of transaction cost theory (TCT). In Section 3, we will state our research hypotheses. In Section 4, we will discuss the methodology of our study, including data sample selection, data collection process, and evaluation indicators. In Section 5, we will provide an in-depth analysis of the experimental results. In Section 6, we will debate our research findings. Finally, in Section 7, we will summarize the core discoveries of this article and share our conclusions.

2. Literature and Theory Framework

In this section, the research literature and theoretical models related to this topic will be introduced. First, the theoretical framework based on the Cost of Transaction Theory will be analyzed. Secondly, the previous research results will be summarized from dimensions such as project performance, contract governance, and relational governance, and a conceptual model for this study will be established.

2.1. *Transaction Cost Economic Theory*

The concept of transaction costs was initially proposed by Coase in 1937 and was later further researched and developed into a comprehensive theoretical framework by Williamson [47]. Bounded rationality and opportunism are the assumptions that must be taken into account when applying TCT. The theory posits that in economic activities, individual behavior is not entirely rational [48]. Due to personal differences, there are limitations in information processing and cognitive abilities. These limitations create barriers for participants when acquiring and processing information, thus affecting their ability to make entirely rational decisions [49]. To compensate for these limitations, participants must invest resources in collecting and processing information to ensure the rationality of their decisions [50].

However, opportunistic behavior refers to the possibility that individuals may deliberately disclose incomplete or misleading information, seeking their own interests through means such as deception, misdirection, or camouflage [25]. This behavior adds complexity to economic organizations. Due to the presence of bounded rationality and opportunism, organizational management becomes more complex, and the uncertainty and risk of transactions also increase [51,52]. To address these challenges, participants need to invest more resources to verify the authenticity of information, assess the risk of being deceived, and increase safeguards to reduce transaction uncertainty and the risks it bears [53].

The TCT has a broad research foundation in the field of project governance. Due to the complexity and dynamism of engineering projects, they are often regarded as temporary organizations [54]. In such organizations, participants are usually temporarily assembled, and there may be new members joining or existing members leaving during the course of the project. Because each participant's interests differ, this dynamism may exacerbate behaviors of bounded rationality and opportunism, which can increase project costs and affect the successful delivery of the project [55].

The use of TCT in the construction sector covers several key areas, such as contract design, risk management, relations with suppliers, and project governance [56]. Firstly, the application of this theory helps create strong contracts that efficiently reduce risks and lower transaction costs [57]. Then, it allows companies to select appropriate governance structures, crucial for additional risk reduction. Moreover, it encourages the development of long-term partnerships with suppliers, which in turn lessens search costs and maintains higher quality standards, ultimately decreasing overall transaction costs [58]. Additionally, the theory supports a fair allocation of contractual responsibilities among all involved parties. This equilibrium, guided by governance frameworks, strengthens cooperation and cuts down the costs related to project execution [59,60]. For a detailed view, please refer to Figure 1.

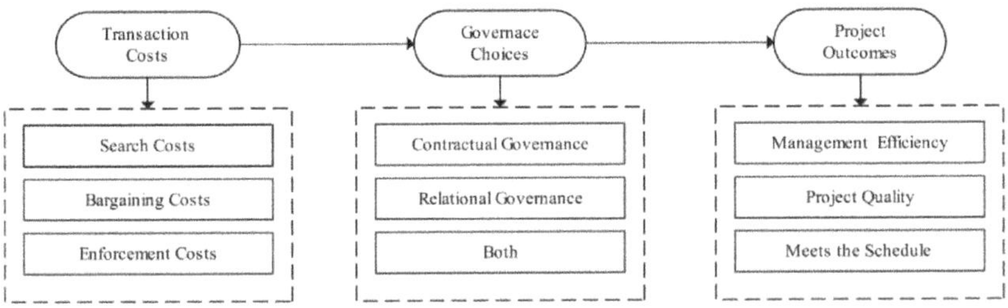

Figure 1. The TCT conceptual figure for the construction application process.

However, Western scholars have predominantly approached trust from an economic perspective, conceptualizing it as an expectation that the other party will not act opportunistically in the face of uncertainty or risk [61]. Trust is thus a latent psychological state resulting from behaviors or choices that subsequently influences future actions and decisions [62]. In construction projects, trust between owners and contractors can be defined as a mutual willingness to believe in and rely on each other, while also accepting potential economic and other losses during the collaboration process [63]. Among the various elements of relational governance, trust is considered a crucial factor influencing contractor behavior and is often viewed as the core element of relational governance [39]. As a state variable, trust has consistently been regarded as the most fundamental relational norm and holds a significant position in the research framework of project governance [11]. Based on this, this study adopts trust as one of the independent variables in its research model.

In addition, the TCT is frequently used to explore project governance mechanisms [64,65]. Wacker et al. [66] pointed out that TCT is a key theoretical framework that can be used to investigate how contract governance can simultaneously enhance financial returns, boost competitiveness, and reduce performance uncertainty. Based on this theory, Young et al. [67] employed quantitative analysis methods to examine the impact of project governance on the successful delivery of IT projects, finding that explicit contract terms, implicit contract terms, reputation, and trust all played a role in mitigating project risks. For instance, Trygges et al. [68] demonstrated that project governance is a crucial factor influencing project success in supply chains. Their research highlighted the significant role of informal relationships in reducing supply chain costs. Zhou et al. [69] employed TCT to delve into the governance mechanisms of food supply chains. Their findings indicated that both contractual and relational governance mechanisms exerted a positive influence on the digitalization of supply chains. The study found that these two governance mechanisms play a positive role in promoting the food industry to achieve large-scale sustainable development.

The aforementioned research findings consistently demonstrate that project governance is a critical factor in enhancing project success and improving overall project performance. Therefore, this study, based on the TCT, proposes that contract governance (including the completeness and flexibility of contracts) and trust in relational governance will play significant roles in constraining bounded rationality and opportunism. The theoretical model of this study is constructed accordingly, based on the above theory. Based on the above theories, the theoretical model of this study is set up in Section 3.

2.2. Project Performance

Project performance is one of the core issues in the field of engineering project research, and its definition varies depending on the research perspective (mainly divided into outcome-oriented and behavior-oriented approaches) [70,71]. From an outcome-oriented perspective, project performance focuses on work results, which include the achievement of organizational goals, stakeholder satisfaction, and return on investment [72]. Therefore, project performance can be defined as the measure of the results produced under specific times and work contents. However, some scholars argue that project performance is not equivalent to outcomes alone, as outcomes are influenced by many external factors [55,73]. They believe that project performance should also include the methods or means of achieving organizational goals. Additionally, some scholars consider project performance to be part of organizational performance, reflecting the overall operation of the enterprise or organization and requiring the establishment of a comprehensive model to evaluate whether the organization and individuals have achieved the planned goals [74].

Existing research indicates that the assessment of performance indicators commonly uses project maturity models, excellence performance evaluation models, and key performance evaluation models [75]. In the Chinese context, the key performance evaluation model is the most widely used method [76]. The key performance evaluation model, based on key success factors, can start from the overall strategic goals of the project, subdividing the project into different levels and stages, thus developing more precise evaluation indicators [71]. The multi-faceted nature of engineering projects necessitates a nuanced approach to performance evaluation. The proposed framework acknowledges the complexity of these projects by recognizing the distinct requirements and objectives associated with each phase of development [72]. This stage-specific and hierarchical goal-setting approach allows for a more accurate and comprehensive assessment of project performance, accounting for the diverse stakeholders involved and their respective targets. Such a tailored evaluation method can potentially lead to more effective project management and improved overall outcomes in complex engineering endeavors [63].

The evaluation model of project performance has shifted from the traditional "iron triangle" of cost, quality, and time to include more success criteria, such as customer satisfaction, stakeholder satisfaction, and knowledge management [77]. This study uses a comprehensive indicator of economy, efficiency, and effectiveness to evaluate project performance, considering not only the impact of individual factors on the project but also the overall impact of the project on society, the environment, and investment efficiency. This evaluation system is more suitable for the current transformation of Chinese engineering projects from extensive to refined development goals.

2.3. Contractual Governance

Contract governance is a formal mechanism in project governance that involves the establishment of a system of legally binding contracts, including clear instructions, regulations, and rules, to define the powers and duties of the parties involved [39]. In engineering project management, formal contract mechanisms are a common practice and serve as an important means for stakeholders in temporary organizations to quickly establish trust. Therefore, for such temporary organizations, contracts play a significant role in constraining the behavior of members, especially in the handover of construction processes [33,78]. Heydari et al. [42] have demonstrated that the application of BIM and block-chain technologies to construction material procurement can facilitate the achievement of contractual governance objectives and enhance decision-making flexibility through effective information management. The TCT suggests that stakeholders can use contract terms to protect their interests and limit the opportunistic behavior of others [79]. At the same time, it can resolve disputes and conflicts, thus improving transaction efficiency.

Therefore, this study will focus on two core principles of contract governance: contract completeness and contract flexibility [80]. Contract completeness refers to whether the contract terms are comprehensive and clearly define the rights and obligations of the parties involved [80]. Contract flexibility, on the other hand, considers the contract's adaptability in the face of changes and instability in the project organization, for example, project adjustments, delays, and price changes; that is, looking at whether the contract can flexibly respond to these changes [50].

Based on previous context, this study endeavors to investigate how contract governance can more effectively restrain the actions of project participants, diminish transaction costs, and enhance the comprehensive performance of the project.

2.4. Relational Governance

Relational governance is an informal mechanism in project governance that complements formal contract governance [69]. It primarily refers to the binding force of social Relational rules and social norms in transaction activities, guiding and adjusting the behavior of both parties through the application of a series of relational norms to coordinate their relationships [81]. There is a complementary relationship between contractual governance and relational governance. Contracts can provide a foundation for building trust and cooperative relationships [82], while relational norms can fill the gaps in incomplete contracts. Strong relationships can facilitate contract negotiations. However, an excessive emphasis on contracts may limit the flexibility of management activities, while over-reliance on relationships can lead to unclear responsibilities [83]. Unlike contract governance, which relies on laws and contract clauses, relational governance relies on informal mechanisms such as mutual dependence and cooperation to respond to environmental changes, coordinate transactions, and ensure the smooth progress of transactions, using relational behaviors such as trust between participants to curb the occurrence of opportunistic behavior [84].

Moreover, trust is an important indicator in the relationships between participants, as it can prompt both parties to address issues in cooperative relationships based on shared values, enhance adaptability to environmental changes, and reduce transaction and communication costs [85,86]. In projects, trust is key to the development of relationships between service providers, suppliers, contractors, and clients [86]. Once a foundation of trust is established between the parties, cooperation becomes smoother [87]. Trust is also regarded as an alternative to control mechanisms, which can reduce agency costs inherent in project delivery methods [88]. Moreover, trust, as a soft constraint, can promote cultural integration among participants, drive different trading parties to form unified values and goals, establish cooperative relationships of shared risks and benefits, and thus enhance supply chain performance [85]. In research on EPC (Engineering, Procurement, and Construction) projects, scholars have found that trust in relational governance has a significant positive impact on the performance of EPC projects [89]. Based on TCT and Social Exchange Theory, some scholars have found, in China's situation-relational society, that trust in the informal relationships between cooperative parties can effectively mitigate the impact of environmental and project uncertainties, thereby improving project performance [27,90].

2.5. BIM Application Level

In recent years, BIM technology has been recognized as an innovative tool and data management solution that is used in all phases of the project lifecycle [91]. BIM technology digitizes building information and integrates the goals and plans of different participants in the project [92]. Therefore, BIM platforms can resolve obstacles arising from differing goals and ideas among project participants, which often become barriers to achieving project objectives [93]. In contract governance, costs can be categorized into contract development costs and contract management costs. BIM technology primarily focuses on reducing contract management costs [94]. When contract changes, unforeseen events, or other unpredictable circumstances arise, the BIM platform can efficiently establish communication channels and rapidly develop feasible implementation plans [42]. Consequently, it minimizes resource waste and delays caused by ineffective communication and coordination, indirectly improving project management efficiency [93]. In other words, the lack of trust and information sharing is a key issue leading to inefficient communication, which in turn affects project performance.

Additionally, the data from the planning, design, construction, and operational aspects of the project will be recorded into the BIM platform [93]. Meanwhile, it can also effectively resolve conflicts and promote collaboration among participants [91]. In other words, the BIM technology can reduce costs, enhance communication, and improve information sharing among stakeholders throughout the project lifecycle. (Figure 2). For example, Marinho et al. [95] argue that BIM applications improve collaboration efficiency by minimizing errors, building trust, and eliminating information asymmetry. In addition, the application of BIM cause a reduction in misunderstandings and conflicts that will improve collaboration and communication efficiency. Rajabi et al. [96] indicate that the key evaluation criteria for BIM capabilities, such as the quality of BIM models, the efficiency of BIM collaboration, and the utilization of BIM data, are significant factors influencing project performance. Additionally, Wu et al. [97] point out that BIM maturity refers to the comprehensive level of an organization or project team in adopting and implementing BIM technology. It reflects the maturity of the organization in terms of processes, technology, personnel, management, and strategic planning in BIM application [98,99]. Sadeghi et al.'s [100] research highlights the potential application of BIM combined with advanced 3D optimization techniques in transportation engineering projects. This integration can significantly reduce the negative environmental impacts of construction, such as by mitigating the effects of vibrations on the surrounding environment during the construction process.

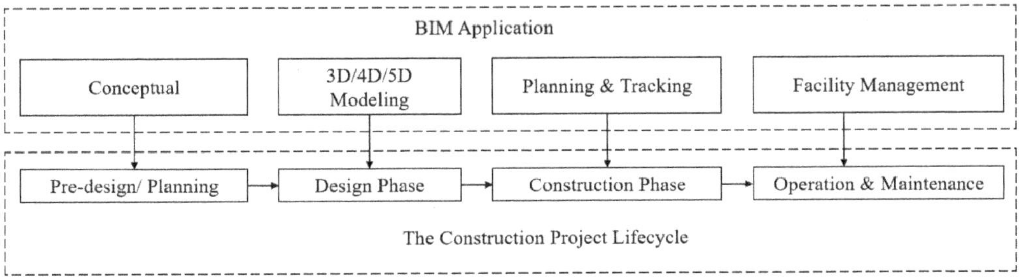

Figure 2. The application of BIM in the whole project life cycle.

BIM technology has demonstrated extensive applicability in healthcare-related engineering projects. Arjanaki et al. [101] investigated the efficacy of BIM in uncertain conditions, revealing its capacity to provide substantial support for real-time control system engineering in the healthcare sector. Their findings indicate that BIM significantly enhances real-time decision-making processes in these complex environments. Complementing this research, El Fathi et al. [102] explored BIM's utility in managing intermittent projects within the healthcare industry. Their study demonstrated that BIM technology facilitates improved data control in such projects, enabling the more effective formulation of strategies for periodic project review and adjustment. Collectively, these research outcomes underscore the potential of BIM technology in addressing the need for adaptive control mechanisms in project management. By enhancing real-time decision-making and improving data control for intermittent projects, BIM offers promising solutions to the challenges of managing complex, dynamic healthcare engineering projects.

Therefore, compared to traditional management models, project governance has achieved more effective management by establishing structured decision-making mechanisms and control systems [36]. BIM technology has significantly impacted project management practices and outcomes by improving collaboration, enhancing communication, and providing better project visualization [45]. Therefore, when examining the mechanisms through which project governance influences project performance, it is essential to consider the supportive role of BIM technology.

While the moderating role of BIM technology in relational governance and knowledge collaboration has been well-established, empirical studies examining how these elements collectively influence project performance from the integrated perspective of contract governance and relational governance remain scarce [38]. Therefore, this study aims to propose and test hypotheses to investigate the mechanisms through which these factors impact project performance [38,98].

3. Hypothesis Development

3.1. Contract Governance and Project Performance

The core of contractual governance lies in balancing contract integrity and flexibility. Integrity emphasizes clarity, comprehensiveness, and the enforceability of terms, while flexibility ensures that the contract can adapt to a constantly changing external environment [103]. Poppo et al. [104] found that more detailed and specific contract terms help to improve the effectiveness of project execution, highlighting the close relationship between contract integrity and project efficiency.

Therefore, flexibility in contracts refers to the adaptability of contractual terms when facing unforeseen events, allowing effective risk control during project execution [105]. When unexpected events occur, project participants can fulfill their responsibilities according to the contract terms rather than focusing on short-term opportunistic gains [106,107]. In other words, contract flexibility creates room for adjustments, allowing for sufficient communication in the face of uncertainties or issues, thereby fostering an effective cooperative environment and enhancing transaction performance. Based on this, the following hypotheses are proposed:

H1: *Contract completeness positively influences project performance in construction projects.*

H2: *Contract flexibility positively influences project performance in construction projects.*

3.2. Relational Governance and Project Performance

Relational governance is an informal governance mechanism grounded in trust and commitment, emphasizing interaction and cooperation between organizations during project execution [95]. Extensive research in the field of engineering projects has demonstrated that trust is founded on positive expectations of the other party's behavior and a willingness to accept associated risks [23]. For instance, trust between the owner and contractor is reflected in their mutual reliance, even in the face of potential economic or other losses during cooperation [108]. Moreover, trust helps establish a sense of identity among project participants, strengthens unity, and promotes information exchange [62], thereby reducing opportunistic tendencies and enhancing the quality and efficiency of cooperation, ultimately improving project performance [66].

Hence, this study proposes the following hypothesis:

H3: *Trust positively influences project performance in construction projects.*

3.3. The Moderating BIM Application Level

Previous studies have shown that the application of BIM technology plays a positive role in the relationship between project governance and performance [109–111]. For instance, in BIM-supported EPC projects, contract completeness effectively facilitates collaboration, thus positively influencing project outcomes [112]. However, in real-world cases such as the Shanghai Expo project, the failure to clearly define the responsibilities of different parties in the contract led to inefficient information transmission, negatively impacting cooperation. This underscores the importance of contract flexibility in a BIM environment [113]. Moreover, studies indicate that in the construction phase of PPP projects, BIM technology helps to maintain a positive correlation between contract flexibility and project performance [114]. Trust significantly enhances team collaboration efficiency, and the level of BIM application serves as a key moderating factor that supports broader social goals, such as promoting transportation equity [115,116].

Therefore, they found that higher levels of BIM application positively affect team cooperation efficiency and contribute to improved project performance. Collectively, these studies suggest that the moderating role of BIM application levels in the relationships between contract completeness, flexibility, trust, and project performance is not only theoretically plausible but also supported by empirical evidence and practical applications.

Therefore, this study proposes the following hypotheses:

H4: *The effect of contract completeness on project performance with the BIM application level is higher in construction projects.*

H5: *The effect of contract flexibility on project performance with the BIM application level is higher in construction projects.*

H6: *The effect of trust on project performance with the BIM application level is higher in construction projects.*

All hypotheses from this study are shown in Figure 3.

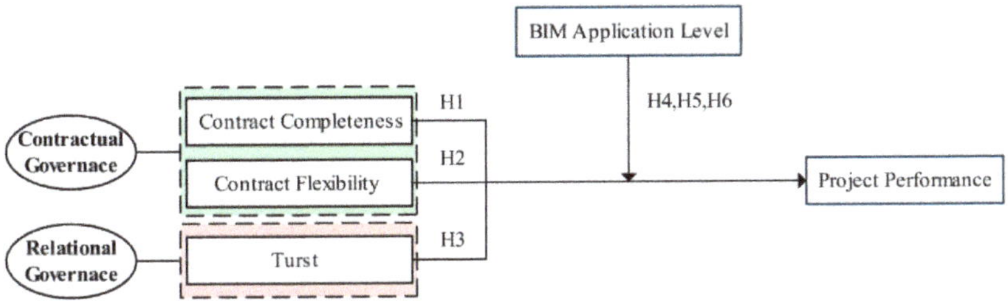

Figure 3. The research framework model.

4. Research Method

The specific research methods for this study are shown in Figure 4.

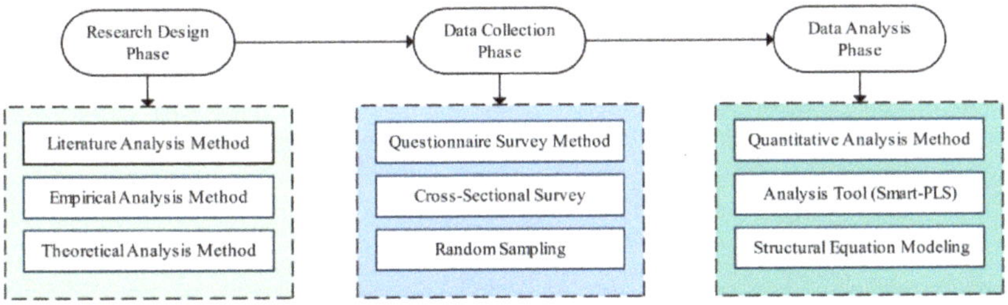

Figure 4. The research method of this research.

4.1. Sampling and Data Collection Procedures

This study adopts a positivist quantitative research method and uses a cross-sectional survey design to collect data. Online survey questionnaires are widely used in project management research because they can reduce costs, save time, and ensure that data collection is based on statistical principles. Given China's extensive infrastructure development, rapid urbanization, diverse regional characteristics, and widespread adoption of emerging technologies, a unique and dynamic context for studying project management practices is provided. This study will focus on middle- and senior-level managers in China's construction industry who have been involved in large-scale projects with investments exceeding RMB 1 billion [9]. Participants will include representatives from owners, contractors, design firms, and cost consulting companies.

To ensure that the target population has a rich project experience, individuals were selected from different construction companies in Sichuan Province, including from economically developed cities such as Chengdu, Mianyang, and Yibin. These companies engage in engineering projects such as subways, housing, water conservancy, and bridges. To ensure the validity and reliability of the survey instrument, three experienced project managers with a minimum of ten years in the construction industry and a proven track record of managing at least three large-scale projects were invited to conduct a rigorous review. Their extensive industry knowledge was instrumental in assessing the survey's alignment with industry practices. To maintain data quality and mitigate bias, the study was confined to professionals within a specific industry sector. Questionnaire items underwent multiple rounds of refinement by experts and scholars with engineering backgrounds, with careful attention paid to consistency between Chinese and English versions. Data cleaning procedures were implemented to identify and remove inconsistencies, errors, and outliers. Common method bias analysis and other statistical techniques were employed to assess data quality and address potential biases.

The data collection process took 2 months and was divided into three stages: first, identifying suitable research subjects; second, sending them emails; and finally, sending the survey questionnaire to those who responded to the email to ensure the quality and response rate of the questionnaire. Based on G-POWER software (version 3.1) testing, the sample size was determined to be 138, which helps to ensure the effectiveness of the sample size. The variables of the survey questionnaire in this study were derived from previous research literature; therefore, they have good reliability and validity.

4.2. Measurement

This study measures 25 indicators, all derived from the existing literature. This ensures that the measurements meet academic requirements. Since the measurements selected for this study come from the English-language literature, the indicators were carefully translated and reviewed to maintain their accuracy and relevance. The first step was to translate them into Chinese, and then to invite experts to review them to make them more understandable to the research subjects and to prevent any ambiguity in their responses. Finally, the Chinese translations were compared with the original English to make appropriate adjustments to the original items and to conduct a pre-test to check the feasibility of the items. Each item in the questionnaire used a five-point Likert scale (1 = Strongly Disagree to 5 = Strongly Agree).

The variables and their measurement indicators in this study are contract completeness ([78]), contract flexibility [31], trust [27,62], BIM application level [38], and project performance [117,118]. Table 1 lists the measurement items.

Table 1. Operating variables and survey questionnaire.

Variables	Operationalization	Survey Questionnaire	References
Project Performance (PP)	Eight items Five-point Likert Scale	1. After the project is completed, the cost target is achieved. 2. After the project is completed, the capital value target is achieved. 3. The important decisions in the project implementation are made in a timely manner. 4. During the project implementation, the relationship between the government and the private sector is harmonious. 5. During the project implementation, the targets of safety and environmental protection are achieved. 6. The important decisions in the project implementation are correct. 7. After the project is completed, all parties are satisfied with the project delivery results, and the quality and schedule targets are achieved. 8. After the project is completed, it receives positive feedback from the public (or users).	[117,118]
Contract Completeness (CC)	Four items Five-point Likert Scale	1. The contract includes detailed specific clauses (project features, rewards, breach resolution methods, etc.). 2. The contract terms are thoroughly detailed. 3. The contract clearly specifies the rights and responsibilities of all parties. 4. The contract provides detailed procedures for conflict resolution.	[78]

Table 1. *Cont.*

Variables	Operationalization	Survey Questionnaire	References
Contract flexibility (CF)	Five items Five-point Likert Scale	1. The contract clauses set a floating range to deal with potential risks or uncertainties. 2. The contract clauses provide corresponding solutions for potential risks or uncertainties. 3. The contract allows us to supplement, adjust or perfect some clauses for certain problems. 4. The renegotiation procedure in the contract is flexible. 5. According to the contract, we can easily apply for reasonable changes.	[31]
Trust (TR)	Five items Five-point Likert Scale	1. We believe that our project partners have sufficient capability to execute their tasks. 2. We believe that our project partners can meet the technical and managerial requirements of the project. 3. We believe that the engineers and other technical personnel involved in the project are competent in their roles. 4. We trust that our project partners can fulfill the contractual agreements. 5. We have confidence that our project partners can uphold their commitments throughout the entire project process.	[27,62]
BIM Application level (BAL)	Three items Five-point Likert Scale	1. Managed 3-D environments are established in a separate discipline BIM model in which data exchange is mainly on the basis of proprietary of exchange formats. 2. The model contains rich data, including program data, cost information and other dimensional data. 3. Fully open process with a unified project model and data integration and exchange among key contracting parties.	[38]

4.3. Data Analysis

In this study, data analysis was primarily undertaken employing SPSS 26 software and PLS-SEM (Partial Least Squares Structural Equation Modeling) techniques. SPSS 26 was used for demographic descriptions and common bias analysis of the sample data, while PLS-SEM assessed the strength of relationships between research variables. Notably, PLS-SEM has been widely used in management research within construction projects over the past few decades, exploring causal relationships among variables and demonstrating its practical validity.

This study chose PLS-SEM over AMOS (Analysis of Moment Structures) for two main reasons. Firstly, the relatively small sample size might not meet the multivariate normal distribution requirements. Secondly, the conceptual model and non-normal variables are better suited to PLS-SEM, which is more adaptable to exploratory models. In contrast, AMOS is more appropriate for confirmatory theoretical research, which demands a more rigorous theoretical foundation. Therefore, Smart-PLS 4.0 software was used for structural modeling and evaluation.

Data analysis in this study was conducted in two phases. In the first phase, SPSS software was used to analyze the data. Initially, the statistical functions were employed to conduct a demographic analysis of the sample data. Subsequently, factor analysis and reliability analysis were performed to assess the validity and reliability of the data. Finally, the IBM SPSS AMOS functionality was utilized to conduct a common method bias analysis. In the second phase, Smart-PLS software was employed to assess the measurement model and structural model, thereby validating the research hypotheses.

5. Results and Findings

5.1. Descriptive Model

A total of 210 samples were collected, and after eliminating incomplete samples, 175 valid samples remained. After the data collection phase was completed, this study conducted a demographic analysis using SPSS software with the objective of verifying whether the sample's demographic distribution was scientifically and reasonably structured. The specific results of the analysis are presented in Table 2.

Table 2. Demographics and profiles of respondents.

Item	Frequency	%	Item	Frequency	%
Gender			Level of Education		
Male	102	58	Below Undergraduate	4	2
Female	73	42	Bachelor's Degree	85	49
Total	175	100	Master's Degree	52	30
			Doctor's Degree	34	19
			Total	175	100
Age			Working Time		
30–39	111	63	Less than 5 years	93	53
40–49	40	23	5–10	34	19
50–59	18	10	10–15	32	18
Over 60	6	3	More than 15 years	16	9
Sector			Project Type		
Leader	30	17	Building Construction Projects	95	54
Project Manager	88	50	Infrastructural Construction Projects	54	34
Others	57	33	Others	26	15

5.2. Common Method Bias

To reduce common method bias in data collected during the data collection process, we employed several strategies. Firstly, the questionnaire content was well designed. Secondly, each question was clearly articulated to prevent confusion. Finally, volunteers' information will be kept confidential. Additionally, the items in the questionnaire were disordered, preventing volunteers from predicting the objectives of the study. Furthermore, statistical methods were used to test for common bias, such as Harman's single-factor test, used to examine common method bias issues. The results indicate that without rotation, five factors had eigenvalues greater than 1, with the first factor explaining 21.344% of the variance, which is less than 50%. Therefore, common method bias will not impact subsequent hypothesis testing.

Additionally, this study employed a Confirmatory Factor Analysis (CFA) model to further investigate the issue of common method bias. The results indicated that the single-factor model had the poorest fit ($\chi 2/df = 3.732$, TLI = 0.438, CFI = 0.485, IFI = 0.494, RMSEA = 0.195), further confirming that common method bias is not a significant concern in this study. At the same time, given the limitations of the Harman single-factor test, such as its lack of sensitivity, we drew on previous research by comparing the model with the method factor included in the original model [115]. Although the five-factor model with the method factor ($\chi 2/df = 1.825$, TLI = 0.921, CFI = 0.957, IFI = 0.937, RMSEA = 0.055) outperformed the four-factor model (original hypothesized model), the CFI of the five-factor model only increased by 0.001 compared to the hypothesized model, which is below the 0.050 threshold. In summary, this study does not exhibit significant common method bias.

5.3. Measurement Model

This study conducted tests on the measurement model using indicators of structural reliability and validity. We utilized SPSS Statistics 26.0 software to perform Cronbach's α coefficients, KMO values, and Bartlett's test of sphericity. The results indicated that all variables surpassed the threshold of 0.7 for Cronbach's α and KMO (Kaiser–Meyer–Olkin) values, and Bartlett's test of sphericity was significant (Table 3), affirming good reliability of the scale. Subsequently, we employed Smart PLS software to further assess the reliability of each variable, analyzing factor loadings, Cronbach's α coefficients, CR (Composite Reliability), and AVE (Average Variance Extracted), with specific data detailed in Tables 4 and 5. Finally, the AVE values were used to test the discriminant validity. As shown in Table 6, the square root of AVE for each diagonal variable is greater than the standardized correlation coefficients of the other variables, indicating good discriminant validity among the variables.

Table 3. Reliability results of variables.

Item	Rotated Factor Loading	KMO
BAL1	0.859	
BAL2	0.864	0.760
BAL3	0.774	
CC1	0.859	
CC2	0.864	0.809
CC3	0.774	
CC4	0.859	
CF1	0.755	
CF2	0.794	
CF3	0.681	0.845
CF4	0.782	
CF5	0.780	
PP1	0.784	
PP2	0.738	
PP3	0.660	
PP4	0.813	0.875
PP5	0.729	
PP6	0.911	
PP7	0.655	
PP8	0.757	
TR1	0.791	
TR2	0.743	
TR3	0.800	0.847
TR4	0.838	
TR5	0.824	

Table 4. Cross-loading.

Item	BAL	CC	CF	PP	TR
BAL1	0.938	0.323	0.291	0.549	0.486
BAL2	0.929	0.209	0.296	0.456	0.478
BAL3	0.929	0.195	0.194	0.576	0.488
CC1	0.262	0.903	0.524	0.510	0.235
CC2	0.068	0.702	0.277	0.271	0.084
CC3	0.271	0.900	0.386	0.626	0.339
CC4	0.218	0.867	0.446	0.469	0.253
CF1	0.263	0.440	0.826	0.353	0.207
CF2	0.208	0.347	0.809	0.301	0.287
CF3	0.235	0.499	0.847	0.438	0.418
CF4	0.140	0.325	0.733	0.243	0.152
CF5	0.237	0.275	0.770	0.271	0.201

Table 4. *Cont.*

Item	BAL	CC	CF	PP	TR
PP1	0.438	0.550	0.379	0.854	0.426
PP2	0.583	0.442	0.363	0.827	0.517
PP3	0.542	0.417	0.435	0.787	0.462
PP4	0.456	0.513	0.286	0.882	0.512
PP5	0.375	0.641	0.374	0.825	0.360
PP6	0.429	0.340	0.273	0.854	0.347
PP7	0.437	0.474	0.374	0.761	0.290
PP8	0.520	0.469	0.283	0.859	0.492
TR1	0.554	0.285	0.339	0.576	0.924
TR2	0.345	0.409	0.265	0.483	0.826
TR3	0.408	0.158	0.284	0.397	0.842
TR4	0.337	0.091	0.263	0.301	0.806
TR5	0.543	0.217	0.261	0.349	0.853

Table 5. Confirmatory factor analysis.

Latent Variable	Cronbach's Alpha	CR	AVE
BAL	0.925	0.935	0.868
CC	0.870	0.925	0.717
CF	0.860	0.890	0.637
PP	0.936	0.938	0.692
TR	0.906	0.942	0.725

Table 6. Discriminate validity (Fornell–Larcker criterion).

Constructs	BAL	CC	CF	PP	TR
BAL	0.932				
CC	0.262	0.847			
CF	0.276	0.489	0.798		
PP	0.571	0.583	0.418	0.832	
TR	0.520	0.292	0.336	0.518	0.851

In this paper, Smart PLS 4.0 was used to fit the main effects of governance mechanism on project performance, and the fit degree of the model was tested according to the GOF (goodness-of-fit) index calculation formula ($GOF = \sqrt{\overline{communality} \times \overline{R^2}}$) (1) proposed by Tenenhaus et al. [119]. After calculation, GOF = 0.579, which is greater than the standard value (0.36) of the complex model.

5.4. Structural Model

This research employed Partial Least Squares Structural Equation Modeling (PLS-SEM) to construct a comprehensive structural model examining hypothesized relationships between contract completeness, contract flexibility, trust, BIM application, and project performance. To assess the explanatory power of the structural model, we employed the coefficient of determination R^2 (R-squared) for endogenous constructs and used the cross-validated redundancy measure Q^2 (cross-validated redundancy measure) to evaluate forecasting capability. According to the results depicted in Figure 5, the R^2 value was 0.604, exceeding the threshold of 0.50 [52]. Concurrently, the Q^2 value was 0.376, greater than 0, indicating that the model achieved an acceptable level of predictive performance. The specific results are shown in Figure 5.

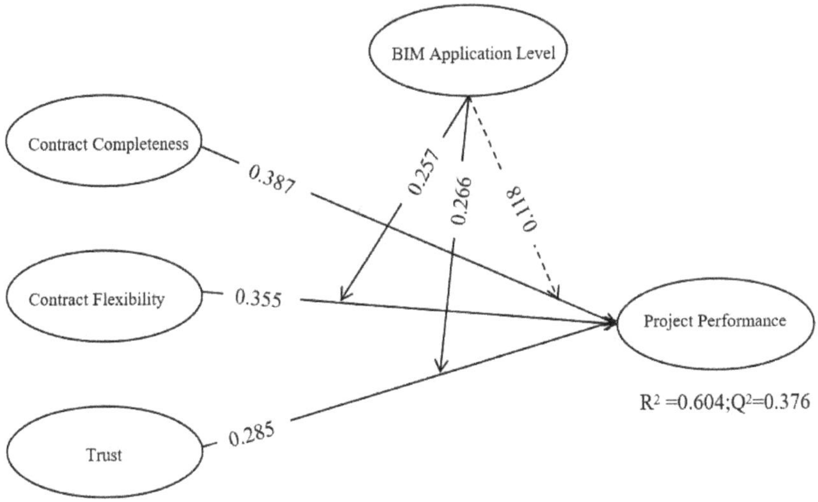

Figure 5. Path analysis by PLS.

The main path indicators and the index values of the adjustment effect in this study are shown in Table 7.

Table 7. Path coefficient results.

Hypothesis	Path	β	T	P	f^2	Effect Size	Result
H1	CC→PP	0.387	2.290	0.022 *	0.178	Large	Support
H2	CF→PP	0.355	2.256	0.010 *	0.156	Large	Support
H3	TR→PP	0.285	2.029	0.000 ***	0.044	Small	Support
H4	BAL × CC →PP	0.118	0.913	0.361	0.020	Small	Reject
H5	BAL × CF →PP	0.257	2.219	0.012 *	0.049	Small	Support
H6	BAL × TR →PP	0.266	2.363	0.015 *	0.061	Small	Support

Notes: *** $p < 0.001$, * $p < 0.05$.

The results from the table suggest that if the *p*-value for H4 is greater than 0.05, this suggests that the moderating effect of BAL adoption between project performance and contract completeness is not valid. However, both H5 (β = 0.257, P = 0.012 *) and H6 (β = 0.266, P = 0.015 *) show positive path coefficients, with *p*-values less than 0.05, indicating that the level of BIM technology application moderates the relationship between contract flexibility and trust.

In order to further analyze the effect of moderating, three moderating effect diagrams were drawn using simple slope analysis method. The results are detailed in Figures 6–8.

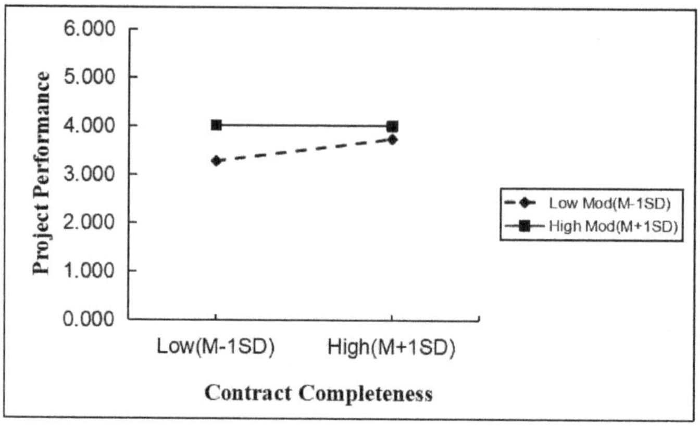

Figure 6. The moderating effect of BIM application level on the relationship between contract completeness and project performance.

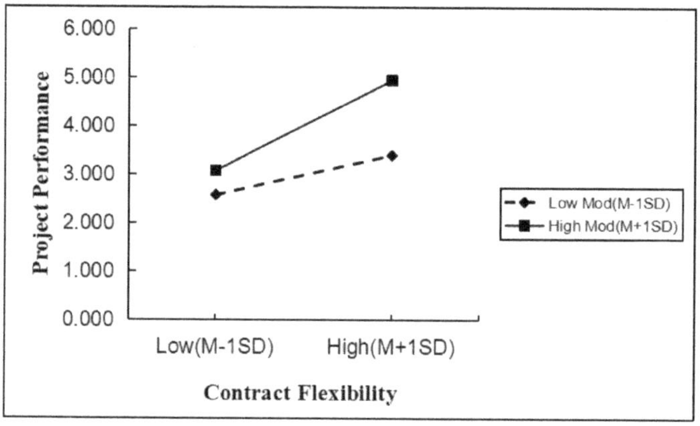

Figure 7. The moderating effect of BIM application level on the relationship between contract flexibility and project performance.

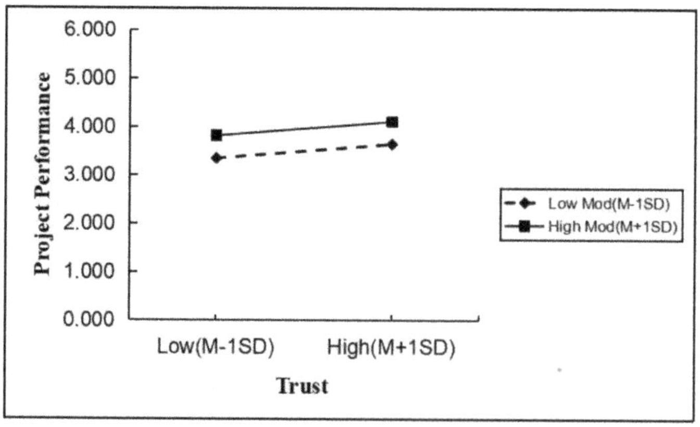

Figure 8. The moderating effect of BIM application level on the relationship between trust and project performance.

6. Discussion

From the perspective of TCT, this study aims to investigate the relationship between project governance mechanisms and project performance. Furthermore, it examines the moderating effect of BIM adoption levels on the relationship between project governance and project performance. The following discussion presents the findings of this research.

1. CC and CF in governance can enhance project performance.

H1 and H2 suggest that contract governance effectively enhances project performance. Firstly, during project implementation, the establishment of contracts enables stakeholders to devise comprehensive terms around project objectives. These terms extensively constrain behaviors, effectively curbing opportunistic actions and thereby facilitating successful project delivery. Secondly, the construction process is inherently dynamic, necessitating contract flexibility beyond rights and obligations to accommodate various engineering changes and swiftly coordinate work processes. Such flexibility helps mitigate dispute costs arising from unforeseen events, thus contributing to cost savings. Lastly, contracts, as formal legal relationships, hold legal validity and enable the effective supervision of participants' conduct. Project stakeholders mutually monitor each other based on contractual provisions to ensure the protection of respective interests, thereby promoting smooth project delivery.

2. TR mechanisms in relational governance can enhance project performance.

H3 indicates that informal relationships play a crucial role in project performance. Despite the ability of formal contractual relationships to constrain participant behavior during project development, their limitations are apparent. In contrast, trust-based informal governance significantly enhances cooperation between organizations and reduces communication and coordination costs, thereby improving project management efficiency. This conclusion aligns with findings by Lee et al. [112].

Furthermore, trust contributes to participants' sense of security within the organization. When projects encounter challenges, trust enables participants to promptly take effective measures and willingly assume risks that benefit the organization's overall interests or improvements, rather than focusing solely on personal gain. These behaviors contribute to enhanced project performance.

Trust dynamics among project partners can be categorized into three phases: (a) an initial trust phase based on reputation, past experience, or contractual agreements; (b) an in-process trust phase shaped by interactions among participants during the project; and (c) a final trust phase determined by project outcomes [28,37]. Initial trust significantly influences project success. High initial trust fosters an open and collaborative environment, laying a solid foundation for project execution. Conversely, low initial trust may lead to excessive reliance on formal contracts, reducing project flexibility [98]. The sustainability of project outcomes influenced by initial trust is contingent upon factors such as project complexity, duration, external environment, organizational culture, and trust mechanisms [99,111].

Moreover, trust is influenced by cultural and organizational differences. First, in highly competitive industries, establishing trust is particularly challenging. Organizations may prioritize short-term gains over long-term relationships, creating an atmosphere of distrust. This competitive environment can hinder open communication and collaboration, both of which are crucial for effective relational governance. Second, diverse organizational cultures may affect how trust is perceived and developed. For instance, in cultures that emphasize hierarchy, the open communication necessary for building trust may be difficult to achieve, whereas more egalitarian cultures may find it easier to establish relational norms. These cultural differences can lead to misunderstandings and conflicts, ultimately impacting project performance. The effectiveness of trust mechanisms may also vary depending on contextual factors such as project complexity and organizational maturity. Therefore, relying solely on trust may be insufficient; formal contracts may also be necessary to mitigate the risks associated with opportunism and ensure accountability.

Therefore, overall, the application of trust mechanisms in relational governance significantly enhances project performance by fostering cooperation, reducing costs, and strengthening participants' commitment to the organization. This effectively promotes successful project implementation and management.

3. BAL adoption plays a moderating role between project governance and project performance.

H4 suggests that the data platform established by BIM technology enhances information transparency, which effectively reduces ambiguities and fosters trust among all stakeholders. Furthermore, the real-time availability of data enables project managers to make informed decisions more quickly, thereby strengthening the trust that various stakeholders place in them.

H5 and H6 indicate that the information-sharing mechanisms based on the BIM platform positively influences the establishment of trust among participants and facilitates contract flexibility. However, in terms of contract integrity, BIM's moderating effect is less pronounced because dynamic information exchange among participants during contract formulation is relatively limited. Trust and contract flexibility are critical factors during project management, influenced by various constraints.

Therefore, effective information exchange and coordination mechanisms are crucial for enhancing inter-organizational workflows, helping to avoid information barriers and making the project management process more efficient and transparent. This suggests that leveraging BIM technology effectively can optimize information flow and improve collaboration efficiency, thereby enhancing the viability of governance mechanisms and improving overall project performance.

7. Conclusions and Implications

This study adopts an inter-organizational cooperation perspective to investigate the moderating effect of BIM technology on the relationship between project governance and project performance. By integrating BIM technology into the relationship model, this research not only expands the theoretical framework in the field of project management but also provides new insights for engineering practice. The findings contribute to a more comprehensive understanding of how technological innovations can influence project outcomes within complex organizational contexts.

Empirical results revealed several significant findings in this study. Firstly, the completeness of contractual governance positively influences project performance; higher levels of contractual completeness enable parties to significantly improve project outcomes through reduced disputes, enhanced risk management, clearer communication, and more efficient project execution. Secondly, contractual flexibility aids project managers and participants in better adapting to unforeseen circumstances or changes in project requirements, fostering a more collaborative environment between owners and contractors, reducing the likelihood of costly disputes, allowing for the effective management of project uncertainties, and providing ample space for decision-making by project managers. Consequently, construction projects are likely to achieve superior performance outcomes. Thirdly, trust, established as a mechanism between participants under common goal requirements, ensures that parties believe their interests will not be compromised and will act proactively for project success out of genuine commitment, significantly enhancing management efficiency and performance. Notably, both contractual flexibility and trust are moderated by the influence of BIM technology on project performance, suggesting that BIM implementation may alter the dynamics of how flexibility and trust impact project outcomes. However, unlike contractual flexibility and trust, contractual completeness is less susceptible to the influence of construction activities, primarily because it involves the formulation and execution of initial terms, which are typically established before the active construction phase.

These findings contribute to the development of a comprehensive framework for project governance, exploring the intricate interrelationships among contractual governance, relational governance, BIM technology, and project performance in project management. The results empirically demonstrate the moderating role of BIM technology in the proposed model. This study not only supports and extends existing research but also provides a systematic explanation of the mechanisms through which BIM technology influences project performance, thereby advancing our understanding of project management dynamics in the context of technological innovation.

7.1. Theoretical Implications

This article delves into the mechanisms by which project governance influences project performance, and conducts empirical research while examining the moderating role of BAL. Firstly, the study adopts a dynamic perspective, using BIM technology as a moderating variable to explore how information dissemination and effective communication influence project performance through contract governance and trust. Previous studies have primarily focused on the positive influence of contract governance and relational governance on project performance. Traditional project governance often adopts a static perspective, viewing projects as linear processes from initiation to closure [120]. This paper proposes a dynamic governance perspective, emphasizing the flexibility and adaptability of governance processes [121]. Conventionally, governance focuses on foundational setup in the initiation phase, detailed planning in the planning phase, process control in the execution phase, and evaluation and knowledge sharing in the closure phase. In contrast, dynamic governance advocates for the continuous adjustment of governance strategies throughout the project lifecycle to respond to evolving internal and external environments. The advantages of dynamic governance lie in its ability to more promptly identify and mitigate project risks, better meet the evolving needs of stakeholders, and ultimately enhance project success.

In addition, drawing insights from the application of BIM technology in project governance, policymakers should focus on the following measures at the contractual governance level: establishing cloud-based BIM platforms to reduce hardware investment for enterprises and improve resource utilization; unifying BIM data standards to promote data interoperability between different software platforms, enhancing data exchange quality and efficiency; establishing a robust data security management system, including implementing strict data security protocols and hierarchical access control to ensure the security and confidentiality of BIM data; and strengthening BIM technology training and improving talent cultivation mechanisms to meet the industry's demand for high-quality BIM professionals. At the relational governance level, emphasis should be placed on establishing effective communication platforms during the construction process to encourage the active participation of all stakeholders in the BIM platform and enhancing the awareness of all participants regarding the significance of BIM platforms in daily work to promote their proactive application.

Thus, this study fills a gap in explaining certain phenomena in engineering practice. Secondly, from an integrative standpoint, the research underscores the importance of BIM technology in facilitating efficient information sharing among project participants to enhance organizational communication and foster trust, thereby ensuring successful project delivery. Lastly, the study employs the lens of TCT to microscopically examine novel approaches to curbing opportunism and reducing costs, offering new theoretical perspectives and practical guidance for project governance mechanisms.

7.2. Managerial Implications

The research suggests several important management perspectives for project managers. Firstly, the impact of contract governance mechanisms on project performance is crucial. Project managers should ensure thorough communication and establish well-defined contract terms to safeguard the rights of all parties and mitigate the negative effects of opportunistic behavior on project performance. Moreover, contract terms should be designed with flexibility to address complexities and external environmental changes during project implementation, thus reducing additional costs and improving execution efficiency.

Secondly, establishing a trust mechanism among project management personnel is paramount. Trust not only fosters collaboration among organizational members but also reduces collaboration risks and accelerates the implementation of project measures and decisions. Project managers should address the work needs and challenges of team members, provide support and advice, and actively build and maintain trust through positive actions to enhance project management efficiency and performance. To address disparities in stakeholder engagement, the following strategies are recommended: (1) Establish a mechanism for incorporating stakeholder input at the decision-making level to enhance the scientific basis of decisions. (2) Strengthen information sharing and collaboration, encouraging active stakeholder participation in project decision-making to optimize BIM information management processes. (3) Develop a robust coordination mechanism to effectively manage conflicts, balance stakeholder interests, and improve overall project performance [12]. (4) Implement a balanced approach that combines relational governance with contractual governance can effectively enhance trust. Relational governance fosters collaboration, while contractual governance provides the necessary framework for resolving disputes and ensuring accountability. (5) Establish feedback mechanisms, which can help organizations assess the effectiveness of trust-building initiatives, and make necessary adjustments. Continuous evaluation ensures that trust mechanisms remain relevant and effective across different cultural contexts.

Lastly, enhancing information exchange mechanisms among project organizations is crucial, and particularly the leveraging of BIM technology platforms to improve the efficiency and quality of information sharing. The project manager should develop a BIM-based information sharing strategy to motivate the project participants to actively share information, taking into account the specific culture of the stakeholders [44]. For example, firstly, the hierarchical structure of rights is rigorous, emphasizing the authority and seniority of the individual. Secondly, there is a collectivist orientation that emphasizes the idea of harmony and the expectation of balancing the interests of different participants, and finally, there is the impact of personal relationships and influence on the implementation of decisions. This approach enhances cooperation and trust among internal organizational members, thereby further improving overall project performance.

In summary, these management insights not only help project managers better understand and address challenges in project governance but also guide them in adopting effective strategies and measures to enhance the capability for successful project delivery and performance.

8. Limitations and Future Research

This study primarily investigates the impact of project governance on performance, focusing on the efficiency of information exchange and collaboration among different participants within an organization. Cooperation processes across different industries and regions are inevitably influenced by individual and organizational behaviors. By establishing contractual rules and relational norms, it is possible to effectively control and regulate behaviors during the cooperation process, thereby fostering collaboration and improving project performance.

However, the study has some limitations. Firstly, the data collection was limited to Sichuan Province, where BIM technology adoption is relatively low, thus limiting the generalizability of the findings. For instance, the low adoption rate might lead to an overestimation of BIM's effects, as early adopters often experience more pronounced benefits. The unique economic and technological landscape of Sichuan Province may not be representative of other regions, potentially skewing the perceived impact of BIM on project performance. Secondly, the study did not adequately account for external factors influencing contract governance and trust mechanisms, such as policy factors and environmental uncertainties. This omission could affect the interpretation of project performance in several ways: Policy factors, like local government initiatives or national regulations, might be the actual drivers behind some of the observed performance improvements, rather than

BIM implementation alone. Environmental uncertainties, such as market fluctuations or supply chain disruptions, could mask or amplify the perceived effects of BIM on project outcomes. Finally, the challenges faced by BIM technology have not been considered, such as high initial investment costs, data interoperability, data security, and the number of personnel, as well as the integration of future technological advances into the project governance framework.

Future research directions should address these shortcomings. Firstly, due to varying levels of development across different regions in China. it is recommended that future studies broaden their sample size to enhance the applicability of research findings in various contexts. For example, the Pearl River Delta region demonstrates a high level of utilization of BIM technology. Therefore, it will provide more robust comparative data for the study. Such an approach would not only improve the generalizability of the research conclusions but also better capture the variations in technological implementation across different regional settings. Secondly, future studies should incorporate policy formulation and project context influences, including policy factors, market transparency, internal and external environmental factors, and collaboration quality as mediators or moderators. Finally, the influence of BIM technology on project performance from alternative theoretical perspectives should be explored, particularly in information integration and sharing.

Furthermore, the role of project governance should be validated in other industries, such as transportation infrastructure projects and the construction industry supply chain, as well as in different countries or regions. Given that relational norms are particularly emphasized in the context of Chinese characteristics, it is necessary to examine how organizations and individuals in other regions perceive relational behavior.

In light of the rapid development of information technology, future research can explore the integration of BIM with emerging technologies such as blockchain, artificial intelligence (AI), and machine learning in the context of project governance. For example, the integration of blockchain technology can facilitate the creation of smart contracts, enabling the automation and verification of project contracts. Meanwhile, the incorporation of AI and machine learning can enhance the automation of project processes and improve decision-making. The integration of these information technologies holds the potential to revolutionize project governance. These information technologies will significantly accelerate the sustainable development of large engineering projects.

Author Contributions: Conceptualization, B.Y. and N.L.S.N.; methodology, B.Y.; software, B.Y.; validation, B.Y.; formal analysis, B.Y.; investigation, B.Y.; resources, N.L.S.N.; data curation, B.Y.; writing—original draft preparation, B.Y.; writing—review and editing, B.Y. and N.L.S.N.; visualization, N.L.S.N.; supervision, N.L.S.N.; project administration, N.L.S.N. All authors have read and agreed to the published version of the manuscript.

Funding: This research received no external funding.

Data Availability Statement: The data that support the findings of this study are available from the corresponding author, N. Nie, upon reasonable request.

Conflicts of Interest: The authors declare no conflicts of interest.

References

1. Pan, Y.; Zhang, L. A BIM-data mining integrated digital twin framework for advanced project management. *Autom. Constr.* **2021**, *124*, 103564. [CrossRef]
2. Wu, G.; Li, H.; Wu, C.; Hu, Z. How different strengths of ties impact project performance in megaprojects: The mediating role of trust. *Int. J. Manag. Proj. Bus.* **2020**, *13*, 889–912. [CrossRef]
3. Demirkesen, S.; Tezel, A. Investigating major challenges for industry 4.0 adoption among construction companies. *Eng. Constr. Archit. Manag.* **2022**, *29*, 1470–1503. [CrossRef]
4. Mellado, F.; Lou, E.C. Building information modelling, lean and sustainability: An integration framework to promote performance improvements in the construction industry. *Sustain. Cities Soc.* **2020**, *61*, 102355. [CrossRef]

5. Gupta, S.K.; Gunasekaran, A.; Antony, J.; Gupta, S.; Bag, S.; Roubaud, D. Systematic literature review of project failures: Current trends and scope for future research. *Comput. Ind. Eng.* **2019**, *127*, 274–285. [CrossRef]
6. Chan, A.P.; Chan, D.W.; Chiang, Y.H.; Tang, B.-S.; Chan, E.H.; Ho, K.S. Exploring critical success factors for partnering in construction projects. *J. Constr. Eng. Manag.* **2004**, *130*, 188–198. [CrossRef]
7. Turner, J.R.; Keegan, A. Mechanisms of governance in the project-based organization: Roles of the broker and steward. *Eur. Manag. J.* **2001**, *19*, 254–267. [CrossRef]
8. Copola Azenha, F.; Aparecida Reis, D.; Leme Fleury, A. The role and characteristics of hybrid approaches to project management in the development of technology-based products and services. *Proj. Manag. J.* **2021**, *52*, 90–110. [CrossRef]
9. Flyvbjerg, B. Introduction: The iron law of megaproject management. In *Bent Flyvbjerg*; Oxford University Press: Oxford, UK, 2017; pp. 1–18.
10. Hadid, M.; Gunduz, M.; Kwak, Y.H. Assessing interorganizational relationships among key supply chain partners (KSCPs) in public construction projects. *IEEE Trans. Eng. Manag.* **2023**, *71*, 8839–8856. [CrossRef]
11. Han, W.J.; Yin, Y.L. Influence of Project Governance on Opportunistic Behavior: Taking a Dynamic Perspective. *Buildings* **2022**, *12*, 1695. [CrossRef]
12. Yang, J.; Cheng, Q. The Conditional Limitation of Relational Governance: The Moderating Role of Project Complexity. *Adv. Civ. Eng.* **2021**, *1*, 886913. [CrossRef]
13. Muller, R.; Pemsel, S.; Shao, J. Organizational enablers for governance and governmentality of projects: A literature review. *Int. J. Proj. Manag.* **2014**, *32*, 1309–1320. [CrossRef]
14. Wang, D.; Wu, Y. Research on the impact of project governance on organizational resilience of megaproject. *Proj. Manag. Technol.* **2022**, *11*, 39–43.
15. Fernandes, E.B.; Wegner, D.; Möllering, G. Governance of Interorganizational Projects: A Process-Based Approach Applied to a Latin American–European Case. *Proj. Manag. J.* **2023**, *54*, 219–234. [CrossRef]
16. Lovallo, D.; Cristofaro, M.; Flyvbjerg, B. Governing large projects: A three-stage process to get it right. *Acad. Manag. Perspect.* **2023**, *37*, 138–156. [CrossRef]
17. Lin, Y.H.; Guo, Y.Z.; Kim, C.J.; Chen, P.H.; Qian, M.W. The impact of relational governance on the adaptability of international contractors: A comparative study between China and Korea. *Eng. Constr. Archit. Manag.* **2020**, *27*, 3235–3259. [CrossRef]
18. Lu, P.; Guo, S.; Qian, L.; He, P.; Xu, X. The effectiveness of contractual and relational governances in construction projects in China. *Int. J. Proj. Manag.* **2015**, *33*, 212–222. [CrossRef]
19. Al-Tmeemy, S.M.H.M.; Abdul-Rahman, H.; Harun, Z. Future criteria for success of building projects in Malaysia. *Int. J. Proj. Manag.* **2011**, *29*, 337–348. [CrossRef]
20. Flyvbjerg, B.; Ansar, A.; Budzier, A.; Buhl, S.; Cantarelli, C.; Garbuio, M.; Glenting, C.; Holm, M.S.; Lovallo, D.; Lunn, D. Five things you should know about cost overrun. *Transp. Res. Part A Policy Pract.* **2018**, *118*, 174–190. [CrossRef]
21. Ferreira, C.; Cardoso, C.; Travassos, M.; Paiva, M.; Pestana, M.; Lopes, J.M.; Oliveira, M. Disorders, vulnerabilities and resilience in the supply chain in pandemic times. *Logistics* **2021**, *5*, 48. [CrossRef]
22. Sharif, N.; Kookhdan, A.; Mardani, A. Enhancing the performance of residential construction project through stakeholder satisfaction: The application of Structural Equation Modelling (SEM). *Transform. Bus. Econ.* **2018**, *2*, 44–46.
23. Sirisomboonsuk, P.; Gu, V.C.; Cao, R.Q.; Burns, J.R. Relationships between project governance and information technology governance and their impact on project performance. *Int. J. Proj. Manag.* **2018**, *36*, 287–300. [CrossRef]
24. Hunsucker, J. The Strategic Management of Large Engineering Projects: Shaping Institutions, Risks, and Governance. *Eng. Manag. J.* **2002**, *14*, 49–53.
25. Durdyev, S. Review of construction journals on causes of project cost overruns. *Eng. Constr. Archit. Manag.* **2021**, *28*, 1241–1260. [CrossRef]
26. Hwang, B.-G.; Shan, M.; Zhu, L.; Lim, W.-C. Cost control in megaprojects: Efficacy, tools and techniques, key knowledge areas and project comparisons. *Int. J. Constr. Manag.* **2020**, *20*, 437–449. [CrossRef]
27. Lu, P.; Qian, L.; Chu, Z.; Xu, X. Role of opportunism and trust in construction projects: Empirical evidence from China. *J. Manag. Eng.* **2016**, *32*, 05015007. [CrossRef]
28. Xue, J.; Shen, G.Q.; Yang, R.J.; Zafar, I.; Ekanayake, E.M.A.C.; Lin, X.; Darko, A. Influence of formal and informal stakeholder relationship on megaproject performance: A case of China. *Eng. Constr. Archit. Manag.* **2020**, *27*, 1505–1531. [CrossRef]
29. Solis-Molina, M.; Hernandez-Espallardo, M.; Rodriguez-Orejuela, A. Governance and performance in co-exploitation and co-exploration projects. *J. Bus. Ind. Mark.* **2020**, *35*, 875–894. [CrossRef]
30. Mueller, R. Governance, governmentality and project performance: The role of sovereignty. *Int. J. Inf. Syst. Proj. Manag.* **2019**, *7*, 5–17. [CrossRef]
31. Song, H.Y.; Hou, J.H.; Tang, S.Q. From Contractual Flexibility to Contractor's Cooperative Behavior in Construction Projects: The Multiple Mediation Effects of Ongoing Trust and Justice Perception. *Sustainability* **2021**, *13*, 13654. [CrossRef]
32. Zheng, J.; Roehrich, J.K.; Lewis, M.A. The dynamics of contractual and relational governance: Evidence from long-term public–private procurement arrangements. *J. Purch. Supply Manag.* **2008**, *14*, 43–54. [CrossRef]
33. Zhao, D.; Gao, A.Y.; Gu, F.F. The use of ambiguous language in contract adjustment: An empirical study of its impact on interfirm relationship performance. *Ind. Mark. Manag.* **2022**, *107*, 368–382. [CrossRef]

34. Fareed, M.Z.; Su, Q. Project Governance and Project Performance: The Moderating Role of Top Management Support. *Sustainability* **2022**, *14*, 2516. [CrossRef]
35. Li, Y.; Lu, Y.; Cui, Q.; Han, Y. Organizational Behavior in Megaprojects: Integrative Review and Directions for Future Research. *J. Manag. Eng.* **2019**, *35*, 04019009. [CrossRef]
36. Wang, D.D.; Fang, S.Z.; Li, K.L. Dynamic changes of governance mechanisms in mega construction projects in China: The mediating role of opportunism. *Constr. Archit. Manag.* **2019**, *26*, 723–735. [CrossRef]
37. Zhang, L.; Fu, Y.; Lai, J.; Chen, Y. Complements or substitutes? Recipes of contract design, contract enforcement, and trust for enhanced project performance. *Int. J. Proj. Manag.* **2024**, *42*, 102587. [CrossRef]
38. Cheng, J.; Huang, L.; Jiang, L.; Chen, J.; Chen, W.; He, Y. Fostering knowledge collaboration in construction projects: The role of BIM applicationn. *Buildings* **2023**, *13*, 812. [CrossRef]
39. Liu, T.; Chong, H.-Y.; Zhang, W.; Lee, C.-Y.; Tang, X. Effects of contractual and relational governances on BIM collaboration and implementation for project performance improvement. *J. Constr. Eng. Manag.* **2022**, *148*, 04022029. [CrossRef]
40. Borkowski, A.S. A Literature Review of BIM Definitions: Narrow and Broad Views. *Technologies* **2023**, *11*, 176. [CrossRef]
41. Tan, J.H.; Loo, S.-C.; Zainon, N.; Aziz, N.M.; Mohd Rahim, F.A. Potential functionality and workability of blockchain within a building information modelling (BIM) environment. *J. Facil. Manag.* **2023**, *21*, 490–510. [CrossRef]
42. Heydari, M.H.; Naderi, H.; Shojaei, A. BIM and Blockchain Integration in Construction Procurement Planning. In Proceedings of the Construction Research Congress, Des Moines, IA, USA, 20–23 March 2024; pp. 787–795.
43. Patacas, J.; Dawood, N.; Kassem, M. BIM for facilities management: A framework and a common data environment using open standards. *Autom. Constr.* **2020**, *120*, 103366. [CrossRef]
44. Porwal, A.; Hewage, K.N. Building Information Modeling (BIM) partnering framework for public construction projects. *Autom. Constr.* **2013**, *31*, 204–214. [CrossRef]
45. Herr, C.M.; Fischer, T. BIM adoption across the Chinese AEC industries: An extended BIM adoption model. *J. Comput. Des. Eng.* **2019**, *6*, 173–178. [CrossRef]
46. Akram, F.; Abrar Ul Haq, M. Integrating agency and resource dependence theories to examine the impact of corporate governance and innovation on firm performance. *Cogent Bus. Manag.* **2022**, *9*, 2152538. [CrossRef]
47. Lin, Y.H.; Zhang, H.J. Impact of contractual governance and guanxi on contractors? environmental behaviors: The mediating role of trust. *J. Clean. Prod.* **2023**, *382*, 135277. [CrossRef]
48. Liu, F.F.; Wang, Y.S.; Li, H.Y.; Zhou, X.W. Influence of cooperative period of municipal PPPs on transaction cost—A case study. *Eng. Constr. Archit. Manag.* **2021**, *28*, 1261–1277. [CrossRef]
49. Ma, H.; Wang, Y.L. Research on the promotion of cooperative relationship among project participants in the context of BIM—Based on fuzzy cognitive map model. *Sci. Technol. Prog. Policy* **2017**, *34*, 52–57.
50. Biesenthal, C.; Wilden, R. Multi-level project governance: Trends and opportunities. *Int. J. Proj. Manag.* **2014**, *32*, 1291–1308. [CrossRef]
51. Zhao, X.; Feng, Y.; Pienaar, J.; O'Brien, D. Modelling paths of risks associated with BIM implementation in architectural, engineering and construction projects. *Archit. Sci. Rev.* **2017**, *60*, 472–482. [CrossRef]
52. Tang, X.; Chong, H.-Y.; Zhang, W. Relationship between BIM implementation and performance of OSM projects. *J. Manag. Eng.* **2019**, *35*, 04019019. [CrossRef]
53. Vasista, T.G.; Abone, A. Benefits, Barriers and Applications of Information Communication Technology in Construction Industry: A Contemporary Study. *Int. J. Eng. Technol.* **2018**, *7*, 492–499. [CrossRef]
54. Um, K.H.; Kim, S.M. The effects of supply chain collaboration on performance and transaction cost advantage: The moderation and nonlinear effects of governance mechanisms. *Int. J. Prod. Econ.* **2019**, *217*, 97–111. [CrossRef]
55. Yan, L.; Zhang, L. Interplay of contractual governance and trust in improving construction project performance: Dynamic perspective. *J. Manag. Eng.* **2020**, *36*, 04020029. [CrossRef]
56. Iyer, K.; Jha, K. Factors affecting cost performance: Evidence from Indian construction projects. *Int. J. Proj. Manag.* **2005**, *23*, 283–295. [CrossRef]
57. Mwesigwa, R.; Bagire, V.; Ntayi, J.; Munene, J. Contract completeness as a foundation to relationship building among stakeholders in public private partnership projects. *Int. J. Public Adm.* **2020**, *43*, 890–899. [CrossRef]
58. Lin, Y.-H.; Zhu, T.; Kim, C.J.; Ho, S.P. How do institutional pressures moderate the impacts of relational governance on the performance of international projects? An empirical assessment. *Int. J. Proj. Manag.* **2021**, *39*, 726–737. [CrossRef]
59. Ceric, A.; Vukomanovic, M.; Ivic, I.; Kolaric, S. Trust in megaprojects: A comprehensive literature review of research trends. *Int. J. Proj. Manag.* **2021**, *39*, 325–338. [CrossRef]
60. Deng, B.C.; Xie, W.W.; Cheng, F.; Deng, J.J.; Long, L. Complexity Relationship between Power and Trust in Hybrid Megaproject Governance: The Structural Equation Modelling Approach. *Complexity* **2021**, *4*, 8814630. [CrossRef]
61. Huo, B.; Ye, Y.; Zhao, X. The impacts of trust and contracts on opportunism in the 3PL industry: The moderating role of demand uncertainty. *Int. J. Prod. Econ.* **2015**, *170*, 160–170. [CrossRef]
62. Pinto, J.K.; Slevin, D.P.; English, B. Trust in projects: An empirical assessment of owner/contractor relationships. *Int. J. Proj. Manag.* **2009**, *27*, 638–648. [CrossRef]

63. Wang, X.q.; Min, Z.; Dan, W. Influence of Trust Between Government and Social Capital on PPP Project Performance: The Mediating Role of Cooperative Behavior. *J. Eng. Manag.* **2019**, *33*, 54–58.
64. Liao, C.; Zhong, J.; Luo, Y.; Hu, J. Research on the Governance of Agent Construction Project from the Perspective of Transaction Cost Theory. In Proceedings of the ICCREM, Beijing, China, 16–17 October 2021; pp. 470–478.
65. Wang, D.; Li, H.; Lu, Y. Factors affecting transaction costs in megaprojects: A qualitative comparative analysis. *Int. J. Manag. Proj. Bus.* **2021**, *14*, 1245–1269. [CrossRef]
66. Wacker, J.G.; Yang, C.; Sheu, C. A transaction cost economics model for estimating performance effectiveness of relational and contractual governance: Theory and statistical results. *Int. J. Oper. Prod. Manag.* **2016**, *36*, 1551–1575. [CrossRef]
67. Young, R.; Chen, W.; Quazi, A.; Parry, W.; Wong, A.; Poon, S.K. The relationship between project governance mechanisms and project success: An international data set. *Int. J. Manag. Proj. Bus.* **2020**, *13*, 1496–1521. [CrossRef]
68. Trygges, L.D.K.; Müller, R.; Martinsuo, M. The impact of relational norms on information technology project success and its moderation through project governance. *Int. J. Manag. Proj. Bus.* **2015**, *8*, 154–176.
69. Zhou, X.Y.; Zhu, Q.H.; Xu, Z.D. The role of contractual and relational governance for the success of digital traceability: Evidence from Chinese food producers. *Int. J. Prod. Econ.* **2023**, *255*, 108659. [CrossRef]
70. Assaad, R.; El-Adaway, I.H.; Abotaleb, I.S. Predicting project performance in the construction industry. *J. Constr. Eng. Manag.* **2020**, *146*, 04020030. [CrossRef]
71. He, Q.; Wang, T.; Chan, A.P.; Xu, J. Developing a list of key performance indicators for benchmarking the success of construction megaprojects. *J. Constr. Eng. Manag.* **2021**, *147*, 04020164. [CrossRef]
72. Alexandrova, M. Evaluation of Project Portfolio Management Performance: Long and Short-Term Perspective. In Proceedings of the 19th International Scientific Conference on Hradec Economic Days, Hradec Kralove, Czech Republic, 10–11 April 2021; pp. 11–21.
73. Gil, N.; Fu, Y. Megaproject performance, value creation, and value distribution: An organizational governance perspective. *Acad. Manag. Discov.* **2022**, *8*, 224–251. [CrossRef]
74. Sun, Y.; Zhang, C.; Kok, R.A. The role of research outcome quality in the relationship between university research collaboration and technology transfer: Empirical results from China. *Scientometrics* **2020**, *122*, 1003–1026. [CrossRef]
75. Denicol, J.; Davies, A.; Krystallis, I. What are the causes and cures of poor megaproject performance? A systematic literature review and research agenda. *Proj. Manag. J.* **2020**, *51*, 328–345. [CrossRef]
76. Wang, G.; Liu, H.; Li, H.; Luo, X.; Liu, J. A building project-based industrialized construction maturity model involving organizational enablers: A Multi-Case Study in China. *Sustainability* **2020**, *12*, 4029. [CrossRef]
77. Bilir, C. Project success criteria, critical success factors (CSF), and agile projects. *Contemp. Chall. Agil. Proj. Manag.* **2022**, *21*, 52–72.
78. Luo, Y. Contract, cooperation, and performance in international joint ventures. *Strateg. Manag. J.* **2002**, *23*, 903–919. [CrossRef]
79. Valentinov, V.; Roth, S. Relationality in transaction cost economics and stakeholder theory: A new conceptual framework. *Bus. Ethics Environ. Responsib.* **2024**, *33*, 535–546. [CrossRef]
80. Lu, W.; Zhang, L.; Zhang, L. Effect of contract completeness on contractors' opportunistic behavior and the moderating role of interdependence. *J. Constr. Eng. Manag.* **2016**, *142*, 04016004. [CrossRef]
81. Wu, A.; Wang, Z.; Chen, S. Impact of specific investments, governance mechanisms and behaviors on the performance of cooperative innovation projects. *Int. J. Proj. Manag.* **2017**, *35*, 504–515. [CrossRef]
82. Young, S.; Nguyen, P.; Macinati, M.S. Dynamic interplay between contractual and relational governance: An empirical study in Australian healthcare outsourcing. *Aust. J. Public Adm.* **2021**, *80*, 261–282. [CrossRef]
83. Gurcaylilar-Yenidogan, T.; Erdogan, D. Opportunism still remained alive: Conditional limits of trust and contract in software projects. *Int. J. Manag. Proj. Bus.* **2023**, *16*, 374–404. [CrossRef]
84. Cheng, J.J.; Zhang, L.Y.; Chen, J.H.; Jiang, L.X.; Lin, Y. How Does Relational Governance Affect Knowledge Collaboration in Integrated Project Delivery? A Transactive Memory System Process. *Eng. Manag. J.* **2023**, *35*, 299–312. [CrossRef]
85. Agarwal, U.A.; Narayana, S.A. Impact of relational communication on buyer–supplier relationship satisfaction: Role of trust and commitment. *Benchmarking Int. J.* **2020**, *27*, 2459–2496. [CrossRef]
86. Blome, C.; Paulraj, A.; Preuss, L.; Roehrich, J.K. Trust and opportunism as paradoxical tension: Implications for achieving sustainability in buyer-supplier relationships. *Ind. Mark. Manag.* **2023**, *108*, 94–107. [CrossRef]
87. Bond-Barnard, T.J.; Fletcher, L.; Steyn, H. Linking trust and collaboration in project teams to project management success. *Int. J. Manag. Proj. Bus.* **2018**, *11*, 432–457. [CrossRef]
88. Feng, G.; Hao, S.; Li, X. Relationship orientation, justice perception, and opportunistic behavior in PPP projects: An empirical study from China. *Front. Psychol.* **2021**, *12*, 635447. [CrossRef] [PubMed]
89. Zhang, S.; Ma, y. An empirical study on relationship governance and cooperation performance in EPC projects. *Commun. Financ. Account.* **2017**, *12*, 4–6.
90. Li, J.; Jiang, W.; Zuo, J. The effects of trust network among project participants on project performance based on SNA approach: A case study in China. *Int. J. Constr. Manag.* **2020**, *20*, 837–847. [CrossRef]
91. Li, Y.; Li, Q. The Application of BIM Technology in Budget Control of Port Construction Cost. *J. Coast. Res.* **2020**, *103*, 644–648. [CrossRef]

92. Volk, R.; Stengel, J.; Schultmann, F. Building Information Modeling (BIM) for existing buildings-literature review and future needs. *Autom. Constr.* **2014**, *38*, 109–127. [CrossRef]
93. Gao, Y.; Liu, X.; Zhang, S.; Zhang, J.; Guo, Q. BIM application and collaboration in construction projects: A perspective of the Chinese construction market. *Constr. Manag. Econ.* **2022**, *40*, 429–441. [CrossRef]
94. Abd Jamil, A.H.; Fathi, M.S. Contractual challenges for BIM-based construction projects: A systematic review. *Built Environ. Proj. Asset Manag.* **2018**, *8*, 372–385. [CrossRef]
95. Marinho, A.; Couto, J.; Teixeira, J. Relational contracting and its combination with the BIM methodology in mitigating asymmetric information problems in construction projects. *J. Civ. Eng. Manag.* **2021**, *27*, 217–229. [CrossRef]
96. Rajabi, M.S.; Radzi, A.R.; Rezaeiashtiani, M.; Famili, A.; Rashidi, M.E.; Rahman, R.A. Key assessment criteria for organizational BIM capabilities: A cross-regional study. *Buildings* **2022**, *12*, 1013. [CrossRef]
97. Rajabi, M.S.; Rezaeiashtiani, M.; Radzi, A.R.; Famili, A.; Rezaeiashtiani, A.; Rahman, R.A. Underlying factors and strategies for organizational BIM capabilities: The case of Iran. *Appl. Syst. Innov.* **2022**, *5*, 109. [CrossRef]
98. Zhang, H.M.; Chong, H.Y.; Zeng, Y.; Zhang, W. The effective mediating role of stakeholder management in the relationship between BIM implementation and project performance. *Eng. Constr. Archit. Manag.* **2023**, *30*, 2503–2522. [CrossRef]
99. Pérez, Y.; Ávila, J.; Sánchez, O. Influence of BIM and Lean on mitigating delay factors in building projects. *Results Eng.* **2024**, *22*, 102236. [CrossRef]
100. Sadeghi, S.; Rafiee-Dehkharghani, R.; Laknejadi, K. Three-dimensional optimization of wave barriers for mitigating ground vibrations induced by underground train. *Environ. Sci. Pollut. Res.* **2024**, *31*, 384–405. [CrossRef]
101. Arjanaki, M.G.; Pourgholi, M. Takagi-Sugeno (TS) Fuzzy model-basedobserver design for glucose-insulin system in diabetes type 1: An LMI approach. In Proceedings of the 2020 28th Iranian Conference on Electrical Engineering (ICEE), Tabriz, Iran, 4–6 August 2020; pp. 1–5.
102. El Fathi, A.; Ganji, M.; Boiroux, D.; Bengtsson, H.; Breton, M.D. Intermittent Control for Safe Long-Acting Insulin Intensification for Type 2 Diabetes: In-Silico Experiments. In Proceedings of the IEEE Conference on Control Technology and Applications (CCTA), Bridgetown, Barbados, 16–18 August 2023; pp. 534–539.
103. Wu, G.; Zhao, X.; Zuo, J.; Zillante, G. Effects of contractual flexibility on conflict and project success in megaprojects. *Int. J. Confl. Manag.* **2018**, *29*, 253–278. [CrossRef]
104. Poppo, L.; Zenger, T. Do formal contracts and relational governance function as substitutes or complements? *Strateg. Manag. J.* **2002**, *23*, 707–725. [CrossRef]
105. Khan, A.; Waris, M.; Panigrahi, S.; Sajid, M.R.; Rana, F. Improving the performance of public sector infrastructure projects: Role of project governance and stakeholder management. *J. Manag. Eng.* **2021**, *37*, 04020112. [CrossRef]
106. Chiara, N.; Kokkaew, N. Risk analysis of contractual flexibility in BOT negotiations: A quantitative approach using risk flexibility theory. *Int. J. Eng. Manag.* **2009**, *1*, 71–79.
107. Poppo, L.; Zhou, K.Z. Managing contracts for fairness in buyer–supplier exchanges. *Strateg. Manag. J.* **2014**, *35*, 1508–1527. [CrossRef]
108. Wang, X.; Yin, Y.L. Structural Dimensions and Measurement of Trust Networks among Construction Project Participants. *Sustainability* **2023**, *15*, 4112. [CrossRef]
109. Yang, Z.; Zhou, C.; Jiang, L. When do formal control and trust matter? A context-based analysis of the effects on marketing channel relationships in China. *Ind. Mark. Manag.* **2011**, *40*, 86–96. [CrossRef]
110. Salzano, A.; Cascone, S.; Zitiello, E.P.; Nicolella, M. Construction Safety and Efficiency: Integrating Building Information Modeling into Risk Management and Project Execution. *Sustainability* **2024**, *16*, 4094. [CrossRef]
111. Chen, G.; Chen, J.; Tang, Y.; Li, Q.; Luo, X. Identifying effective collaborative behaviors in building information modeling–enabled construction projects. *J. Constr. Eng. Manag.* **2022**, *148*, 04022026. [CrossRef]
112. Lee, C.-Y.; Chong, H.-Y.; Li, Q.; Wang, X. Joint contract–function effects on BIM-enabled EPC project performance. *J. Constr. Eng. Manag.* **2020**, *146*, 04020008. [CrossRef]
113. Tu, B.; Zuo, J.; Chang, R.D.; Webber, R.J.; Xiong, F.; Dong, N. A system dynamic model for assessing the level of BIM implementation in construction phase: A China case study. *Eng. Constr. Archit. Manag.* **2023**, *30*, 1321–1343. [CrossRef]
114. Love, P.E.D.; Liu, J.; Matthews, J.; Sing, C.-P.; Smith, J. Future proofing PPPs: Life-cycle performance measurement and Building Information Modelling. *Autom. Constr.* **2015**, *56*, 26–35. [CrossRef]
115. Xu, Y.; Chong, H.-Y.; Chi, M. Impact of contractual flexibility on BIM-enabled PPP project performance during the construction phase. *J. Infrastruct. Syst.* **2022**, *28*, 04021057. [CrossRef]
116. Halimi, Z.; Bavafa, A.; Cui, Q. Barriers to Community Connectivity: An Assessment of Reconnecting Communities Pilot Program. In Proceedings of the International Conference on Transportation and Development 2024, Atlanta, GA, USA, 15–18 June 2024; pp. 83–91.
117. Atkinson, R. Project management: Cost, time and quality, two best guesses and a phenomenon, its time to accept other success criteria. *Int. J. Proj. Manag.* **1999**, *17*, 337–342. [CrossRef]
118. Sun, Y.G. Research on PPP Project Governance Performance Based on Contract Relationship Dual Governance. Ph.D. Thesis, Beijing Jiaotong University, Beijing, China, 2021.
119. Tenenhaus, M.; Vinzi, V.E.; Chatelin, Y.-M.; Lauro, C. PLS path modeling. *Comput. Stat. Data Anal.* **2005**, *48*, 159–205. [CrossRef]

120. Zhang, S.B.; Chen, J.; Fu, Y. Contract complexity and trust in construction project subcontracting. *Eng. Constr. Archit. Manag.* **2020**, *27*, 2477–2500. [CrossRef]
121. Shi, Q.; Chen, X.; Xiao, C.; Han, Y. Network perspective in megaproject management: A systematic review. *J. Constr. Eng. Management* **2022**, *148*, 03122002. [CrossRef]

MDPI AG
Grosspeteranlage 5
4052 Basel
Switzerland
Tel.: +41 61 683 77 34

Buildings Editorial Office
E-mail: buildings@mdpi.com
www.mdpi.com/journal/buildings

www.ingramcontent.com/pod-product-compliance
Lightning Source LLC
Chambersburg PA
CBHW040820250926
51850CB00007BA/1219
* 9 7 8 3 7 2 5 8 7 1 1 6 2 *